SCIENCE, SOCIETY AND RELIGION



# SCIENCE FOR LIVING- 5 SCIENCE TOPICS OF COMMON INTEREST TO RELIGION AND SOCIETY

# SCIENCE, SOCIETY AND RELIGION

Additional books in this series can be found on Nova's website under the Series tab.

Additional e-books in this series can be found on Nova's website under the e-book tab.

SCIENCE, SOCIETY AND RELIGION

# SCIENCE FOR LIVING- 5 SCIENCE TOPICS OF COMMON INTEREST TO RELIGION AND SOCIETY

RAGHAVAN JAYAKUMAR

NOVINKA

*New York*

**Library of Congress Cataloging-in-Publication Data**

Jayakumar, Raghavan, author.
Science for living : 5 science topics of common interest to religion and society / Raghavan Jayakumar (Lawrence Livermore National Lab, CA, USA).
pages cm. -- (Science, evolution and creationism)
Includes bibliographical references and indexes.
ISBN 978-1-63483-806-1 (softcover)
1. Science--Philosophy. 2. Life (Biology) 3. Space and time. 4. Cosmology. I. Title.
Q175.J348 2014
500--dc23
2014014972

*Published by Nova Science Publishers, Inc. † New York*

# CONTENTS

# PREFACE

The motivation for this book rose from my experiences with people's attitudes in different countries. In northern European countries, and to some extent Japan, people, on average, tend to think of religion as one of the facets of life and culture and science is acknowledged as a real force that shapes lives and their society. In contrast, in countries like the United States, India, Mexico, Latin America and Middle Eastern countries, religion plays a major role in daily life, social activities and in politics and science is assigned as an activity for a group of people who are supposed to just provide answers to questions of science and technology raised by the Government. The exception is Medicine because it serves a need. The fact is that science can provide reliable answers in the same areas that religious guidance is sought and in this author's opinion, science needs to be trusted for providing true solutions for most of the questions and issues of daily and social life.

Like most people who were involved with science from an early age and continually engaged in learning several aspects of science, I have realized the value of science and the life lessons it has taught. Growing up in a strongly religious family and at the same time being exposed to the beauty and demanding rigor of science, I have seen the contributions of both science and religion. While, it is not my place to tell readers what to and how to believe, the fact remains that most occasions of our daily life require a scientific thinking and objectivity. In many cases, scientific knowledge comes into play as analogy or even to point directly to a solution. After observing number of people, I have come to the conclusion that the primary reason why average people trust science and develop a scientific attitude is because they know something about the history of Modern Science and have also taken the trouble to absorb some scientific knowledge, even if it is at a basic level. However, this attitude does not develop when people get involved with technology or highly applied sciences, without appreciating basic science. Therefore, the more people understand the fundamental concepts and methods in science, the more they trust science and seek guidance from it.

In countries with religion in the middle of everything (strong examples are Middle Eastern and South Asian countries, United States, Italy, Spain and Mexico), daily life is imbued with rational and irrational religious practices and beliefs, God or gods are never far away from influence and trust in religion is supreme. Americans pleading that God Bless America or Indians starting some simple work only on religiously auspicious days and times or people believing mythological stories as revealed truths in many countries, are but few examples of the indelible influence of religion. In US and India, for example, few have the patience and the ability to appreciate a perfectly scientific and rational explanation for something they have decided is supernatural. I believe that Science deserves to have this type of access and trust and if that is given, societies would transform into more efficient, just and prosperous ones.

The goal of this book is to provide this acquisition of basic scientific knowledge. Interestingly but not surprisingly, these are the fundamental areas which religion is also interested in. One gets a sense that people have absorbed the religious worldview over millennia and have come to accept it and that is why they are comfortable with them. The same way, if people could acquire this familiarity and knowledge with science, they would be able to thoughtfully choose between scientific and religious guidance, when there is conflict. We are already seeing this in the field of medicine even in countries like the United States, where more modern medical information is available. In the same way, it is hoped that the fundamental topics and discussions covered in this book will enable the readers to have a fuller participation in life, through observing, understanding, analysis and finding solutions that are scientifically sound.

I had to struggle to determine the level at which to write this book- from the level of lay people with no background in science to people who are well versed in science. I decided to straddle the line, to maximize the readership of an important demography- the change makers. The book is intended for those who care about what makes science a viable force and how it is to be compared with other influences. Their choice to include science in their worldview would convince others to share in that view, because scientific knowledge is as infectious as religious belief and more so, because science is more observable for the people who look for it. In straddling the line, I have provided for readers who would rather know concepts than detailed mathematics in the two physics topics of space, time and matter, and cosmology, by separating the mathematics from the main body of the text so that it can be looked up if so desired. I hope this works for the readers.

Readers would not fail to notice that considerable mining of quotes has been done in a few chapters. While that would be a negative for books on a technical or specific subjective topic, in this case, the historical descriptions and the demonstration of the dialog that goes on in society requires it and makes people comprehend the spectrum of opinions on science and religion and therefore they can make up their own minds on the issue, but by employing the scientific approach. I confess that I have not fully read several books from which quotes have been drawn, but have made sure that these are not quoted out of context.

I would like to gratefully acknowledge the advice and comments of my daughter and biologist Dr. Prerana Jayakumar, on the chapter on Evolution. I am very thankful to my daughter and medical doctor, Dr. Archana Jayakumar for advice and comments on the chapter on Medicine. I would like to express my gratitude to Professors Jesse Thomas and Prof. Khaleel Mohammed of State University of San Diego and Prof. Babak Rahimi of University of California, San Diego for their valuable opinions on matters of science and religion and the references they provided. I would like to thank my wife Suhasini Jayakumar for her continued support and many discussions that sharpened my arguments in the book. I am immensely thankful to Dr. Mallika Vasugi Govindarajoo for ably editing the varied text.

*Dr. Raghavan Jayakumar*
*Tel: 858-566-4322*
*E-mail: raghavan.jayakumar@gmail.com*

*Chapter 1*

# INTRODUCTION

"To modern educated people, it seems obvious that matters of fact are to be ascertained by observation, not by consulting ancient authorities. But this is an entirely modern conception, which hardly existed before the seventeenth century. Aristotle maintained that women have fewer teeth than men; although he was twice married, it never occurred to him to verify this statement by examining his wives' mouths."- Bertrand Russell (Russell 1968).

In Gobekli Tepe' near Urfa, an ancient city in southeastern Turkey, stand 11,000 year old concentric rings of massive T shaped pillars. The carvings by the prehistoric, pre-agricultural people emphasize predators and, vultures in particular are even more prominent. Archaeologists and anthropologists believe that the vultures represented the aspiration of humans to reach high into the heavens, even so long ago. The pursuit of this aspiration has always been obviously evident in religion. But, if scientific thought and technological output are measures, then science has been in the same pursuit along with religion or perhaps before it, even in Gobekli Tepe'. Though it may not be evident, the broader aspirations of science and religion are similar and ultimately approached with the intellect as well as the heart and human spirit. With steady and long strides, science has now become a guide to living and to the organizing of society, as religions have been. To many, scientific knowledge and efforts are inspirational and give them a sense of awe when they begin to understand Nature and its mechanisms. However, science takes a path that is paved with physical evidence and establishes milestones of achievements that become the markers for its next steps. The path itself gets wider, well-worn and finds practical use. With the guidance of science, not only have societies come a long way in technologies, but the illumination of the mind by the processes of science has clarified many areas of social responsibility, morality and harmony. Science, which is acknowledged to be a pillar of society in technological and sociological structure, can also be a powerful guide to understanding personal and social options and our ability to control and direct our lives. Like religion, association with science and learning from science requires an intentional engagement. But, unlike religion, sciences are not intended to be handed down as fixed dogmas and this makes science somewhat inaccessible and difficult for many people. Science also proceeds through a rigorous process of rational inquiry, which borders on skepticism. This scientific approach, as much as scientific knowledge, forms the foundation of a scientific attitude that informs the decisions and actions of a person.

## A Brief History of the Origin and History of Science

We can only speculate on the process of the invention of fire or the wheel and wonder how early humans viewed the phenomena behind these inventions. But it is reasonable to think that early humans had not yet developed a rigorous method of analyzing and understanding nature. In the absence of that methodology, it is perhaps understandable that causes and effects of natural phenomena would have either not been contemplated at all and accepted without question or attributed to the intent and power of a Being, God or gods who are beyond human understanding. This attitude might even have led to a fear that such phenomena should never be looked into, so as not to offend those gods. (Some religions still continue to advise people to accept God's will without inquiry, both to abide by the laws of God unquestioningly and also in order to promote peaceful acceptance of one's circumstances). Whatever the circumstances of early inventions of fire, wheel, hunting tools, agriculture, animal husbandry, construction of dwellings, healing potions etc. were, there is no question that these inventions led to the formation of communities and societies that cooperated and derived the maximum benefit out of these inventions. The desire to improve these inventions, increase the benefits and scale up the production must have led to the inquiry into the processes, and subsequently to the discovery of empirical rules. Regardless of whether we call this science in the sense of modern terminology, these inventions clearly liberated societies from superstitions and improved their knowledge, thus laying the foundations for an intellectual inquiry and formalizing various systems of knowledge. The development of languages, writing and methods of communicating and disseminating knowledge continuously changed the status of science.

In early history, science had two disparate components, one related to applications in one's life such as that related to the development of trade and weapon technologies, like metallurgy, chemical compounds, mechanical devices etc. and the other the inquiry into celestial objects and their motion in a spirit of pure curiosity as well as to forecast the weather, climate and so on. The latter interest was also strongly motivated by faith in the astrological influence on the earthly events. For example, many civilizations including Mayan and Indian civilizations built observatories to mark seasons for agricultural and cultural purposes, and also for astrological predictions. Careful astronomical observations were carried out by Babylonians and their method and the results formed the basis of many astronomical systems of calculations in India, Egypt and other places. In parallel efforts, systems of medicine and surgery were developed in India, Greece, Egypt and China. All these scientific endeavors, though drawing influences from each other, still maintained an independent approach, appropriate to local culture and belief system. However, the concerted and unified effort at inquiring into the fundamental nature of matter and forces owes its origin to Greece.

Modern science comprises empirical and theoretical knowledge and observations of natural phenomena that are further supported by experimental discovery and confirmation. This 2600 year old definition was founded in Greek philosophy, before Socrates, by great thinkers such as Thales who was one of the first to propose that natural phenomena do not have supernatural causes. Around the same time, philosophers like Leucippus and Democritus proposed the revolutionary atomic nature of matter, rather than the five-elemental nature proposed by Aristotle. Pythagoras came up with the idea that the Earth was not flat and was, in fact, spherical in shape. Aristotle conceived the principle of empiricism, which argued that sensory experience provides the knowledge of nature. Now, this empiricism includes not only sensors on one's body but also sensors such as observing instruments and other methods of measuring natural phenomena. The idea that natural knowledge can be obtained through studies, observations and experimentation and also by a process of logical deduction, is now an established concept because of these great minds. The record of discoveries and advancement of knowledge was so great during this period and this philosophy of scientific pursuit and the methodology was so instrumental in this success that even now the basis of scientific research and discoveries is founded on principles of this era. The Greek period is famous for scientific advancement through theories and observations on life sciences, astronomy, mathematics and medicine. The Greeks also understood the required rigor of scientific inquiry and the need for integration with logic. Concepts and standards for proof and verification were also developed during this period. Many Greeks also realized the limitations that their instruments and observations posed and kept questions open, rather than declare unconfirmed discoveries. In the second century C.E. Claudius Ptolemy was one of the first modern thinkers in the Roman era, who theorized on astronomy and geography.

Between the $5^{th}$ and $11^{th}$ century CE, many philosophers around the world developed mathematics, astronomy and medicine. The famous Hindu philosophers among these were Bhaskara, Aryabhatta and Sushruta. Their *Vedic* mathematics and medical methods were original and brilliant in their own right. These scientist-philosophers carried out precise calculations of planetary transits, eclipses etc. and developed a thorough knowledge of anatomy, surgical tools and methods, while discovering herbal cures. The Indian philosophers were some of the first modern thinkers to wonder about the origins of the Universe and to propose theories. Since the early $4^{th}$ century, China has had an unbroken history of scientific development and accumulation of scientific knowledge that was bolstered by technological innovations. In some areas of mathematics, such as geometry, Chinese philosophers were ahead of the Europeans by as much as 1200 years. The Chinese developed seismometers, the magnetic compass, gun powder and many such tools both for daily use and for scientific purposes. The Chinese record of inventions is breathtaking in its scope.

The Nobel Laureate Abdus Salam (Salam 1994) points out that between 750 and 1100 C.E., starting a little over a hundred years after the death of Islam's Prophet, Islamic scholars studied, absorbed and advanced all branches of science known to mankind. (It is not coincidental that this period also corresponds to Hindu development of science). For example, Al-Khwarizmi, considered to be the father of Algebra, translated the Indian mathematical numbering and introduced the decimal system to the western world. He adapted Indian astronomical calculations to create a compendium of transit tables. Apart from this, he also wrote an amazing book on the appearance of earth which described the geography and weather zones of the earth, correcting Greco-Roman philosopher Ptolemy's calculations. The $8^{th}$ century scholar Jabir Ibn Hayyan's treatise covered, in addition to alchemical texts, inquiries into subjects ranging from cosmology to music. He developed methods of distillation, crystallization etc. as well as instruments for chemistry in making new minerals and acids. Starting from Indian and Chinese knowledge, Abu Raihan Al-Baruni ($10^{th}$ century) continued previous work on all fields of science and mathematics and contributed to physical geography among other subjects. A great physicist of his time, Ibn Al Haitham ($10^{th}$ century) was the first to declare that light traveled from an object as rays to the eye. Al Haitham anticipated Newton in the concept of inertia and was perhaps well versed in optics. The famous $11^{th}$ century scientist-poet Omar Khayyam developed solar calendars, medical techniques and astronomy instruments. The $13^{th}$ century scientist Nasir Al-Din Tusi propounded the first statement of conservation of mass.

## THE BIRTH OF MODERN SCIENCE

Modern science owes its discipline and methods to the Greek thought, logic and desire to explain nature with universal laws that are discovered through human understanding, rather than postulated as inexplicable phenomena created by Gods. Although there was a strict desire to develop science (philosophy) with fewest assumptions and considerable verification, the amazing intellectual influence of Aristotle that loomed over the community for close to a millennium and even after the demise of Greek civilization, inhibited this scientific approach. Aristotle applied his view which is that one can trust one's senses and logical powers in understanding nature, rather than rely mainly on observation and verification. This allowed science to advance in some cases and yet held it back in many important areas. Examples of the wrong ideas are, (a) the natural state of all objects is to be at the center of the earth and therefore they fall and (b) all earthly things are formed by a different mixture of four elements: fire, earth, water and air, and celestial objects are made of ether. Aristotle's theories were like an edifice built together providing cogent, albeit, often incorrect or false concepts. Therefore, for a long time, his theories could not be brought down without destroying the whole system of thought. One could almost say that it was a different system of science, from which the present sciences arose. The following gives a sparse description of discoveries and progress of science with a few examples to orient the readers. The history of science is so vast that even when approached by individual disciplines, it would take a very large number of volumes to contain it.

The Renaissance Period brought a systemic change in this situation, particularly in the $16^{th}$ century. In the 12th century, Western philosophers came in contact with thoughts from the Eastern and Islamic scholars. As a result, there was an increased flow of alternative thoughts and philosophers saw a diversity of ideas. Institutions were built to translate foreign texts and study these ideas and to codify them into different disciplines. The works of the Islamic scholars described above, became known and the seeds of change were thus planted. The Renaissance Period is speckled with important events in Europe and the cultural landscape assisted the vigorous growth of modern scientific thought. In the meanwhile, scientific thought languished in other regions of the world under the weight of military occupations, foreign colonial rules and religious uprisings and dominance. Science, fed from the intellectual streams from the East, started growing in the West. Therefore, in a way, Western science is not totally Western in its origins and this also explains the fact that Eastern and Middle Eastern cultures see many foundational commonalities with modern science. In the West, with a humanistic (rather than religious) viewpoint gaining popularity, the study of natural sciences gained prominence and prestige, to the extent that the Renaissance Period was also a period of scientific revolution. The discovery of Americas by Columbus and Copernican ideas of a heliocentric Solar system with the Sun rather than the earth being at the center of the Universe, catalyzed revolutionary changes in scientific thought. In England, Francis Bacon inspired a scientific thinking for the benefit of humankind. His statement about scientific discoveries being *"... a light in nature ... spreading further and further should presently disclose and bring into sight all that is most hidden and secret in the world that man would be the benefactor indeed of the human race-the propagator of man's empire over the universe, the champion of liberty, the conqueror and subduer of necessities."* (Bacon 1915), inspired generations of scientific community.

In the $14^{th}$ century, Jean Buridan, reinterpreting Aristotle, stated that objects moved because of an impetus, an idea which would later be the basis for the concept of inertia of an object. In the $16^{th}$ century, William Harvey showed that the venous and arterial systems belonged to the same system of blood circulation in a body. In the $16^{th}$ and $17^{th}$ centuries Pierre Vernier, Evangelista Torricelli, Blaise Pascal, Gottfried Leibniz and many others perfected instruments, calculators and devices to enable precision calculations, measurements and accurate experimentation. The German physician Leonhart Fuchs established the discipline of Botany and Carl von Linne (Carlos Linnaeus) catalogued all known 10,000 livings things into a single taxonomic system in *Systema Naturae*. Georg Agricola, often known as the father of metallurgy, developed processes for ore dressing and metal extraction and treatment of metals. Rene Descartes exhorted fellow scientists to adopt the mechanical philosophy to quantify properties of matter.

$17^{th}$ century science saw the relatively fast appearance of scientific knowledge and adoption of modern scientific methods. Using Tyco Brahe's precision measurements, the accomplished mathematician Johannes Kepler calculated and established the pattern of orbits of planets. When Robert Boyle, perhaps the first modern chemist, separated chemistry from alchemy, he laid the foundations of physical chemistry. Galileo Galilei was the first to describe the properties of gravity on earth. (Figure 1.1). He acquired the telescopes invented by Flemish scientists and made detailed observations on planets and stars confirming some of Kepler's calculations and also establishing that the earth went around the Sun. The discipline of science matured beyond the patronage of the Catholic Church. Galileo also established the use of mathematical methods as tools of science and declared that without them, *"it is humanly impossible to understand a single word of it (the language of the Universe); without these, one is wandering around in a dark labyrinth."* In the year 1600, William Gilbert studied electricity and magnetism and found that these were distinctly new phenomena. In 1687, Isaac Newton published his magnificent work '*Principia Mathematica'*, covering laws of motion and laws of Gravitation, which as we all know, is perhaps the most important scientific success. Thus, for the first time, the theory of motion of celestial objects, once an exotic field of study, was unified with that of the day to day objects, into a single mechanical theory. Benoit de Maillet proved that the earth was older than the 6000 years that the Bible declared. Robert Hooke proposed a theory of earthquakes. Thomas Burnet inspired people to start thinking of earth as not static but ever changing and encouraged secular thought about earth's history.

Figure 1.1. Gravity is a great teacher. Above: Galileo at the Tower of Pisa, showing that both heavy and light objects fall at the same rate. Below- Story of Newton understanding the laws of motion after being hit by a falling apple.

18th century, the age of enlightenment, saw the full blooming of rational inquiry and lost many vestiges of the Aristotelian era. Charles de Coulomb discovered the law that governed the force between two electric charges. Antoine Lavoisier proved that burning of objects was caused by combination with oxygen, thereby disproving the so called 'Phlogiston" theory. In 1774, he showed that the total mass of matter remains the same in every chemical reaction.

Georges-Louis Leclerc, Comte de Buffon, wrote a 36-volume book on the history of natural science, expanding on the state of discoveries and knowledge in biology and geology. Since his history of the Earth bore no resemblance to church doctrines, he was roundly condemned by theosophists. He was the first to note that the prevalence of biological species depended on climate and geography. The development of powerful optical telescopes in this century, would lay the foundations of serious observational astronomy.

These fundamental principles and developments in science made it possible for science to explode in explaining nature in detailed ways and also helped to usher in the Industrial Revolution that changed human culture, economics and living conditions. Alongside the scientific enlightenment, competing approaches of capitalism, socialism and communism were replacing feudalism. Colonial powers began to establish far away presences and Europe benefited from this greatly. European culture and technology influenced the colonial world and changed them forever. One of the key achievements of Colonialism is that it also brought the scientific knowledge and methods of modern science to the colonies, where they have taken hold. As stated before, the history of science is much too vast in scope and depth to adequately describe in this book. It is sufficient to state that these discoveries have made the world more or less completely dependent on science and its fruits in every conceivable aspect, and in doing so, have also impacted the spiritual and religious life of people.

## Modern Science and Its Impact

Compared to other areas of human enterprise, modern science and its children, technology and information, have grown at fantastic rates beginning from the heady days of the 19th century. Fed by the symbiotic relationship between science and technology, the rate of acquisition of scientific knowledge has been exponential in the visceral and literal sense. The actual growth cannot really be quantified, since knowledge of science defies a proper measure. There have been two types of development and the following is just a mention of some aspects and a broad summary:

### Development in the Civilian Sector

The early 1900s, particularly the 1930s, saw an explosion of discoveries in physics and it is said that it is the period when even a third rate physicist could do first-rate science. Discoveries of theory of relativity, the full understanding of electromagnetic phenomena, development of quantum theories and the advance of mathematical physics were phenomenal achievements. Since the 1950s, physics has advanced far beyond the understanding of even many professional researchers in the field and years of training in a specific field of physics or even a specific topic would be needed to get a clear grasp. Starting from thinking of matter as atoms and forces as disparate, physicists now understand that the cosmic beginnings are related to the model of particles (quarks, leptons and neutrinos) and know how these particles get their properties. They have identified the nature of fundamental forces and know that these are but the manifestation of a single force that existed at the dawn of the Universe. The understanding of cosmology, galaxies, stars and black holes has developed rather quickly over the recent decades and there is unprecedented confidence that we are closing in on some basic truths about the origins of this Universe. This has been made possible by fantastic developments in spacebased and terrestrial telescopes, space missions and major advances in theory. Except for the specific uses of nuclear technology in science and medicine, most of today's modern technology is reaping the benefits of physics discoveries (mostly in electromagnetism) from the early 1900s, only now. One can barely imagine what impact present day physics discoveries will have in the future. For example, it is expected that the exotic aspects of quantum mechanics, such as quantum entanglement, are likely to revolutionize information technology. Physics, like other fields, has branched off into many subfields; condensed matter, high energy, nuclear, astronomy, astrophysics, plasma physics, surface physics, biophysics and so on. While discovery of nuclear energy as a viable source was hailed as a breakthrough in the 1940s, it has now become highly controversial because of lack of scientific solutions to radioactive waste and also the guarantee of accident-free operation. Radioactive isotopes, on the other hand, have contributed greatly to medicine and material testing. The contribution of physics to diagnostic modalities in medicine, such as Magnetic Resonance Imaging (MRI), X-ray CAT scans and Positron Emission Tomography (PET) have revolutionized the medical field.

Chemistry and biochemistry kept pace with developments in physics and the field became prolific in its impactful discoveries. Chemistry has spawned its own branches with organic and polymer chemistry changing the face of technology. Physics and chemistry, in their applied form, are more closely linked to real world applications. Condensed matter and surface physics and polymer science have given rise to a rich field of material science where every day new materials with amazing properties are discovered. Semiconductors, nanotechnology and modern day ceramics are some examples. Superconducting materials are making their mark in various domains including medicine and electricity generation and transmission. Plasma physics forms the base of the nuclear fusion energy research and lasers are finding broad applications in medicine, manufacturing and information technology.

Biology has made the maximum progress in the late 20$^{th}$ century and the revolution is ongoing. Advances in biology and biochemistry have provided a deep and unparalleled understanding of living beings and their environment. We understand animal and our own bodily processes in great depth and the diagnostic modalities have enabled an unprecedented view of the functioning or malfunctioning of our bodies. 50 years since the identification of DNA as the code for constructing life forms, genetics and molecular biology have become the bread and butter activity, similar to what mechanics is to physics. Biology and biotechnology are now enjoying the boom that physics experienced in the early 1900s. While fundamental research does go on in biology, such as on the origin of viruses and their impact on human evolution, the relationship between particular sets of genes with traits and inherited diseases, neurobiology, theoretical biology etc., much of the research emphasis, to the disappointment of some, is on finding applications. This has resulted in a contrast between basic research in physics and biology, with physics discoveries exceeding far beyond any applications on the horizon. While its impact is mostly in medicine and preservation of species, it is expected that there will come a time when biotechnology plays an important role in information technology.

Large science projects such as particle accelerators and colliders, space missions and space stations, nuclear fusion experiments, the human genome project and brain research initiatives have been enabled by the technologies described above. These are intimately connected to advancements in engineering design and fabrication with high precision and quality. Massive amounts of data that are crucial for large projects, are now routinely handled, stored, processed, retrieved, searched and analyzed in such projects and this too has been made possible both due to developments in materials and in mathematics and software. The inventions in information generation, transmission and processing needs of the physics community, combined with inventions in computing technology enabled by condensed matter physics, have given rise to what is one of the greatest scientific/technological revolutions- the internet, which has become THE medium for families, workplace and for operating any system. Internet has become so ubiquitous that children from the age of 3 are engaged in it and older people depend on it for everyday life. This facilitation of daily life and increase in productivity are because of initiatives in basic sciences.

What has inspired the common public of all ages is science fiction. Space science and technology has grown in the fertile imagination of the public and so, the race to the moon, mission to Mars and pictures from far away planets bring the public closer to exotic places which they might have thought and dreamed about. Although space related research and development have culled out their own fields, these are actually systematic applications of basic sciences and are technology based. On the other hand, it is also true that space research and various space missions are often for the purpose of advancing basic research.

Today, high technology combines science in its core and it is hard to tell where the science behind the technology begins and ends. It is safe to say that the scientific world looks very different from the one that was experienced in the early 1900. Though it is impossible to determine which discovery in which era is more important, recent progress has made the scientific quest very worthwhile and has been of great benefit to society. But, like in all human efforts, there is also a dark side.

## Development in the Military Sector

There is no doubt that several inventions such as gun powder and cannons were meant for military purposes. But, more often than not, science discoveries are quickly co-opted for use as weapons and for purposes of war. While many nations and populations would consider this application of science as legitimate and investments in science and technology for the sole development of weapons as necessary, it is fair to say that science is not originally intended for this purpose. The inhuman weapons of mass destruction, though enabled by science, are here out of the fears and ambitions of the rulers and it is evident that despite the recognition of the nature of these weapons, it is not only hard to eliminate them, but it is even harder to disengage from research on them. Even scientists engaged in weapons research would consider this research as an evil necessity. Since insecurity of people and propensity of empires and governments to dominate their region or the world, make it easier to get funding for weapons and instruments for war, scientists may instigate and engage in this research, either sharing in the insecurity or the ambition. But, often, scientists justify this research because they see this as advancing overall science, albeit at the price of developing inhuman instruments of war. The societies and governments carry out this work as preventive and defensive, partly because it is true and partly because it is more palatable. In any case, there is a positive feedback between military application and science, with military applications arising out of scientific discoveries and at the same time motivating scientific research.

Whether justified or not, science has been behind every step of the way in military developments. It is no secret that advanced military technology is more or less a guarantee of dominance. This dominance has changed balances of power and the course of history in many regions. As shown in the opening scenes of the film, "2001 Space Odyssey", human ability to convert ordinary items into weapons is an almost natural sign of intelligence. But humans went on to develop sharpened stones and sticks, poisons to be placed at the tip of darts, clubs and projectiles for aggression. During the development of military equipment, the science of metallurgy was at the forefront with copper, iron and then bronze providing increasingly effective weapons in strength and lightness. This was evident in weapons such as bows and arrows with improving shapes and efficacy, swords, spears, tridents, maces, shields and armors, flame throwers and catapults and in the use of horses and elephants as rides and for chariots. Though, it had earlier existed in Greece and Rome, the sophisticated Greek Fire, an incendiary mixture of resins, sulfur and perhaps saltpeter or quicklime, was developed and used in flamethrowers, grenades or just plainly poured on, in the 7$^{th}$ century by the Constantinople based Byzantine empire with devastating effect. The Greek Fire was so feared that it was linked to heavenly punishment and its formula was a state secret for over a century. Chinese developed the gun powder for non- military purposes and this was used by them and the Arabs for warfare, only later. For a long time, gun powder was used in ceramic pots, and guns became cheaper for common use only in the 15$^{th}$ century. The Mongols and Ming warriors developed cannons which brought them many victories and changed the history of the world.

The conversion of cargo ships into naval vessels had happened long ago with the first naval war being between the Hittites, an eastern Mediterranean empire stretching over most of Asia Minor, and the Cypriots. During the middle ages, tough and large ships which could carry over 1000 tons and capable of long voyages were developed. Being able to accommodate cannons, these became formidable instruments of war, occupation and colonization. At the beginning of the 20$^{th}$ century, after the invention of steam engines, steam ships provided the basis for armored steel construction and heavy cannons. Later ships using internal combustion engines ushered in the era of modern vessels, the first of the "motor ships" being the Russian vessel Vandal, built in 1903. The first submarine to be developed was in the service of James I by a Dutch engineer in 1620, but the fore runner to modern day submarines was built by the French physicist Denis Papin. In 1776, an oar powered- one-man submarine was deployed in America and various designs of submarines were developed relatively slowly. The first engine propelled submarine using compressed air was built by the French in 1863 and a 62 ton steam powered submarine in 1885 by collaboration between an English and a Swede engineer. The first fully equipped electrically powered submarine was the French Gymnote in 1888. Development of larger and more advanced submarines armed with more sophisticated sonar and torpedo missiles have continued as with all other military transports.

Though balloons had been used by China as early as the 3$^{rd}$ century, and in the American civil war, the first use of airplanes for battle was in an Italian- Turkish war in 1911 and soon after that Bulgarians bombed Turks from the air in 1912. United Kingdom and Germany were well ahead in air force technology in early 1900 and it was only later (around 1918) that US would catch up. Initially most of the airplanes were wooden biplanes, but by 1939 metal frame planes had become available. After that, the superiority of air power became extremely important and so, research in aeronautics and missile technology became a high priority. The needs of World War I and II, on both sides of the wars, were central to the development of aeronautics and airplanes and the speed and later transition of airplanes to civilian transportation. The need for maneuverability of the planes for bombing missions also spurred the development of nimble planes. The naval forces utilized air power when aircraft carrier vessels were developed and specialized as torpedo bombers and divers during World War II.

While gun powder based explosives were used for warfare as early as 1221 in China, low explosives using charcoal or aluminum powder and potassium nitrate are an example of early explosives used in fireworks and in warfare. The development of nitroglycerine in 1843 by an Italian chemist marks the beginnings of high explosives with large destructive power. The powerful trinitrotoluene (TNT), invented by the German chemist Joseph Wilbrand, forms the standard for explosive power. The invention of the blasting cap by Alfred Nobel enabled the use of these high explosives in a safe way for both peaceful uses and for attacks. In 1867, Nobel also invented the dynamite, which is nitroglycerine bound in earth or sawdust for safe handling and which requires an external igniter such as the blasting cap. Modern day bombs are made with TNT or RDX (an organic compound). Their shapes, the shells, the detonator etc. are specially designed for specific purposes.

The testing of the atom bomb and its use by the U.S. military to bomb Hiroshima and Nagasaki in Japan, mark sharp turning points in the use of science ~~to~~ for weapons. As Robert Oppenheimer quoted from Hindu scriptures, the atom bomb represented God-like power: *"I am become death, destroyer of worlds"*. (Winkler1993). But today, the atom bomb and its successor the hydrogen bomb are key parts of the arsenal of advanced countries and the proliferation of nuclear weapons has not stemmed because of the extraordinary power and (seeming?) security they bestow. The arsenal is justified by the powers as a deterrent. Nuclear energy is much less controversial for military than for civilian use and is used to power battleships, aircraft carriers and submarines. Clearly military technology changed immensely and the world became a much more dangerous place with the invention of nuclear power. Again to quote Robert Oppenheimer, *"In some sort of crude sense, which no vulgarity, no humor and no over-statement can quite extinguish, the physicists have known sin and this is a knowledge they cannot lose."* and *"The atomic bomb made the prospect of future war unendurable."* (Winkler 1993). In a rueful statement, Albert Einstein said, *"Today the atomic bomb has altered profoundly the nature of the world as we know it, and the human race consequently finds itself in a new habitat to which it must adapt its thinking."* (Einstein 1946). His expression of regret was clear when he said, *"The release of atomic power has changed everything except our way of thinking ... the solution to this problem lies in the heart of mankind."* (Einstein1956).

One great benefit that came out of military application is the science and technology of radio and radar communication. The growth of the entertainment and communication industry and society's ability to deal in information transmission in such volume and the growth of the electronic industry is due to the devices and methods developed during World War I and II. Many of the newer discoveries such as lasers and fiber optics also rose in tandem with civilian and military applications. The other military need that spurred scientific and technology development, is aerospace applications for intercontinental ballistic missiles and satellite observation and communication.

Key developments in physics, chemistry and material science have been behind each of these technologies. Examples are advances in fluid dynamics, mathematical physics, neutron physics, chemical analysis and the development of new chemical compounds. Modern science has brought in an era of global knowledge, global technologies and global co-thinking. But, also because of the scientific nature of this era, societies are changing rapidly. While one value system would call this good and another bad, the fact remains that this impact on society will inexorably continue. As the late Czech President Vaclav Havel said,

*"The dizzying development of this science, with its unconditional faith in objective reality and its complete dependency on general and rationally knowable laws, led to the birth of modern technological civilization. It is the first civilization in the history of the human race that spans the entire globe and firmly binds together all human societies, submitting them to a common global destiny. It was this science that enabled man, for the first time, to see Earth from space with his own eyes; that is, to see it as another star in the sky."* (Havel 1994). But he also laments that, *".. the relationship is missing something. It fails to connect with the most intrinsic nature of reality and with natural human experience. It is now more of a source of disintegration and doubt than a source of integration and meaning. It produces what amounts to a state of schizophrenia: Man as an observer is becoming completely alienated from himself as a being."* Nevertheless, Havel remained correctly hopeful when he went on to say *"...Paradoxically, inspiration for the renewal of this lost integrity can once again be found in science...a science producing ideas that in a certain sense allow it to transcend its own limits....."* This statement points to the fact that science, at the hands of a wise and knowledgeable society, can transcend limitations of science, religion, language and even basic human needs to achieve a harmony that eludes us today.

## NATURE OF MODERN SCIENCE AND ITS METHODOLOGY

Science is the field nurtured both by nature that is astoundingly open to inquiry and human curiosity that is astoundingly eager to inquire into nature. As Einstein said, *""The most incomprehensible thing about the world is that it is comprehensible."* (Polkinghorne 2011), "*...why is science possible at all in the deep way..? Of course, evolutionary survival necessity can be expected to have moulded our brains ...to make sense of the world of very day experience. However, our human ability to understand the subatomic quantum world, totally different from the macroscopic world of everyday happening and requiring counterintuitive ways of thinking for its understanding is another matter altogether...The fact is that the universe has proved to be rationally open to human enquiry to a very remarkable degree.*". Science along with mathematics has shown that the Universe and the order that it incorporates, are intellectually elegant, visually and viscerally beautiful and spiritually awe inspiring. To quote John Polkinghorne,"*The universe has proved to be not only rationally transparent, making deep science possible, but also rationally beautiful, affording scientists the reward of wonder for all the labours of their research.*" The openness of the universe and physical phenomena invites humans to the field of science. In paying homage to this awesome nature, the human ideal of scientific inquiry incorporates curiosity and a reverence to the truth behind objects and phenomena. But, human weaknesses, imperfection and incompetence make for a flawed pursuit. Therefore, the discipline of science has been developed and continues to be perfected by humans, almost by a process of evolution, to be as close to the truths as possible and to remain in the realm of what is rationally open about the universe. The Renaissance period, frustrated with many conflicting ideas, with no way to select between various notions and theories, developed the process that modern science uses today. While the process, methodology and mental approach of the practitioners are discussed more in detail in a later chapter of this book, here is a brief description to orient the readers and hopefully the chapters illustrate this methodology.

All science derives its primary motivation from curiosity. But the ideal character of scientific inquiry is marked by the discipline of inquiring into it, while making a minimum number of assumptions, presumptions, premises and hypotheses and while rejecting a "received" knowledge, or dogmatic ideas that are just passed on without questioning. The general marks of a good scientific discovery are as follows. Reproducibility, so that any competent person wishing to do so can reproduce the results, verifiability, so that a concept, an idea or a theory can be verified by observations and mathematical derivations, and consistency so that no observed physical phenomenon or theoretical foundation contradicts it. The process of discovery is rarely this ordered, but once a discovery or scientific knowledge is declared by the originating scientist to be ready for dissemination, it has to meet these criteria. It is accepted that objectivity, lack of bias towards a particular outcome and watchfulness to detect and prevent any errors are characteristics of a good scientist. While these principles are observed in both, fundamental research and applied research take somewhat different routes. The scientist working in fundamental research has the charge of discovering a new thing or phenomenon, explaining an unexplained phenomenon or how a thing behaves. The applied scientist goes forward from the fundamental discoveries and applies it to different needs to come up with a package of knowledge that can be used, for example, in engineering. The line between fundamental and applied research is blurred but each has its own clear domains too. While particle physics belongs in the fundamental category, physics of particle accelerators belongs in the applied physics. While molecular biology is a fundamental field, the application of molecular biology to food, agriculture and medicine is applied biology. Sergey Kapitza states in his presentation at the UNESCO World Conference on Science, *"In fundamental research the motivation is the search for the unknown, for new knowledge, to reach an understanding of nature, of the human being, for that matter. This is the main stimulus, a powerful and very personal force, driving the scientist against all the odds of nature and the established wisdom. In other words, the scientist is always a dissident and often his dissidence goes beyond science. The applied scientist, the engineer, is in general a master of compromise, a member of a system, rather than a lone individual, the majority of one, who is right."* (Kapitza 2000). A practicing researcher usually prepares for a scientific career for over 25 years, including years of doctoral and post doctoral work. This is an investment of time and dedicated effort with no guarantee of fame or fortune. While Kapitza's statement is by and large true, scientists are also human beings subject to the forces of their upbringing, pressures of their living and society and have individual perspectives and therefore, there is a continuum of variation in how a scientist works. Though a scientist ought to be focused on knowledge, modern scientific race pits one scientist or science team against another and therefore there is always competition, because the only prize a scientist seeks is a high recognition for the discovery. The higher the recognition by the general and scientific community, the higher is the competition. The positive side of this competition (and this is why the system survives) is that this has created a peer review system to weed out erroneous ideas. Although imperfect, unwieldy and haughty in the sense that the review can and may reject some good ideas based on establishment prejudices, the review ensures that a lineage of sound and successful scientific ideas is established. The peer review is like a trial by a jury of the peers, but is not very good at detecting scientific fraud. (Rennie and Gunslaus 2008). But the continued application of a given discovery or a concept is guaranteed to uncover any flaws, and this, in turn, is brought on by the emphasis on consistency between various results. The lineage of scientific concepts ensures that an entrant into a research field does not have to (but can if he or she wants to) go back and reconfirm all the previous findings and the foundation of the field. For example, if a physicist is looking into how a gas mixture flows through a certain system, he does not have to re-derive all the fluid equations that were derived in the last several centuries. Every scientist (including Newton, Einstein and Darwin) stands on the shoulders of scientists who came before them. But in the subject of science, every element of science is perched on all of the science and therefore, the soundness of the enterprise is ensured by the high level of quality of what is universally accepted. Even when accepted, a scientific idea or discovery is subject to rejection, if convincing and opposing evidence comes along. The scientific enterprise depends upon this since, otherwise ideas would become dogmatic and doctrinaire and this would be totally against the spirit of modern science. Aristotle and the later physician Galen's ideas, which reigned for over a millennium are examples, which were weeded out by the process of scientific selection.

Scientific research on any particular sub field is divided into theoretical and experimental areas. The preparation of the scientist is different in each, the theoretical emphasizing areas of mathematics, statistics, conceptual thinking and modeling and the experimental research emphasizing instruments, observations, statistics and understanding of the environment of the experiment in detail. While theoretical work is intensive in its own subject and would only borrow analogous ideas from other fields, a typical experimentalist would have to be well versed in a variety of physics and engineering topics, for example, vacuum, optics, electronics, structural details etc. in addition to having expertise in his/her own field. But both theoreticians and experimentalists have to be completely aware of the status and progress of each other's domains, at least in their common field. Contrary to the incorrect colloquial usage, "it is only a theory", a sound theory satisfies accepted norms of science, including consistency with observations and experiments.

This arrangement of training, methods, materials, approaches and philosophy of work has withstood the rigors of scientific pursuit and has unquestionably yielded many answers and solutions that society and scientists seek. The solutions have been satisfactory and useful for the most part and the fact that science returns to correct itself when necessary, makes it a reliable field for society to invest in. Most of the scientists would refuse to pursue anything other than science, if they have reasonable freedom in their work. The thrill of discovery is not restricted to the great minds that make great discoveries, but is felt by every scientist. Even the news of a discovery by someone else creates great excitement and stimulates every one. The joy and awe felt by scientists is comparable to the descriptions and promises in religion and the fact that this joy can be felt during their lifetime is, indeed, a blessing for the scientists.

## PUBLIC PERCEPTION VERSUS INFORMED OPINION ON SCIENCE

Surveys indicate that public perception of science is predominantly positive. For example, the Pew Research Poll of July 2009 finds that *'Americans believe overwhelmingly that science has benefited society and has helped make life easier for most people.*' Among broad segments of Americans, more than 8 people in 10 say that science has a positive impact. Even more than science, scientists are highly regarded and are believed to contribute a lot to society. (No similar survey from an Eastern country is available to the author, but it is likely that public sentiments are not dissimilar). Despite the positivity, one has to be concerned about how much of this view is based on their actual knowledge of science. It is not surprising that with the technology surrounding them, and their health and even life in the hands of modern scientific medicine, people in advanced countries see science as a positive element in the society. While space explorations have sparked the imagination of people, particularly children as they dream and think about new worlds beyond the frontiers of earth's sky and cheer space exploration, one has to ask if they truly understand the limitations on space voyages and the sacrifices and choices the society has to make for a truly momentous space mission. The same is true for the support for large science projects. The recent discovery of the Higgs Boson, the so called 'God Particle', in the European Large Hadron Collider (LHC), inspired broad interest in science among people. (Figure 1.2). Many believe that this has to do merely with the association of this research with the word "God". This rings true because the forerunner of the LHC, the prestigious Supercollider project in the US, which could have made this discovery sooner, was abandoned and very few in the public lamented that or even took notice. There are several instances where the larger public shows approval or disapproval after the fact but do not show intellectual interest or civic engagement in science endeavors.

Figure 1.2. Mass for the mass.

As Bruce Lewenstein states in his presentation at the above cited UNESCO World Conference on Science, *"Some analyses of large quantitative surveys suggest that, although scientists perceive science as open investigation of nature, public perceptions of science have more practical images in mind; public perceptions view science like medicine, an applied field that uses knowledge of the natural world to yield specific practical benefits."* (Lewenstein 2000). While, this may be good news for applied science, the broad field of science suffers from the fact that people don't know that the pursuit of science is not primarily for discovering better tools for life. Most scientists enter into the field because of a relatively pure interest in the science itself. (The subject of the book is, of course, that science and its approaches themselves are great tools for life).

There is agreement among scientists that they enjoy the admiration and respect of general public and note that the public have come a long way from the image of the "mad scientist" of the 19$^{th}$ century, fostered by Mary Shelley's Frankenstein. Yet the very same scientists see that trust in science is in serious decline and the public ambivalence about science is at an all-time high. There may be several reasons for this. One counterintuitive reason is the success of science in generating practical applications and technology. Familiarity breeds contempt. As people become too familiar with day to day technology they stop wondering about its discovery and the science that brought it. When we flip the light switch and the room is flooded with light from a bulb, the number of physical phenomena, the surface physics of the contacts of the switch, the electrical and solid state properties of the wire material, the heating of the filament or the ionization of the gas in a fluorescent bulb and the lattice or atomic processes that lead to the emission of light, the physics of the light itself and finally the science of vision, are enormous and barely muster notice of the regular user of the device. For the lay public, an item developed from a scientific discovery is at first an object of excited curiosity, a brief item of wonder and when the workings of it exceeds the immediate understanding, it becomes just a useful object for routine use and not worthy of inquiry. The television and the mobile phone are indispensible items for a modern day person. But an average user is not very interested in learning the workings of it, let alone the science discoveries that lie behind the parts, because he or she is not interested enough to spend the required time.

Sergey Kapitza notes that "*...the rate of growth of the scientific endeavour has gone far beyond the level of understanding not only of the content of science but of its deeper message as a contribution to modern culture. Mesmerized by the immediate usefulness of science, under the influence of the short-range pressure of the market, of advertising salesmanship of the media and of the pragmatic trends in modern education, most people are lost in the brave new world of science and technology."* This statement summarizes the state of public involvement with science, very well. While the public is attracted by the glamour and becomes habituated to the comforts from technology, the technological product becomes raison d'être' for science in society and science becomes the job of trained professionals, much like any other professionals.(To draw a contrast, no average believing person thinks that religion is only a job for the religious professionals).

The lack of training in scientific thinking during school years and particularly during college and the absence of easy engagement in natural sciences, leave people far behind in inquiring about the technology they use. A child in the elementary school is extraordinarily interested in scientific topics. But the system of education and the societal values are such that by the time the children are in the middle school years, many of them see science and mathematics as difficult topics and eventually come to fear them. This persists as people grow into adulthood to the extent that people believe that science can only be understood by the "brilliant" people, which is far from the truth. Of course, later this fear of scientific knowledge leads to distrust of science, in the absence of a broad understanding of science.

What is even more frustrating is that applied scientists do such a good job of creating packaged information that technologists can create products without understanding the science. The magnetic resonance imaging often uses superconducting magnets, a device with high physics pedigree. The engineers that work in the superconducting magnet factory, know nothing about superconductivity, a wonderful field of physics, and can still do a great job of designing and fabricating it. One can see that today's information technology, which has earned the arguable title of high technology, has engineers and technologists who know little of the discoveries in electromagnetism, particle physics, quantum mechanics, material science and mathematics, that are behind every nook and corner of the devices they work on or use. This lamentable lack of scientific knowledge in technologists is symptomatic of the greater problem for science in society.

Daniel Yankelovich, referring to the most important issue in this situation states that during the middle of the 20$^{th}$ century, in the U.S., a social contract was developed between science and society creating a separateness and autonomy for science. (This autonomy was more or less set in place by the U.S. Government in the middle of the cold war competition). His belief is that this social contract has enabled science to pursue long term research and build up new knowledge base gradually. He states, "*.But this same social contract is responsible for the widening disparity between the sophistication of our science and the relatively primitive state of our social and political relationships.... Now, 20 years later, both the successes and the price tag of this social contract have grown. Science has reached greater heights of sophistication and productivity, while the gap between science and public life has grown ever larger and more dangerous, to an extent that it now poses a serious threat to our future."* He states further that *"In today's public domain, scientists are highly respected but not nearly as influential as they should be. In the arena of public policy, their voices are mostly marginalized. They do not have the influence due to them by virtue of the importance and relevance of their work and of the promises and dangers it poses for our communal life...".* According to him, the reason is that*"....rest of society operates out of vastly different worldviews, especially in relation to assumptions about what constitutes knowledge and how to deal with it. Scientists share a worldview that presupposes rationality, lawfulness, and orderliness. They believe that answers to most empirical problems are ultimately obtainable if one poses the right questions and approaches them scientifically...*" (Yankelovich, 2003).

Yankelovich's statement is as much about science as it is about scientists. The lack of influence is also due to a lack of trust in many sections of the society. As the gap between science and public knowledge has grown, the distrust has also grown. But this is, at least an undesirable, and, at worst, a dangerous trend because as Kapitza says, *"When the gap is so great, people not only cease to understand any more what is happening, they lose trust, then get afraid and reject much that comes with science. Finally, the void left is filled with pseudoscientific, mystic or fundamentalist ideas, taking us far back in time and history."* It is also true that the previously stated autonomy has, in a way, spoiled some in the academic community, who would rather be left alone and not be bothered to bring scientific information to the public. So, it has become fashionable to write publications in complicated and jargon filled prose that is deliberately made inaccessible to the public. (See later discussion). In this context, scientists like Jacque Cousteau, David Attenborough, Carl Sagan, Stephen Hawking and Brian Green are to be applauded. There are many more scientists interested in informing the public about the science, but few get the opportunity, mainly because of public apathy.

Increasingly, societal solutions are based on science and range from choices in medical treatments to energy production to dealing with climate change to dealing with demand for food and water. In addition, the pursuit of scientific goals for fundamental and applied research requires increasingly larger public expenditures and the public must weigh in on these knowledgeably rather than allow politics to guide such decisions. Examples of these abound: The public was not engaged in the decision to cancel the American Supercollider project, but the European public was, and despite delays and overspending, the LHC is a great success and brings prestige. The Human Genome Project and the Hubble telescope, generally unknown to public, were funded and have been a great success. A past political climate practically stopped stem cell research. (In the absence of a public engagement on the topic, the stem cell research was hijacked by the above quoted religious fundamentalists). Today, decisions, such as whether to send a mission to Mars or to link rivers in India for preventing floods and drought, choosing whether to use nuclear energy for reducing greenhouse gases, or whether fracking for oil is desirable, require engagement by a knowledgeable public.

There are also considerable differences in perceptions of science, in different parts of the world. This indicates that appreciation and perception of science has local context. In traditionally rooted countries like India, science is dichotomized- science that is not concerned with personal life, such as basic physics, chemistry and microbiology, and science that requires one to make a choice in attitudes and practices. The former is accepted largely and even respected and admired by many. But when it comes to understanding and choosing "Western" medicine (medicine that has been discovered by the methods of modern science), there is often a conflict with traditional medicine which has not met the standards of science. Western medicine might be distrusted because "they have side effects" or "many Western medicines are recalled", while home or traditional remedies that have only anecdotal or hearsay efficacy might be accepted unquestioningly. In the case of the recall of a drug, the public in these countries fail to recognize that recall proves that the modern medical system works, even if in some cases, the recall happens a little later than needed. A similar distrust of science can be seen wherever science comes into competition with traditional practice, say in nutrition or social practices that are scientifically unhealthy or discriminatory or raising children. This distrust of the "Western" science can spill over to philosophical debate about how scientists cannot know what they claim to know. Further, since Eastern cultures emphasize holistic learning, realization of truth through meditation and mind control over body, the highly categorized, classified, specialized and impersonal system of modern science is resisted.

Brian Green stated in a New York Times article, "*It's striking that science is still widely viewed as merely a subject one studies in the classroom or an isolated body of largely esoteric knowledge that sometimes shows up in the "real" world in the form of technological or medical advances. In reality, science is a language of hope and inspiration, providing discoveries that fire the imagination and instill a sense of connection to our lives and our world.*" (Green 2008). Science is not only for the classrooms and also not only for scientific professionals who gaze into their telescopes or microscopes or their navel. Science enriches the otherwise everyday life. Again as Brian Green stated, *"Like a life without music, art or literature, a life without science is bereft of something that gives experience a rich and otherwise inaccessible dimension."*

Figure 1.3.Scientist's love of science.

## COMPARISON WITH PUBLIC RESPONSE TO RELIGION

A good comparison can be made with public response to religion. Except in some European countries and some far Eastern countries, religion is a dominating influence among a majority of the public and they in turn form a majority of the people of the world. The unquestioning adherence to religious practices, faith in the will of God, trust in the counsel of priests and scriptural guidance in these regions of the world, point to a situation in which science might want to be. This is illustrated by a humorous quote by the Nobel laureate Leon Lederman: *"Physics is not a religion. If it were, we'd have much easier time raising money."* (Lederman and Teresi 1993). Science and religion are both engaged in the goal of enlightenment and harmony that comes from understanding. But sometimes science and religion stand in the way of each other. While many people do see a serious and fundamental conflict between theology and scientific understanding, the conflict that matters for our discussion arises out of the poor accessibility of science and its demand of unconditional rationality with few promises for the future, compared to the gentle (a more recent adaption of religion) and humane persuasions of religion and its promise of salvation, demanding in return, only unquestioning faith. Science's demand for a certain amount of scientific knowledge in the average person contrasts with religion's allowance of easy association requiring no training or 'do what you can' theme, but pointing to peace and other worldly returns. For the pure scientist, there is not much leeway in belief. The best a scientist can do is to say 'I don't know but I will start at this unknown point to proceed and see if it leads to a rational result confirmable by observation or analytical means.' A religious person can declare that it is the will of God and proceed from any point. For example, they can say that God created my house and He wills me to make alterations to it. This would seem to be totally consistent for them. This religious allowance makes for easier living and the attitude worked for people before science had permeated lives. As stated before, this attitude when combined with ignorance or ignoring of scientific information, leads to incorrect and harmful personal and social decisions.

Faith is the antithesis of rational (skeptical) inquiry. Even though daily material needs are the primary preoccupation and science overwhelmingly caters to them, spiritual needs have a higher value once material needs are satisfied. (Figure 1.3). Therefore, even as the public satisfies their basic material needs, their spiritual needs take them to religion which asks them to have faith. Faith offers easy access and a peaceful and harmonious break from the daily grind. Even the discipline and routine that religion demands have a calming effect and promote a socially friendly environment. For most, religious pursuits do not cost any money and may even offer financial incentives. Science, on the other hand, offers answers only through the difficult and time consuming process of learning complex concepts, mathematics and discipline of inquiry. Often it is a lonely quest, since it is difficult or expensive to find teachers who will teach with patience. One can see then why people, with little time left from their hard lives for any serious learning, choose religion over science in obtaining answers. There is, however, a parallel to science in religion, particularly in eastern religions. Realization in these religions comes through years of meditation and practice of exercises, chants etc. and abjuring of worldly desires. In Christian, Islamic and Jewish faiths, the rigorous study of scriptures and strict adherence to a religious discipline is required for the priestly class. In such cases, exactly as in science, religion draws a very small number of people. The penalties for non-observance or ignorance of religious matters are also easy to understand and are themselves supernatural. For human beings, who would like to believe in immortality or second chances through reincarnation or existence beyond the body, rewards and penalties are great- heaven and hell, rebirth as a superior or inferior being, consorting with gods and angels or being with God Himself or Eternal Fire and Damnation. The acquisition or lack of knowledge of science does not offer any such extraordinary rewards or punishments, but its rewards are wisdom, spiritual satisfaction and harmony when combined with right attitudes and respect from others. (In some ways, the rewards of scientific learning are like what Zen Buddhist practices promise). Failure of learning science is failure of leading a life without these rewards and perhaps being a cause or part of a failed society or civilization.

So, perhaps one can just emulate religion's approach and offer scientific concepts on faith and have scientific congregations. (There are science fiction movies such as Planet of the Apes and other television episodes where a bomb or computer is worshipped). But, this is against the creed of science itself. Trust in science should not be handed over blindly and should come with the knowledge of what is good science. Today, the media may state two sides of a scientific issue as if they are equal even when there is a logical side with preponderance of supporting evidence on one side. In the absence of the hard work of verification and analysis of the information by the media, the public has to undertake this effort and force politicians to make the right decisions. Science gives only probabilistic results and cannot decide what is good for the society. In the democratic society that the world has largely become, the responsibility of making choices for the society that are increasingly scientific, lies with the public and scientists can only assist in that effort.

***

A person with little knowledge of the origins of science might actually approve of today's situation where the public is relatively disengaged with science. But, in reality, science is a human quest which is part of everyone, in lesser or greater form. It is cultural in the deeper meaning of it, like art and spirituality are. It is well known that even as the lay public shies away from advanced knowledge in science because of the difficulty in learning, it yearns for that knowledge and respects those who have it. Though the educational system and the job market do not demand it, every person needs to have a good sense of science and a scientific attitude in order to serve and be part of the society, especially in current times where science is needed in all walks of life. But science has advanced very far and a person, with familiarity with science at high school level, would not know where to start, if he or she wants to know more. This book is written on the premise that people will trust science if they know the important foundational material that informs them in their deeper questions. These questions are the same whether they are asked by a person, religion or science. Once aware of these foundations, an average citizen can make rational choices. Also, once these basic concepts of science are understood, the excitement of science and thirst for more knowledge are guaranteed to be felt by the reader. The book is also written with the assumption that such foundational material can be made accessible to the reader.

With the unique approach of describing the brief history, the methods and the content in each topic and then describing its value in religious or social perspectives, the book is a guide to understand the importance of science in one's personal life and how integral it is to that life. This way, the book attempts to achieve parity of science with religion which, in many countries and societies, has a tremendous influence on people's thinking and their daily life. The book demonstrates to the readers that science progresses with a great deal of integrity in methodology and arrives at conclusions that are robust and yet open to modification, if it is called for. By familiarizing the readers on the detailed science of major discoveries in major areas, the book is intended to persuade the reader that science takes the high road to finding knowledge and therefore is to be trusted in its conclusions, when one is confronted by apparent contradictions or denials from personal, cultural and religious perspectives. Even for the section of people who are very religious, it is shown that, in most areas, religion has updated itself so as not to contradict science.

The main body of the book takes the reader through 5 science topics of interest to society and religion: evolution of species, medicine and healing, order and disorder, mechanics and space-time, and cosmology. The book presents the historical background and introduces the fundamental scientific concepts involved, with details to get a good understanding of the topic without overwhelming the reader. The material covered is adequate for the reader to understand where the science came from and where it is now and to be able to ask intelligent questions for further learning. This way, the book gives a starting point for the reader who might otherwise be bewildered by the complexity of the topics.

The book is meant for readers with a sound high school education. The content is not written as in most popular science books, which more often than not, give an incorrect picture of the scientific detail or oversimplify with improper analogies. This book gives not only an overview but examples of detailed science to get a clear understanding. For a beginner in science, the book will be a powerful start but will demand studious attention. Readers with college level science education will be able to better access the two topics of mechanics and cosmology because of the mathematics in them. However, readers, who are intimidated by the mathematics, may read the text and skip the mathematical steps given in the boxes without much loss of broad understanding. (It is strongly suggested that readers without a firm background in mathematical concepts, first go through the primer on the calculus, given in an Appendix). Readers with good knowledge in a topic will get a good view of the science in other topics. With each topic, the conflict and agreement with Christianity, Hindu and Islamic viewpoints, are discussed. The book will only note or summarize scholarly discussions and will present the much simpler public discourse and opinion influenced by religious belief and any scholarly influence. The final two chapters discuss the agreements and conflicts between science and religion and how science can be applied in daily life and in social issues. It is shown that science has its own spirituality and deserves greater influence in society, at least, at par with religion. The last chapter advocates the adoption of the scientific attitude in individual, public and corporate matters, since this leads to the way of conducting one's life with integrity and respect. Furthermore, despite polarizing debates in the media, a scientific attitude is, more or less, in agreement with religious attitudes.

Francis Bacon's philosophical work "*Novum Organum*" was the revision to Aristotle's "*Organum*" and detailed the new scientific method, which emphasized the importance of experimental science. It rejected the un-argued doctrinaire knowledge as non scientific. His following statement on "idols" is a forerunner of the present thinking on how scientific knowledge is impeded:

*"Idols of the Tribe are rooted in human nature itself and in the very tribe or race of men. For people falsely claim that human sense is the measure of things, whereas in fact all perceptions of sense and mind are built to the scale of man and not the universe..... Idols of the Cave belong to the particular individual. For everyone has (besides vagaries of human nature in general) his own special cave or den which scatters and discolours the light of nature. Now this comes either of his own unique and singular nature; or his education and association with others, or the books he reads and the several authorities of those whom he cultivates and admires, or the different impressions as they meet in the soul, be the soul possessed and prejudiced, or steady and settled, or the like; so that the human spirit (as it is allotted to particular individuals) is evidently a variable thing, all muddled, and so to speak a creature of chance. There are also Idols, derived as if from the mutual agreement and association of the human race, which I call Idols of the Market on account of men's commerce and partnerships. For men associate through conversation, but words are applied according to the capacity of ordinary people. Therefore shoddy and inept application of words lays siege to the intellect in wondrous ways. Lastly, there are the Idols of Theatre which have misguided into men's souls from the dogmas of the philosophers and misguided laws of demonstration as well; I call these Idols of the Theatre, for in my eyes the philosophies received and discovered are so many stories made up and acted out stories which have created sham worlds worthy of the stage."*

In modern science, overcoming these idols is perhaps the first step to becoming scientifically aware. The worldwide correlation of independent and cooperative development with advances in science, demonstrates that advancement of a civilization, in its culture and prosperity is closely related to the advancement of scientific knowledge. It is for the people to be aware of this deep influence and know enough so that they can guide the progress of society. This book provides a very good foundation for this awareness.

## REFERENCES

Bacon, F.; *"New Atlantis",* Editor: Alfred B. Gough, Oxford University Press, London, U.K. (1915), p.xii.

Einstein, A.; *A Documentary Biography* by Carl Seelig, the English translation by Mervyn Savill, Staples Press (Indiana University), Bloomington, IN (1956), p.223.

Einstein, A.; "*Atomic Education Urged by Einstein*", *New York Times,* New York, NY (23 June 1946).

Green, B., *Put a Little Science in Your Life,* New York Times, New York, NY (June 1, 2008).

Havel, V.; Physics and Society Newsletter, Americal Physical Society, Vol 23, Number 4, Oct 1994.) (Etext: http://www.aps.org/units/fps/newsletters/1995/october/coct95.html#a1), Speech when receiving the Philadelphia Liberty medal, Philadelphia, PA (July 1994),.

Kapitza, S.; Proc. World Conference on Science, Editor: Ana Maria Cetto, Banson, London, UK (2000), p. 287.

Lederman, L. and Teresi, R.; *"The God Particle: If the Universe is the Answer, What is the Question"*, Mariner Books, New York, NY (1993).

Lewenstein, B.; Proc. World Conference on Science, Editor: Ana Maria Cetto, Banson, London, UK (2000), p. 287.

Polkinghorne, J. Science and Religion in Quest of Truth, Yale University Press, New haven, CT (2011),pp. 71-73.

Rennie D. and Gunsalus C.K.; *"Fraud and Misconduct in Biomedical Research.",* Edited by Frank Wells and Michael Farthing, Royal Society of Medicine Press, London, U.K. (2008)), pp.13-31.

*Russell B.; The Impact of Science on Society*, AMS Press New York, NY (1968), p.7. Also, Russell, B.; *The Impact of Science on Society - B. Russell", (eBooks and Texts) https://archive.org/details/TheImpactOfScienceOnSociety-B.Russell.*

Salam, A; *Renaissance of Science in Arab and Islamic Countries*, Editors: Dalafi, H.R and Hassan., M.H.A.; World Scientific Press, Singapore (1994), p. 14-17.

Winkler, A.M., *Life under a Cloud,* Oxford University Press, New York, NY (1993), p. 38,.

Yankelovich, D.; *Winning Greater Influence in Science*, Issues in Science and Technology, Richardson, TX, (Summer 2003), Etext: http://issues.org/19.4/yankelovich.html.

*Chapter 2*

# EVOLUTION: HOW LIFE AND WE CAME ABOUT

"As buds give rise by growth to fresh buds, and these, if vigorous, branch out and overtop on all sides many a feebler branch, so by generation I believe it has been with the great Tree of Life, which fills with its dead and broken branches the crust of the earth, and covers the surface with its ever-branching and beautiful ramifications." - (Darwin 1872)

"Evolution could so easily be disproved if just a single fossil turned up in the wrong date order. Evolution has passed this test with flying colours." – (Dawkins 2009)

Few accounts on the history of evolutionary theories can surpass that of John W. Judd's "*The coming of evolution: The story of a great revolution in science*" (Judd 1910a). In the early passage in his book, Judd states,

> "It will be seen from these considerations that in attempting to decide between the two hypotheses of the origin of species—the only ones ever suggested—namely the fashioning of them out of dead matter, or their descent with modification from pre-existing forms, we are dealing with a problem of much greater complexity than could possibly have been imagined by the early speculators on the subject.
>
> The two strongly contrasted hypotheses to which we have referred are often spoken of as 'creation' and 'evolution.' But this is an altogether illegitimate use of these terms. By whatever method species of plants or animals come into existence, they may be rightly said to be 'created.' We speak of the existing plants and animals as having been created, although we well know them to have been 'evolved' from seeds, eggs and other 'germs'—and indeed from those excessively minute and simple structures known as 'cells.' Lyell and Darwin, as we shall presently see, though they were firmly convinced that species of plants and animals were slowly developed and not suddenly manufactured, wrote constantly and correctly of the 'creation' of new forms of life."

The classic yet simple language of the early $20^{th}$ century befits the grandeur of this subject. Readers would do well to read this beautiful narration of the history of the Theory of

Evolution. Below is a much briefer and hopefully simpler account, written primarily for the purpose of explaining the basics of the development of natural selection, followed by sections on the Theory, the evidence and consequences of natural selection, and the modern theories of evolution. Hopefully, this will elucidate the scientific process which, under the pressure of various non-scientific theories and doctrines, won the acceptance by the world, except in some more dogmatic sections of society. The history of the progress of the theory of evolution also is an example of the vigorous and rigorous process of science.

The 17$^{th}$ century Archbishop of Armagh, James Ussher declared that the earth was created at 9:00 AM on October 23$^{rd}$, 4004 B.C. This confident proclamation came from his study of the Bible and counting of generations from Adam and Eve to his time. Needless to say, in his time, there were considerable numbers of voices critical of it, but unfortunately, they were about his calculations and not his premise. Even more importantly, this declaration of the Archbishop was because of the Man-centric assumption that Man was created directly and hence the Universe more or less at the same time as humans (or according to teleologists, for humans). This is to be expected because, in those days, life and its source were strictly interpreted by Judeo Christian religion in the western world, according to the "*Scala Natura*-The Great Chain of Being", a construct of Plato and Aristotle. The "chained" people believed that God created life forms starting from the simplest insect going to the most complex- The Man, nearly simultaneously, and that for 6000 years those life forms had been unchanging and as naked, as furred, as beaked or as finned as the day God created them. This was consistent with the established order in their faith and literature, like the Christian hierarchy of angels and other beings in the heaven.

The 18$^{th}$ century biologist, Carolus Linnaeus was one of the first in the Western world to create taxonomy (a system to document and classify animal and plant species) diligently, but he did not analyze their origins and history. He classified forms as kingdoms, *Regnum Animale*, *Regnum Vegetabile* and *Regnum Lapideum*, based on whether it was animal or vegetable or a mineral. He further classified these into subgroups. Today, the naming of an organism by its "genus" and the "species" follows his method, derived from the Anglican Priest John Ray's proposal. The classification is not based on any ad hoc arrangement, but is based on the detailed knowledge of anatomy of many species and their related characteristics and behavior. The classification of Linnaeus is shown in figure 2.1. In parallel, another approach, not inconsistent with Linnaeus taxonomy, an Evolutionary Tree and, in principle, rooted in the Chain of Being, was proposed by Edward Hitchcock in 1840. (Archibald 2009). This Tree of Life or phylogeny described the relationship between species. Charles Darwin also used an evolutionary tree in his Origin of Species to explain speciation and this approach is used in limited ways. Linnaeus was seriously troubled that new plant species (hybrids) could be produced by cross pollination, because this did not fit the prevailing ideas of simultaneous creation of all life forms. But he appeared to have reconciled these forms as potential organisms in the Garden of Eden, which come into existence later in time. Linnaeus anticipated the concepts of the struggle of life forms for survival and nature's way of weeding out the unfit life forms. Yet, he believed that all this was according to the divine order.

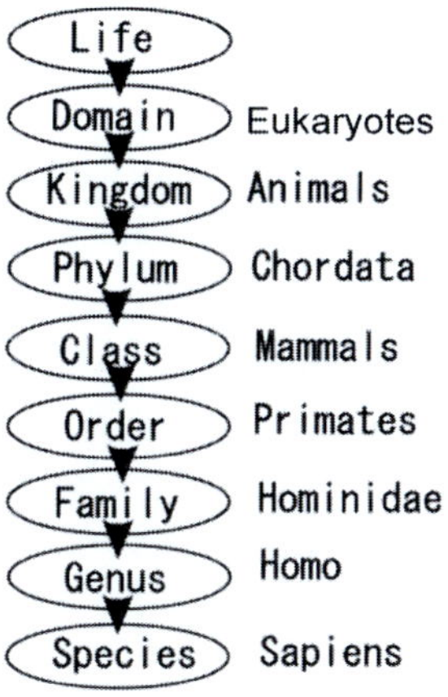


Figure 2.1. Classification system of Linnaeus and John Ray. (Chordata refers to the type of spinal cord of nerves and backbone and tailbone etc.).

## EARLY EVOLUTIONARY CONCEPTS

In the second half of the 18th century, much against this established order of biology that was dictated by theosophical considerations and orthodoxy of the prevailing science, a number of scientists started to advocate that things were different and living forms changed over time. Foremost among them was the Comte de Buffon, George Louise Leclerc, a naturalist, mathematician, a cosmologist par excellence and an accomplished writer. 100 years before Charles Darwin, Leclerc wrote his brilliant 37 volume treatise *Histoire Naturelle,* questioning everything about the prevailing notions in Biology, including Biblical statements. He bobbed and weaved against church and establishment opposition and continued his writings. His works were very widely read and followed to such an extent that he is sometimes called the father of Natural History (Mayr 1982). He was, according to Darwin, *"the first author who has treated evolution in a scientific spirit"* (Darwin 1866). Leclerc speculated in the then frontiers of human evolution, such as thinking that humans originated in Asian Caspian region, 6000 years back, and that the earth was a lot older, perhaps 75000 years. Leclerc stated that species can change over time, but not into other species. While he was willing to admit that an ape and man could be of the same family, he did not countenance the idea that humans descended from apes or the other way round. (Figure 2.2).He was the first "transformist" before Darwin, because he stated that all the quadrupeds of the world descended from an original set of 38 quadrupeds.

In the early 19th century, Jean-Baptiste Chevalier de Lamarck and George Cuvier clearly delineated the branch of biology and were responsible for initiating ideas on inherited characteristics. Lamarck's ideas were new and radical but many were wrong, such as the idea that an animal gradually acquired traits, such as the long neck of a giraffe for better survival, and passed on its improved trait to its offspring. Couvier corrected many of these ideas. Couvier was a first rate scientist and an expert on dinosaurs and held the now proved opinion that, the dinosaurs died out because of catastrophes and new species emerged in their place. Charles Lyell was a famous geologist, a lawyer and some of his ideas anticipated Darwin and Wallace's theory of evolution. After examining European fossils, he concluded that, contrary to previous ideas, most changes in species were gradual and slow. He drew the same conclusion for the geology of the land in his three volume *Principles of Geology* (1830-1833), stating that all changes were gradual due to the wearing by air and water movements, in addition to a few catastrophic events like earthquakes etc. The “Uniformitarianistic” idea of James Hutton that geological changes that occurred in the past are the same that are happening now, along with this concept of species evolution, was placed on a firm footing by Lyell. Though both Couvier and Lyell implied that there was a close similarity between changes in geology and biology, neither of them expressed their belief in evolution of species into another. While Lyell did believe in the evolution of species, he was reluctant to express it, as shown in his statement in a letter,

> “If I had stated….the possibility of the introduction or origination of fresh species being a natural, in contradistinction to a miraculous process, I should have raised a host of prejudices against me, which are unfortunately opposed at every step to any philosopher who attempts to address the public on these mysterious subjects” Lyell 1881( quoted in (Judd 1910b)).

Lyell would change his mind later after Charles Darwin himself was inspired by Lyell, as he acknowledged it to Lyell's father-in-law,:

Figure 2.2. “Human” Evolution.

> "I always feel as if my books came half out of Lyell's brains and that I never acknowledge this sufficiently, nor do I know how I can, without saying so in so many words—for I have always thought that the great merit of the Principles, was that it altered the whole tone of one's mind and therefore that when seeing a thing never seen by Lyell, one yet saw it partially through his eyes." (Darwin 1844).

## Charles Darwin, Theory of Natural Selection and Evolution

Charles Darwin was very much the grandson of the creative and intellectual poet Erasmus Darwin, who himself believed in the theory of Lamarck, probably because it was radical. But unlike his grandfather, Charles had to prove things to himself and was therefore, foremost, a scientist. Neglecting his medical education, he developed a passion for natural science and its methodology. In 1831, at the young age of 22, Darwin set sail on the ship Beagle to map out the geology of different places. (We notice, like Charles Lyell, how Darwin too was connected to the study of life forms through his interest in geology). He visited ports in nearly all the countries of South America, Brazil, Argentina, Patagonia, Falkland islands. Uruguay, Chile, Peru etc. During his voyage, Darwin studied marine vertebrates and their behavior, the distribution of different species and the relationship of various species to their environment. He dissected specimens and studied their anatomy and the distribution of characteristics of different species. After nearly 9 months of travel, the ship Beagle arrived in Galapagos Island, where, over 5 weeks, Darwin saw, observed and wondered about the variety of species in this small island, such as giant tortoises and "clumsy and ugly" iguanas. Because of the rather special nature of the island and its relatively large distance from other lands, Darwin could assess the effect of the local environment on the development of species. He was to come to the realization that the environment had a strong connection to the type of species present there. He had discovered that, in this small island, there were 13 species of finches whereas, on the mainland, there was only one. Each of the island's species was differentiated by the beak. (See figure 2.3). Though he knew that was significant, he would figure out the reasons only later as adaptations to occupy specific ecological niches and to maximize survival, in a process called Adaptive Radiation. He was puzzled by the fact that the mockingbirds (he called them mocking-thrushes) on Charles Island, on Albemarle (now called Isabela), and those on James and Chatham Islands were all different.

Darwin agreed to the request of the Beagle captain Robert FitzRoy to write part of the captain's book detailing the Beagle's two voyages. He had enthusiastically made copious field notes and had been sending many letters and a sort of journal to his family, in order to give an account of his travels. So, it was not a difficult task to write the journal. The 3rd volume of the 4 volume book, *The Narrative of the Surveying Voyages of His Majesty's Ships Adventure and Beagle* carried Darwin's "*Journal and Remarks, 1832—1835"*. An example of his writings is below, (It is on the subject of giant tortoises in the Galapagos Islands);

> "The tortoises ...During the breeding season, when the male and female are together, the male utters a hoarse roar or bellowing, which, it is said, can be heard at the distance of more than a hundred yards. The female never uses her voice, and the male only at these times; so that when the people hear this noise, they know that the two are together. They were at this time (October) laying their eggs. The female, where the soil is sandy, deposits them together, and covers them up with sand; but where the ground is rocky she drops them indiscriminately in any hole: Mr. Bynoe found seven placed in a fissure. The egg is white and spherical; one which I measured was seven inches and three-eighths in circumference, and therefore larger than a hen's egg. The young tortoises, as soon as they are hatched, fall prey in great numbers to the carrion- feeding buzzard. The old ones seem generally to die from accidents, as from

falling down precipices: at least, several of the inhabitants told me, that they never found one dead without some evident cause." (Darwin 1839a)

Figure 2.3. Darwin's finches drawn by the ornithologist John Gould (Illustration from Voyage of the Beagle). Source: Wikipedia Free Public Image.

On his findings about the finches, he wrote, *"It is very remarkable that a nearly perfect gradation of structure in this one group can be traced in the form of the beak, from one exceeding in dimensions that of the largest gros-beak, to another differing but little from that of a warbler."*(Darwin 1839b).

This book became extremely popular and Darwin became well known among his peers, almost a celebrity. Thanks to Darwin's later fame, the Voyage of the Beagle would be a historically famous document and a popular read even to this day. Darwin's father organized his funds and Darwin became a self -financed scientist and rose among the ranks very quickly. With the help of Charles Lyell, Darwin worked with various institutions and his peers like Richard Owen at the Royal College, who studied the fossils collected by Darwin. All these associations and results of the studies gave Darwin insights that he could not have got by himself. In February he was elected to the Council of the Geological Society of London. By March 1837, Darwin was well along into thinking that one species may change into a different one-transmutation of species. He reasoned that this must be the reason for the geographical distribution of rheas, guanacos (wild relative of the llama) and the mockingbirds in South America.

Darwin presented his observations during his voyage, to the Geological Society of London. He showed the animal and bird specimens and the puzzle of the finches. An ornithologist John Gould, fascinated by Darwin's account, stopped all his other work, collected the bird specimens and analyzed them and made the famous declaration that the blackbirds and finches were actually ground finches of a distinct group of 12 birds. He also found that the mockingbirds were each of a different species. In 1838, Darwin read Thomas

Malthus's essay called *"Principles of population",* in which he observed that human population doubled every 25 years and that it would continue to do so. The only things that could curtail the population growth would be disease, war and especially a lack of sufficient food supply. Darwin came to realize that this thesis was applicable to all life forms. He surmised that the species that survives would be the one that can nourish itself from what is available from the environment, feed their offsprings and protect themselves and the offsprings from elements and predators. He reasoned that species that developed traits which helped them to make good use of the available environment, would be the fittest to survive in large numbers. A favorable variation of traits would make organisms survive better and if this variation could be passed along to their offspring and at the same time unfavorable variations are lost, then the species as a whole would thrive in the specific environment. This adaptation to specific environment, he called, "wedging", referring to the fact that the species inserted itself into the wedge of a survival zone. Later, he would identify the pressure to adapt, as the "pressure for natural selection". He wrote that the *"final cause of all this wedging, must be to sort out proper structure, & adapt it to changes",* so that *"One may say there is a force like a hundred thousand wedges trying force into every kind of adapted structure into the gaps in the economy of nature, or rather forming gaps by thrusting out weaker ones."* (Darwin 1838, quoted in (Levine 2011)).

For the next 15 years, Darwin continued his quest and research on the origin, distribution, variations, reproductions, inheritable characteristics and other aspects of life forms that make them distinct species. He had befriended the botanist Joseph Dalton Hooker who became his close ally and consultant. At one stage, Darwin was distracted from his work on the evolution of species as he worked feverishly on barnacles and found that their "homology" reminded him of the Galapagos finches and saw similar patterns in distribution, based on environmental pressures. The sexual development of barnacles also gave him insights into sexual selection. In 1853, The Royal Society of London awarded him a medal for his work on barnacles. Much to the annoyance of his friends, he guardedly published his findings on the evolution of species and even then in bits and pieces, although he wrote copious letters to his friends like Joseph Hooker. To some extent, his secrecy had to do with his thoroughness, although the fact that he published all his work as soon as he realized that he had competition, might indicate a personal motive. In April 1856, for the first time and to his first audience, Charles Lyell, Darwin explained the theory of natural selection, during Lyell's visit. Later Lyell exhorted Darwin not to delay and to publish all his results. In May 1856 Darwin wrote an outline of his ideas and in October, he started to write his major work, "The Origin of Species".

## ALFRED RUSSELL WALLACE AND THEORY OF NATURAL SELECTION

Unlike the independently wealthy Darwin, Wallace was born in 1923 in a poor family in Wales, descended from the famous Scottish rebel William Wallace, the Braveheart. His family circumstances prevented him from getting much formal education. Despite that, his passion for natural sciences kept him learning and exploring. In 1845, Wallace moved to Neath, Wales where he avidly read the book *Vestiges of the Natural History of Creation* by Robert Chambers (the authorship was revealed only in the 12$^{th}$ edition of 1884), a radical theory of evolution, much against church teaching. The following is a quote from this book, insightful for that period.

> "Not one species of any creature which flourished before the tertiary (Ehrenberg's infusoria excepted) now exists; and of the mammalia which arose during that series, many forms are altogether gone, while of others we have now only kindred species. Thus to find not only frequent additions to the previous existing forms, but frequent withdrawals of forms which had apparently become inappropriate — a constant shifting as well as advance — is a fact calculated very forcibly to arrest attention. A candid consideration of all these circumstances can scarcely fail to introduce into our minds a somewhat different idea of organic creation from what has hitherto been generally entertained". (Chambers 1884).

The book even connected human intellect as an evolutionary advancement of animal instincts. Wallace then came to consider evolution as a serious natural phenomenon.

Wallace had a friend Henry Bates, who had interests and a background in education which were very similar to Wallace's and both were inspired by the work of Alexander von Humboldt and Thomas Robert Malthus. Bates had introduced Wallace to entomology- study of insects. A person with an adventurous spirit, Wallace was inspired by W. Edward's book on a voyage along the Amazon River and suggested to Bates that they should embark on a trip to the Amazons to collect specimens of various kinds that were in great demand for purchase by rich collectors. Unemployed and penniless, both boarded a ship in April 1848, from Liverpool, England. After a year of exploring the region of Belem near the mouth of the Amazon River together, the two parted ways to explore on their own. For another 3 years, Wallace collected specimens of insects and butterflies and birds. Unfortunately, much of his specimens that he sent back, were destroyed in a ship fire. In 1852, because of poor health, he returned to Britain, but on the way, he was almost lost to the sea and survived on leaky boats until his group was rescued by a cargo ship. Fortunately, his notes and a few specimens he had sent separately, survived the journey and these became the source for his two books and several publications after his return. (Figure 2.4 shows a butterfly, an example of Wallace's findings). The insurance payment he received was helpful in continuing his work and to pay for his living expenses. The Royal Geographical Society recognized his work and with this recognition, Wallace was able to collect funds to pursue his interest in the study of species and geology.

In 1854, he went on an 8 year long study in the Malay Archipelago (present Indonesia and Malaysia). Here he collected a very large number of specimens and at the same time started publishing a number of articles on zoology.

In 1855, several months before Darwin started his writings, Wallace had already written about his discovery that every species had come into being from a closely allied species. On January 8, 1858, his thirty-fifth birthday, Wallace arrived on an island Ternate' in the Maluku group of islands and would stay there for 3 years. His momentous participation in the discovery of evolution would happen here. Soon after his arrival here, Wallace was struck with the idea of Natural Selection that would help him understand the phenomenon of speciation- development of a new species from a preceding one. He was suffering from a rather severe bout of fevers due to a resistant strain of Malaria and perhaps this helped him in this discovery. He wrote that this origin of species came about by a progression and divergence of a species from its predecessor through struggle for existence. This discovery of "divergence" was a key finding that Darwin had missed. Considering Wallace's integrity as a scientist, this astounding discovery was a tremendous insight rooted in the analysis of his specimens. Whatever it was, it was certainly a much quicker acquisition of knowledge than Charles Darwin had, and was from an original and lonely quest. But it is a matter of great joy and affirmation that two scientific theories from two independent minds and approaches, Darwin's, through slow, methodical, cautious and extensive studies using many other people's contributions, and Wallace's, through a single handed and focused analysis, using self taught science, resulted in the same discovery. In 1858, learning of similarity between his ideas and Charles Darwin's, Wallace sent a letter which stated the summary of his thoughts and findings and immediately Darwin saw the conjunction of ideas and also realized that he himself could not delay the publication of his work. Darwin, with the knowledge of Lyell and Hooker acknowledged Wallace and sent a set of 3 papers to the Linnean Society with both Darwin and Wallace as authors. These were published as "*On the Tendency of Species to Form Varieties; And On the Perpetuation of Varieties and Species by Natural Means of Selection*", By Charles Darwin, Esq., F.R.S., F.L.S., & F.G.S., and Alfred Wallace, Esq. Communicated by Sir Charles Lyell, F.R.S., F.L.S., and J. D. HOOKER, Esq., M.D., V.P.R.S., F.L.S, &co. in the Proceedings of the Linnean Society issue of August 1858. It would seem that Wallace was not consulted in the sending of this publication and Wallace criticized that action. Following that, instead of waiting to complete the voluminous book that he was planning, Darwin published his so called "Abstract" (not in the meaning it carries at present), of the Origin of Species.

Figure 2.4. Wallace's Golden Birdwing Butterfly (Ornithoptera Croesus). Similar to the one that Wallace cataloged. Source: Wikipedia Free Public Image.

Wallace, on his part, recorded his thoughts in his autobiography and the quote below shows his intelligent quest.

"My paper written at Sarawak rendered it certain to my mind that the change had taken place by natural succession and descent - one species becoming changed either slowly or rapidly into another. But the exact process of the change and the causes which led to it were absolutely unknown and appeared almost inconceivable. The great difficulty was to understand how, if one species was gradually changed into another, there continued to be so many quite distinct species, so many which differed from their nearest allies by slight yet perfectly definite and constant characters. One would expect that if it was a law of nature that species were continually changing so as to become in time new and distinct species, the world would be full of an inextricable mixture of various slightly different forms, so that the well-defined and constant species we see would not exist... The problem then was, not only how and why do species change, but how and why do they change into new and well-defined species, distinguished from each other in so many ways; why and how do they become so exactly adapted to distinct modes of life; and why do all the intermediate grades die out... and leave only clearly defined and well-marked species, genera, and higher groups of animals... And the answer was clearly, that on the whole the best fitted live. From the effects of disease the most healthy escaped; from enemies, the strongest, the swiftest, or the most cunning; from famine, the best hunters or those with the best digestion; and so on. Then it suddenly flashed upon me that this self-acting process would necessarily improve the race, because in every generation the inferior would inevitably be killed off and the superior would remain - that is, the fittest would survive." (Wallace 1905). (Not in copyright).

Wallace continued his work in Malay Archipelago, resulting in one of his other contributions, "the Wallace Line", the geologic divide that runs between Bali and Lombok islands, and Borneo and Sulawesi islands and forms the eastern edge of a sudden disappearance of Asian species and correspondingly the western edge of the appearance of Australasian species. On his return, Wallace published the book, The Malay Archipelago, a celebrated travelogue, ranked as one of the best.

## THEORY OF NATURAL SELECTION

The evolution of the theory of evolution and natural selection is a continuing one and it mirrors the evolution of life forms it attempts to describe. The survival of the ideas (and "memes", using the terminology of Richard Dawkins) in the environment of rigorous and strenuous examination, the ending of one "species" of ideas giving rise to a new branch of the knowledge tree, the divergence of ideas into many branches, while still somehow connected to the trunk of the tree, the sudden innovations and discoveries that give rise to new ideas and explanations etc., are astonishingly similar to the evolution of life and the mechanism of natural selection. One sees that this is also true for many branches of life and science. One then comes to believe that Life imitates Science.

The biographical narration given above was intended to convey to the reader, how the path for this major theory was laid out. Personal efforts, collaborations, contributions from different fields of science had to come together to form this theory. There had been individual ideas like that of Linnaeus, Leclerc, Lamarck, Lyell and others, but the theory of Natural Selection as embodied in the work of Darwin and Wallace, marked a paradigm shift, based on a large body of accurate and convincing evidence in the form of specimens and observations and an analysis that was undeniably logical and irresistible. This type of work, though supported by statistics, is very different from that of discoveries in physics, where the tool of mathematics stands along with concepts and provides the methodology for proof. Here, as in general biology, the empirical evidence and a unique explanation that explains widely, consistently and logically, form the basis of acceptance. Like in physics theories such as Special Theory of Relativity, the strength of the theory remains in the fact that no single incontrovertible evidence has been found to disprove the philosophical and scientific foundations of this theory. Again, this is not for lack of trying by the members of the field as well as opponents of the theory.

Darwin and Wallace understood that Natural Selection is the operating mechanism of evolution. They knew that, while breeding (mostly by sexual mating) would increase the population with traits of the parents, environmental factors, competition, predation and catastrophes were primarily responsible for reduction and control of the population. Taking the example of a housefly, a fly lives about 30 days and a female can lay 500 eggs every 3 days, after mating once. If all eggs were to hatch and the flies were to live to full life, without predation and environmental factors there would be something like a billion trillion flies in a matter of few months. While Darwin knew that Natural Selection was operating and Wallace identified the source as divergent variations in traits, they did not exactly know how. Now we know that it is through the selection of genes that determine the traits.

Ernst Mayr has summarized Darwin and Wallace's theory of evolution and these are as follows (Mayr 1985):

- If a species is fertile enough, it will produce offsprings and the population will grow.
- Resources such as food are limited but relatively stable over time.
- A struggle for survival follows in presence of competition and when resources diminish.
- Individuals in a population vary significantly from one another and provide advantages and disadvantages in a given environment of resource and predation.
- This variation is inherited.
- When there is a struggle to survive because of resources or domination of predators, the process of natural selection is this – individuals that are more suited (with suitable variation), survive and reproduce and the less suited (with unsuitable variations) do not survive and do not reproduce.
- The variations accumulate over generations and over a period a new species forms.

In the following, an abbreviated description of the theory of evolution is given, some following Darwin and Wallace, but mostly, that which has emerged over the last 150 years. The theory now integrally involves the advances in biology, particularly the discovery of cellular mechanism, the DNA/RNA, the gene, the aspect of how proteins are the intermediaries in the mechanism of life processes and how a process, initiated by matching pair of genes called alleles, determine traits. Therefore, first a description of these biological subjects is in order.

## CELLS, CHROMOSOME AND DNA

Proteins in the body of a life form are the sources of many things that happen in the body and how the body appears and functions. They also mediate the replication of genetic material and catalyze metabolic processes occurring in the body. So proteins are the biochemicals that interactively function to confer traits to a living form, as it is conceived and grows, with each protein being triggered to be formed and in turn, invoking the necessary biochemical and biological processes to create functions and organs. These proteins are large molecular chains of amino acids with thousands of types in a single body. There are structural proteins for making keratin, collagen and elastin, which strengthen and protect hair, quills, feather, nail and beak, hormonal proteins that help make hormones like insulin which regulates glucose absorption and somatotropin which regulates bodily growth. There are proteins to make enzymes that help in process of digestion, and catalysis of body metabolism. Then there are proteins to make antibodies, proteins to allow muscle function and so on. The code for the assembly (synthesis) of various proteins is written in the strands of deoxyribonucleic acid (DNA) (see below). So, it is the DNA that determines the traits.

## Biology of a Cell

Going from macroscopic to microscopic, animal organs are made from cells. The cell was discovered by Robert Hooke in 1665, facilitated by the invention of the compound microscope. He called it the cell, because under the microscope, the structure of cells looked like compartments. (See figure 2.5). He did not know these were living things. When a microscope with a magnification of 300 became available, Anton van Leeuwenhoek discovered the cell to be living by seeing a single cell organism. He called them animalcules. The cell theory took hold after many biologists found further structures in the cell and understood various functions of the cell. In 1839, Theodor Schwann, Matthias Jakob Schleiden, and Rudolf Virchow made a comprehensive description of the cell. In 1855, Rudolf Virchow declared that the cell is also a reproductive unit and cells are daughters of parent cells.

The cell is the basic structure of an animal, with specific characteristics, analogous to elements on a periodic table. All cells are living things (except those of virus), since they metabolize food to provide energy, can move and reproduce (multiply). There are two kinds of cells- The prokaryote cell of bacteria and archaea, and the eukaryotic cell – that of animals and plants;

We will take the Eukaryotic cell for illustration. The cell (see figure 2.6) consists of 3 major parts, the cell membrane, which is the wall of the cell to keep the contents of the cell together and separate from other cells, the nucleus which is the organizing part of the cell or the command center containing the DNA, and the cytoplasm in between. The cytoplasm consists of the liquid cytosol and the cell's organelles or the little machines that consume the food and manufacture proteins, keeping the cell alive. The cytoskeleton keeps the cell shape, the mitochondria converts food into the chemical ATP that provides the energy to the cells, the ribosomes are the protein factories, endoplasmic reticulum makes the fat cells and other proteins, hosts some of the ribosomes and transports materials through the cells, and the Golgi apparatus sorts and packages different materials for distribution. The Lysosome organelle destroys undesirable materials and protects the cell and the centrosomes are involved in the process of cell division and reproduction. The cytoskeleton consists of the so called fiber-like microtubules that not only help to keep the cellular structure but are also involved in transporting intercellular materials. In cell division the cytoskeleton helps to hold on to the daughter cells, orienting them in the right way by lengthening and shortening.

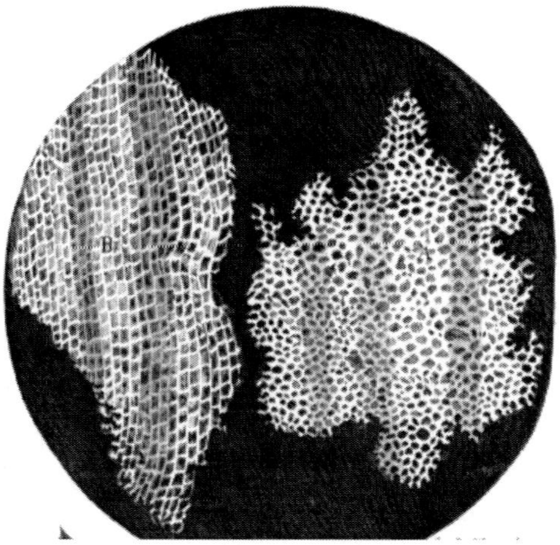

Figure 2.5. Drawing of the structure of a slice of cork (which is from the cork tree) under a microscope by Robert Hooke. His book, Micrographia, contained these and other amazing drawings detailing structure of plants, insects etc. Source: Wikimedia Commons Free Public Image.

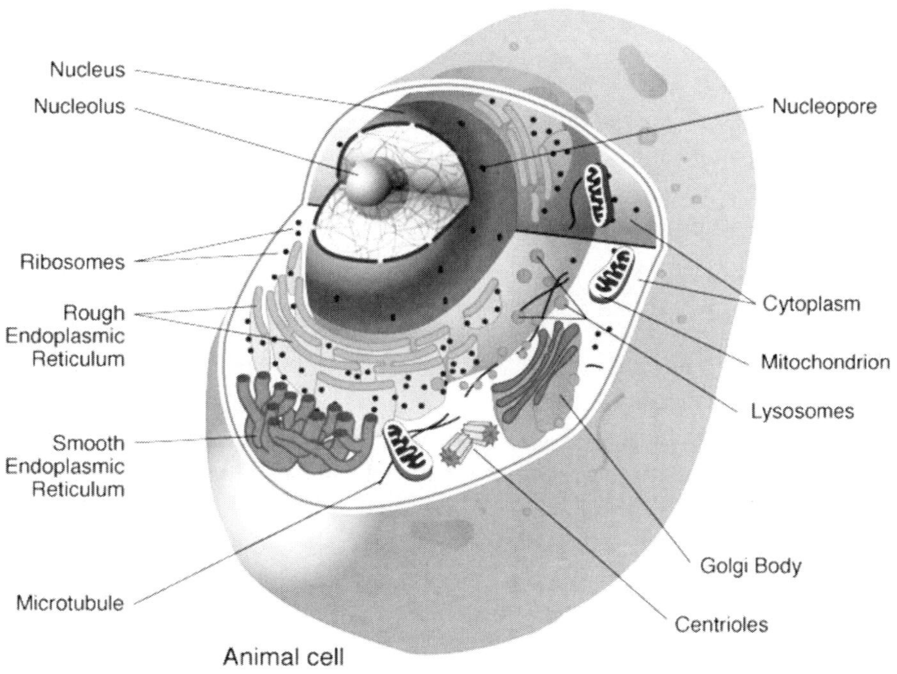


Figure 2.6. Animal cell and components. (The materials in the cytoplasm are called organelles). (Credit-US national Library of medicine. Free Use).

The materials needed for the maintenance and multiplication of cells are proteins of different kinds. For producing the proteins, the DNA in the nucleus has the instructions to determine the nature of these proteins. At first, a copy of the DNA is made as a messenger RNA (ribonucleic acid) in the nucleus. The nucleus which also has a membrane, allows the RNA to leave and carry the instructions to the ribosomes, accompanied by other RNAs that assist and carry the needed materials for the fabrication of the proteins. The ribosomes build the protein and send it to the Golgi apparatus which adds some sugar molecules, packages the proteins in a bubble sac and sends the package to a specific destination (storage or to the membrane for exiting or to the lysosome). If there is a mistake in the process, the lysosome recognizes it and destroys the newly formed protein and sends it out.

For the purposes of our discussion on evolution, the DNA in the flagellum of the prokaryotic cells and in the nucleus of the eukaryotic (plant and animal) cells are only relevant, since they are the ones that create the characteristics of the cell itself and therefore the specific traits. The cell contains two types of genetic materials- The deoxyribonucleic acid (DNA), in which the code for biological information (potential characteristics of the animals, inherited and environmental) is stored and ribonucleic acid (RNA) which can be the messenger mRNA for carrying information from the DNA to the ribosome or the transport tRNA that carries the material (amino acids) to the ribosome for the fabrication of the protein. The mitochondrial DNA is inherited from the mother while, the other DNA is a mix of the father's and mother's DNA. The mitochondrial DNA (mtDNA) is involved in coding for the above mentioned energy production and is suspected to be involved in a few diseases and the ageing process. Much of the discussion below does not apply to the mtDNA.

## The DNA

While in prokaryotic cells, genetic material is found in the cytoplasm in the form of a simple circular DNA molecule, in eukaryotic cell, the genetic material is divided into different molecular units called chromosomes inside the cell nucleus, along with additional genetic material in mitochondria, and chloroplasts in the case of plant cells. In humans, who have eukaryotic cell nucleus, the package of (nuclear) chromosomes called chromatin, contains 46 chromosomes, 23 inherited from the father and 23 from the mother. (The adders tongue fern has 1260 chromosomes. Other animal and plant species have different numbers, smaller and larger). A chromosome is a coiled structure made of DNA containing many genes and other nucleotide sequences. The packaging of the DNA into a chromosome is done by special proteins. A chromosome carries a large number of nucleotides, 100,000 to 3.75 billion depending on the species. A nucleotide is a molecule made with certain types of sugar molecules, phosphates and nitrogenous bases called nucleobases, and the DNA is strung from the nucleotides.

In 1869, soon after the recognition of the cell as the unit of life, Swiss biochemist Friedrich Miescher decided to find the biochemicals within the cell. For this he procured used bandages from a local clinic, because these contained the large pus (white blood) cells that could be easily seen under the microscope. When an alkali is added to the cell collection, the cells would burst open and spill their contents. One of the chemicals he obtained was an acid he called nuclein, because he identified it as coming from the cell nucleus. He found that, in addition to the carbon, hydrogen and nitrogen, which are common to many biochemicals, the acid contained phosphorus. He estimated it to be a compound having 29 carbon, 49 hydrogen, 22 oxygen, 9 nitrogen and 3 phosphorus molecules. (This is a gross underestimate). The importance of this acid, later called nucleic acid was not recognized for a considerable length of time. In 1879, Walter Flemming discovered that the nucleus contained material that has fiber-like structures, later called chromosomes, because they could be stained with colors. He also identified this as the material for inheritance, as he traced the process of cell division.

When it was found that the chromosomes contained the nuclein, the German biologist declared that the chemical nucleic acid (by then identified as DNA) was responsible for inherited characteristics. Most did not accept this possibility, since the chromosomes also contained the large molecule proteins, which were clearly involved in all the biological processes.

Considerable work after that revealed that the DNA consisted of 4 types of bases, one type of sugar and a phosphate. (Many molecules of these are in a DNA). Phoebus Levene of Rockefeller Institute discovered that the DNA is a linked chain of what he called nucleotides, with sugar-phosphate backbones, connected together by the bases. But no one had a clue that the DNA was a very long molecule (The human DNA molecule is 1 meter long when stretched, but is coiled to a very small size in the cell nucleus).

In 1944, Oswald Theodore Avery, Colin McLeod and Maclyn McCarty discovered that the biological molecule that was responsible for genetic transformation in bacteria was the deoxyribonucleic acid (DNA), a molecule that encodes genetic instructions used in generating proteins that in turn control the development of traits and functioning of organisms and viruses. They were experimenting on the pneumonia causing bacteria pneumococci and found that the mixing of a dead harmful variety with the harmless variety caused some genetic information to pass from the dead cells to the living bacteria and the harmless colony became virulent. By the regular scientific process, they isolated the component that made the bacteria virulent and found that this was not a protein, but the DNA. DNA thus acquired the right place in science and determination of its structure became an urgent topic of research.

Erwin Chargaff had shown that the DNAs of all species contained the same proportions of the adenine (A) to thymine (T) and the proportions of cytosine (C) to guanine (G) bases. The famous physicist Erwin Schrödinger proposed that a material that would confer traits would have to have a regular structure such as a crystal and suggested that the DNA was an aperiodic crystal. Linus Pauling, another famous chemist and biochemist with keen insights into molecular behaviors, had theorized that certain proteins and biological macromolecules would have helical structure, like many proteins that he had already proven to have helical structure. He went on to propose a structure of the DNA, which though incorrect, was amazing for its insight. The data for the discovery of the actual DNA structure was from the work of Rosalind Franklin and Maurice Wilkins, who built a dedicated X-ray lab for this purpose. Their X-ray images showed that the DNA had a helical structure. Franklin was the first to clearly understand the structure. (Pauling was denied an opportunity to visit Wilkins and Franklin's lab in England, because of his anti-war views. If he had seen their early images of the DNA, he would have made the correct discovery). Unfortunately, Franklin died before the Nobel Prize was awarded for this discovery.

James D. Watson, the biologist and Francis Crick, the physicist, joined forces to take on the challenge of determining the DNA structure accurately. They built models and looked for functions that would correspond to the known characteristics of the DNA. Using Franklin and Wilkins images and using Jerry Donohue's findings on how hydrogen chemical bonds would exist in these bases, they identified the structure of DNA in 1953.

The DNA contains the primary code of life. The genetic information is coded in the DNA as a sequence of the above quoted 4 nucleobases A,C,G and T. The DNA is a double helix of backbones, made of sugars and the A,C,G, and T nucleobases are attached to these sugars like rungs of a ladder. (See figure 2.7). A and T are complementary and C and G are complementary bases, that is, for example, A and C or A and G pair cannot form. A gene is a section or a stretch of the DNA, identifiable with the function of instigating (coding for) the fabrication of a specific protein. Therefore the gene is the unit (section of DNA) that codes for and causes a particular trait in the life form or is a "unit of inheritance". The arrangement or the sequence of the nucleobases is called the genome. The genome is therefore like a book with the letters A,C,G and T in different sequence. The gene then is like a sentence. The set of genes corresponding to characteristics of living organisms (black eyes, blue feather, short stature etc.) are called alleles.

The sequence of these 4 nucleobases along the helix forms the code of the gene which is a specific characteristic of the life form. These sequences are genes that have been inherited from parents and determine the life form's life cycle along with environmental factors.

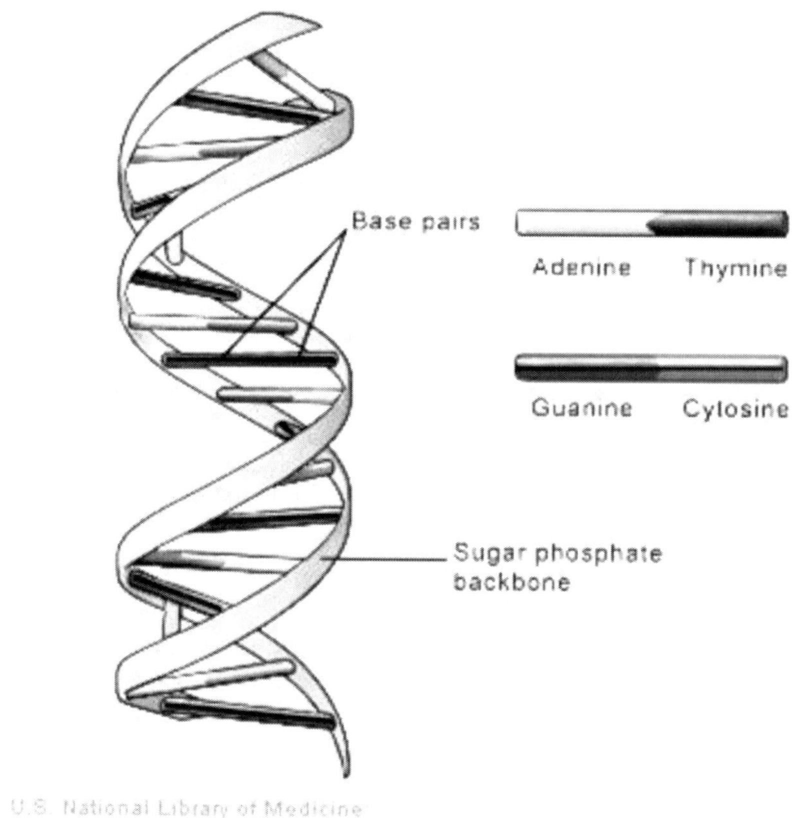


Figure 2.7. DNA double helix. Credit – U.S. National Library of Medicine, Genetics Home Reference Handbook, Cells and DNA, Dec 2013. Free Use.

In an approximate summary, a cell carries the DNA which codes for proteins. The RNA carries the message of DNA to the ribosomes of the cell, the protein factory and also carries amino acids that are required to build the proteins. The proteins manufactured in the ribosomes, give specific characteristics and traits to the organism, and also cause the functioning of the organism. Stretches of DNA which are coded for specific proteins are called genes. The role of genes is intricately connected with interspecies genetic influence. One such interaction between genes is disease. Correspondingly, the present disease model has broken away from the previous empirical concept based descriptions to structure based description. In this new paradigm, the molecular structure confers a specific mechanism to a strand of DNA or a protein. The first important step in understanding inheritance and infection, for example, is the understanding of DNA replication.

## DNA Replication

If a new cell is to be produced, host bodies like that of animals and humans do have the resources to make any kind of new cell, perhaps even that of a dinosaur. Of course, what cell is made and how it is made is determined by the DNA. Therefore, if a compatible DNA of a disease organism is introduced into the host cell and this new DNA can overtake the host DNA, it would commandeer the resources of the host cell and produce the cells of the disease organism. How this disease organism affects the host body is a matter of the property of the microorganism. Since the key to replication of lots of cells of the disease organism is the replication of its DNA, it is worthwhile to illustrate DNA replication, which is a very elegant and yet somewhat complex process. The following are the steps in DNA replication (See figure 2.8): (For a better understanding, see (Campbell and Reece 2002)).

1. The cell division process requires copying of the double helix DNA. A "helicase" protein unwinds the coiled DNA.
2. The replication starts at the so called "origins of replication", which is a single site in circular bacteria but are many in most other long chain DNA. Proteins that recognize these locations are "single strand binding proteins", in addition to the helicase. The process of copying is carried out in several of these locations simultaneously. At these locations, the helicase unzips the double helix into two strands with one backbone each and one set of bases. The binding proteins hold up the two strands preventing them from reattaching.
3. Each set of the strand has the information on what its complementary base pair is and therefore a copying mechanism will know what new complementary DNA strand is required and this is produced at every base pair. The new strand is built by building each sugar and each phosphate (of the backbone) in sequence and the corresponding base. DNA polymerase enzymes, using the triphosphate nucleotides, collect the resources and build the backbones and the bases. On one of the strands, the copying is carried out continuously towards the replication fork, but at the other strand it progresses in the opposite direction, piece by piece of the DNA called "Okazaki" fragments. An RNA primer which is an analog of the DNA is first created and put in place by an RNA primase enzyme and then this RNA primer is replaced by a corresponding DNA Okazaki fragment. This reversal of direction is required on the second strand because the sugar molecules along the backbone are in the same direction on both strands. As shown in the figure, the sugar molecule has 5 carbon locations marked 1' to 5'. (See figure 2.9). New backbones can only be added to the 3' side and this requires that the copying be in antiparallel direction in the second strand.

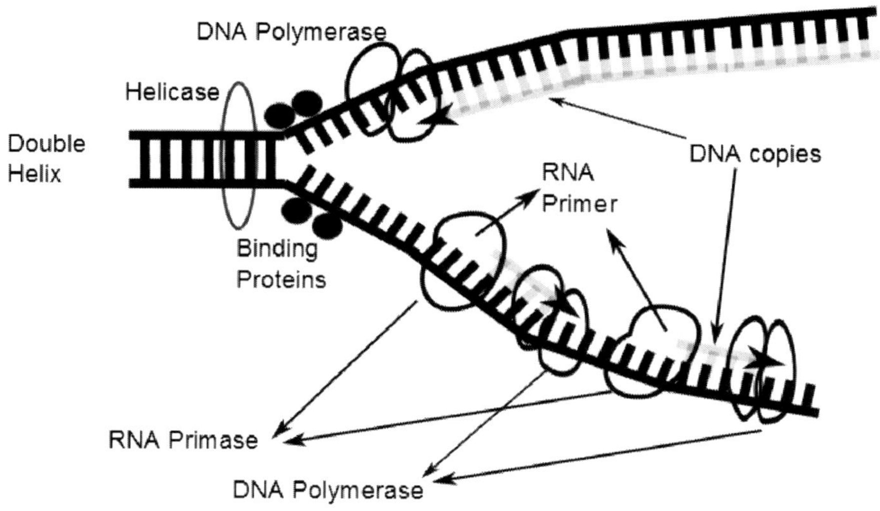


Figure 2.8. DNA Replication process.

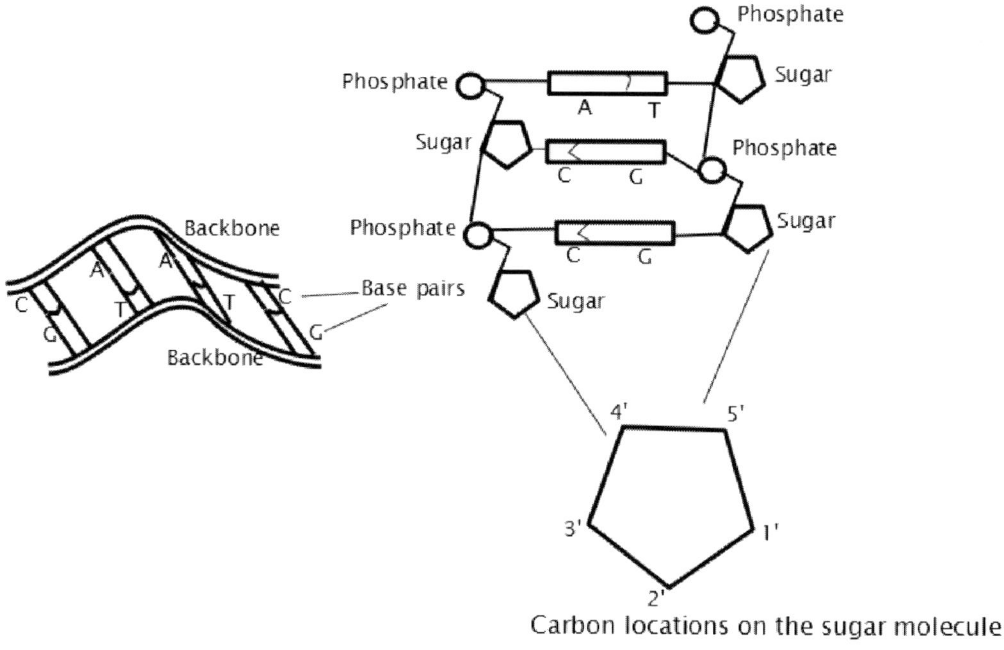


Figure 2.9. Backbone of the double helix and carbon location on the sugar molecule of the backbone.

4. The DNA polymerase proofread the copy that has been made and mistakes are cut out by an enzyme nuclease and redone by a ligase and DNA primase.
5. Since the DNA replication begins at various points in the DNA, the ends are reached last and these are finished by a separate process to create sequences called telomeres.

### Chromosomes

Chromosomes are the structures of DNA chains coiled around proteins called histones. A typical human DNA of about 1.8 meter is wound around the histone to give about 90 microns thickness (thickness of hair). A centromere divides the chromosome into two arms, one short p arm and the other q arm. The location of the centromere is a characteristic of each chromosome and plays a key role in cell division and reproduction. (Figure 2.10). Chromosomes that can pair for meiosis (see below) are called homologous. The pair has the same length and has the same location of the centromere. It also has the loci of alleles (groups of genes conferring traits) at the same location. The ends of the chromosomes have what are called telomeres and are related to cell death.

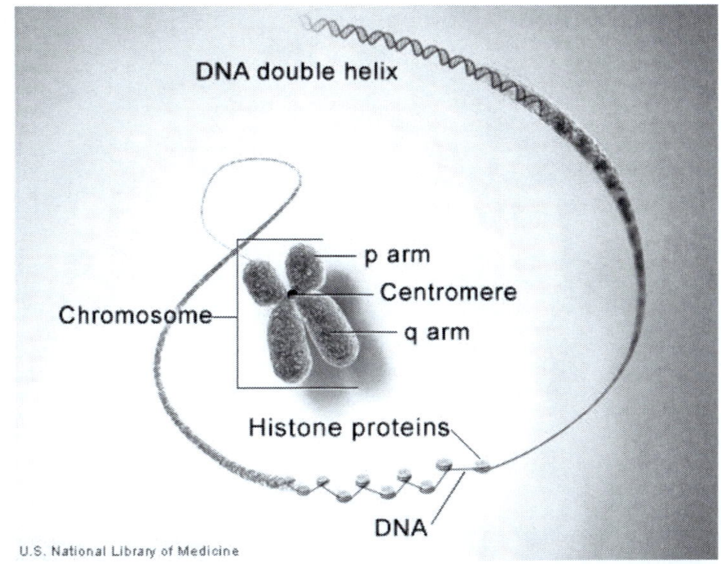


Figure 2.10. Rendering of a chromosomal pair prior to mitosis or meiosis (after replication and tying together by a centromere). The DNA and the histones are also shown. Credit – U.S. National Library of Medicine, Genetics Home Reference Handbook, Cells and DNA, Dec 2013. Free Use.

## Mendel and Genetics

Gregor Mendel, an Austrian-German scientist, became the father of a new field called genetics, when, in 1865, he demonstrated that the inheritance of specific characteristics by pea pods followed certain rules. The importance of his work was not fully appreciated until the 20th century and its impact has been accumulating ever since. Around the age of 20, Mendel joined the Department of Natural History and Agriculture at the University of Olomouc, where scientists carried out research on inherited traits in plants and sheep. In 1843, Mendel trained to become a priest and while at the Augustinian Abbey of St Thomas in Brno, he taught physics and eventually headed the abbey as the Abbott. While at the abbey, Mendel cultivated, among other things, a large number of pea plants on 2 hectares of land and studied them. He published a seminal work on his findings and most of his contemporaries believed the papers to be just on the production of hybrid varieties and did not take much notice. Mendel's work started to be recognized only in the 1900s, after high resolution microscopes were invented and the field of genetics was born. Though Mendel's work was on plants, his findings and their codification into principles of genetics are applicable to all living organisms. It is a pity that Mendel's work was not recognized for what it was, because Mendel came up with his profound findings, just when Darwin was struggling to explain how Natural Selection could explain origin of new species and their prevalence or otherwise. If Darwin had paid attention, there is no doubt that there would have been tremendous collaboration between the two. The following were Mendel's findings on the pea plants:

Mendel found that there were 7 characteristics (traits) of pea plants with two variations in each trait that were replicated from the parent plants without any plant (pea) having an intermediate characteristic. For example, when pea plants with purple and white flowers were cross pollinated, there were either only white or purple colored flowers in offspring plants and no intermediate color was observable on any plant. The following lists the exclusivity of the traits (see figure 2.11):

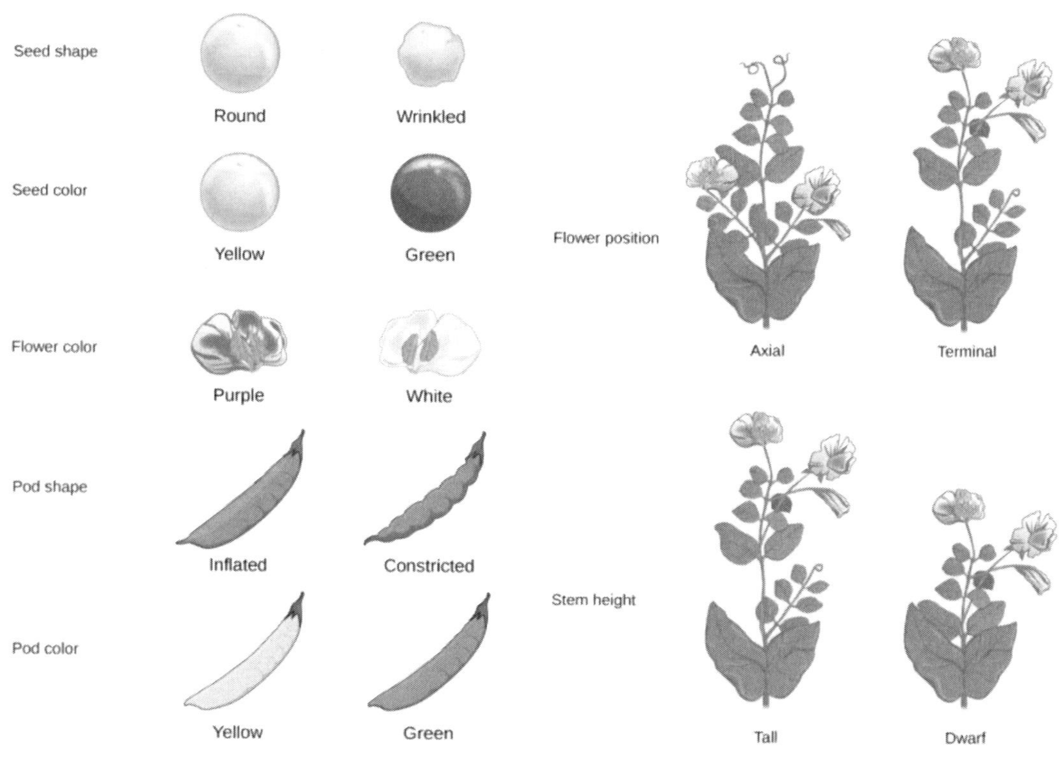


Figure 2.11. Mendel's experiment and pea characteristics. (Source: OpenStax College. Mendel's Experiments, Connexions Web site. http://cnx.org/content/m45469/1.3/, Mar 21, 2013).

- Purple or white colored flowers
- Flowers were axial (on the stem) or terminal (at the end of the stem)
- Short or long stems
- Seeds are round or wrinkled
- Peas are green or yellow

- Pods are yellow or green
- Pods are puffy (full) or tight around the peas (constricted)

This finding that no intermediate trait appears was, at the time of Mendel's discovery, unknown and extrapolation from the nature of other mixed traits in people and animals had led people to believe that intermediate traits always existed. Mendel was careful in breeding between genetically similar plants (pure or true bred as opposed to hybrid bred) and therefore, could observe the principles of inheritance. He found a very interesting fact when he cross-pollinated a plant and observed generations of offsprings on the above 7 exclusive traits (see figure 2.12):

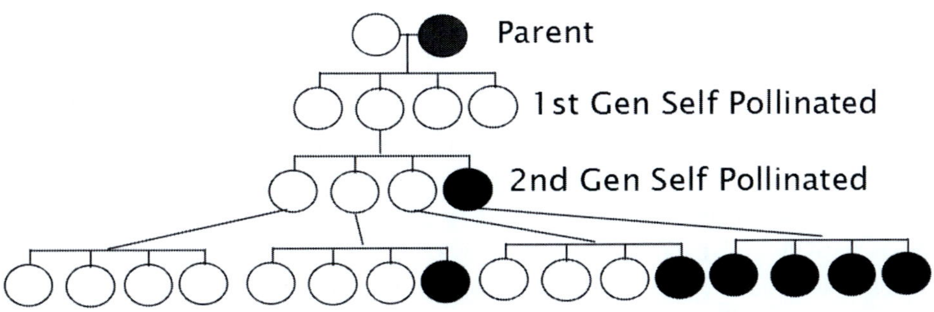


Figure 2.12. Experimental results on multiple generations of Mendel's green (black here) and yellow (white here) peas.

Taking the case of yellow and green peas, when the two varieties are cross pollinated and next generations are self pollinated, the first generation did not have any green peas, but in the second generation one out of four (or two out of eight etc.) peas were green and this ratio of green to yellow persisted in further generations, while the pure breeds continued to produce only green or only yellow peas. This is related to the dominant and recessive alleles described in the later section.

Another of his important findings was that the showing up of a trait of one kind does not mean that it is always accompanied by the simultaneous appearance of another trait that co existed in the original plant that was cross pollinated. The appearance of yellow or green peas did not have anything to do with the 'wrinkleness' or fullness of the pea. The latter characteristic was inherited independent of the color of the pea. (This is similar to inheriting the light colored eyes of the mother who has straight hair and the curly hair of the father who has dark eyes). Mendel dared to generalize these findings into the following principles of inheritance

***Principle of Segregation***

The inherited trait is determined by some inherent unit characteristic of the parent plant. (Now we know this unit to be alleles or groups of genes. The first idea that the unit was a gene on a chromosome was proposed by Thomas Morgan in 1911, based on, what is now the most common instructional tool, study of the eye of a fruit fly).

- The off spring inherits one unit of this characteristic from each parent (and therefore, the offspring has two units corresponding to a characteristic).
- One type of unit expresses itself more readily than another of the same type.
- Therefore, a trait corresponding to one of the units may not show up but will be passed on to the next generation.

Principle of independent assortment:

- A plant may inherit one type of trait independent of the inheritance of another trait.
- Therefore, the units of the different seven traits must be independent of each other.

## Cell Division and Genetic Inheritance

We can understand how fantastically important and insightful these principles are, from the present status of our knowledge on biological mechanism of inheritance. The readers are encouraged to return to the above description of Mendel's laws after going through the paragraphs below and the description of natural selection.

We take the example of humans, because it would be viscerally easier to the reader. In humans, each of the somatic cells (that is cells that do not constitute a woman's egg or a man's sperm) that are involved in our body organs, has 46 chromosomes (bundles of DNA-protein complex), which in turn, contain the genetic information in the form of the sequence of nucleobases in the DNA double helix chain. A specific gene for a specific trait is said to be located at the Gene Locus. When chromosomes are colored by specific dyes, one can see, under an optical microscope, a specific colored band for a particular segment of a chromosome. (Figure 2.13).

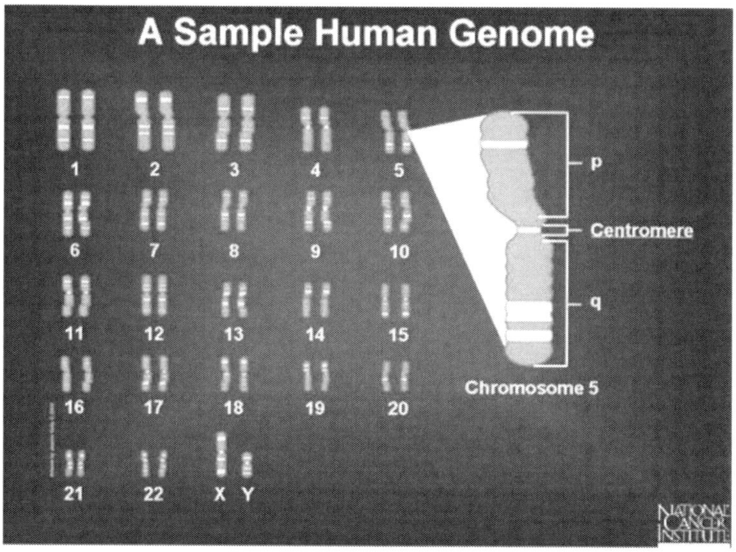


Figure 2.13. Human Genome and Chromosome pairs (male) (Source – U.S. National Cancer Institute, Free Use).

We can see that the 46 chromosomes are actually 23 pairs of strands of DNA, some larger in size than others. In all chromosomes, except the sex chromosome of the male (X and Y in the above figure), one chromosome leg of a pair is homologous (see above) with the other chromosome leg, that is they are similar in length in the two arms and when carefully seen in color bands, have the same gene locus for the same trait (although each leg of the pair may code for different variations of that trait, for example, brown vs. blue eye color). For the male, the 23$^{rd}$ chromosome pair is non-homologous and consists of X and Y chromosomes while the female has a homologous pair of two X chromosomes.

## Mitosis

This is the process when a cell divides and multiplies for purposes of growth or maintenance of the organs and body. (The creation of sperm and egg cells occur by the process of meiosis, given below). In creating two cells from the first cell (in cell division), the cell does not pinch off like a soap bubble, but has to first duplicate itself in terms of the structure within it and then only break off into two. (See figure 2.14). The cell spends 90% of its life time during this interphase when the duplication is done. During this interphase, the cell accumulates resources of proteins and grows (G1 phase) creating enough material for duplicating chromosomes and doubling the cytoplasm and ribosome content. In the S phase of the interphase, the cell duplicates the 46 chromosomes that is, each of the 23 homologous pairs of a female and, if it is a male's cell, 22 chromosomes and the XY pair of chromosome. The G2 phase of the interphase not only grows the cell further but is also a cell check point for DNA damage. (Cells progressing to division without the G2 phase are either cancer cells or becoming cancer cells). At this point the large cell contains a nucleus with duplicate pairs of chromosomes in the form of loosely packed fibers and a pair of, what are called, centrosomes that control the fiber-like microtubules. (See figure 2.14).

The mitotic M phase then is the process of division which is divided into prophase, prometaphase, metaphase, anaphase and telophase. During the prophase, the loose fibers of the duplicate chromosomes (pairs of 46, 92 in all) start becoming more and more tightly coiled. The tightly wound pairs are then joined at the “waist” by the protein called centromere, to make what are known as sister chromatins. Microtubules or tubular strings start extending from the centrosome and as they elongate, the centrosomes move part. As the sister chromatids tighten further, the microtubules extend more and more and eventually, in the metaphase, the sister chromatids are lined up along a central plane and a microtubule passes through each of the waists. In the anaphase, as the tubules extend further, the chromosomes break at the waist, one going to one side of the cell and the other to the other side. Now each half of the cut chromosomes has the full set of 46 chromosomes. During the telophase, the chromatids unwind into loose DNA fibers and nuclear envelops are formed to enclose the fibers. Cytokinesis, formation of the cytoplasm and ribosomes starts and generation of two cells is completed with two cell envelopes.

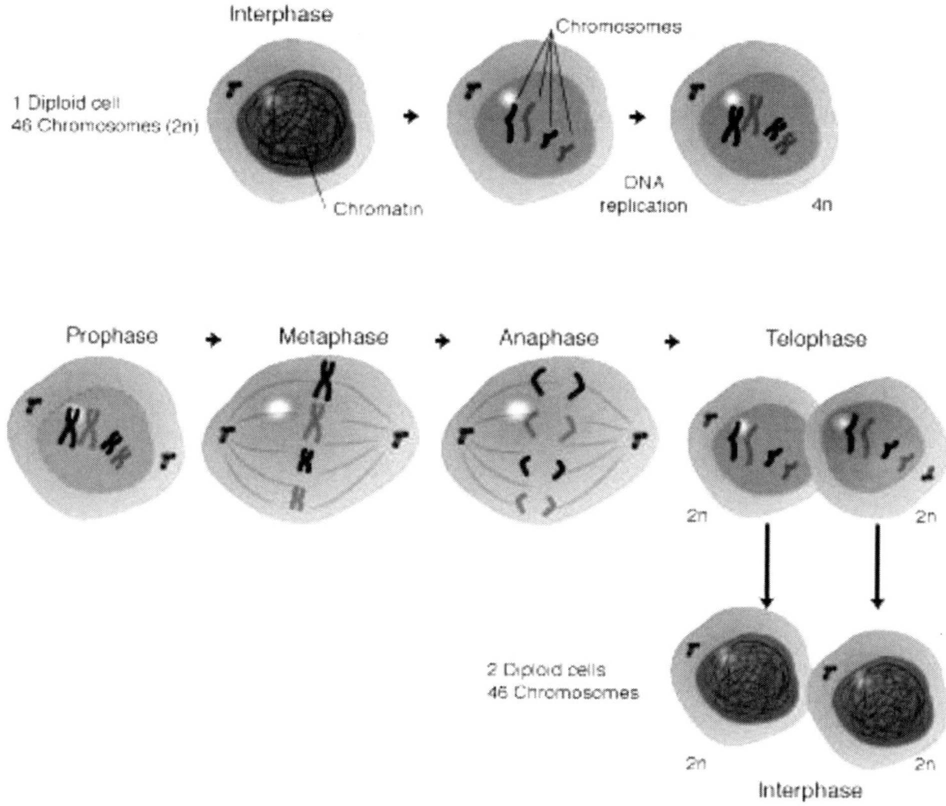


Figure 2.14. Steps in mitosis (cell reproduction) (Source: National Library of Medicine, National Institute of Health, U.S. Government. (GeneED, Cell Biology, Mitosis). Free Use).

## Meiosis and Sexual Reproduction

The sex chromosomes required for sexual reproduction are different from the somatic chromosomes (used in growing organs and general growth) in that while the 22 chromosomes have homologous pairs, for the female, the sex chromosome pair termed XX is also a homologous pair, but the sex chromosome pair XY of a male is not homologous. When the male sperm fertilizes the egg, the result is that there is a 50% chance of XX and 50% chance of a XY or in other terms, an equal chance for the off spring to be a female or a male.

The sex cells (sperms and ova) are also different from the somatic cells. These sex cells (gametes) have only 23 chromosomes and are therefore called haploid cells (Greek word *haploos* meaning single), as opposed to a full set of 46 chromosomes which is for a diploid cell. The sperm has 22 autosomes (not sex chromosomes) and one X or one Y chromosome. When the egg is fertilized by the sperm, the resulting fertilized egg (zygote) has 23 chromosomes from the mother and 23 from the father, a set with 46 chromosomes and hence a diploid cell. If the fertilizing sperm carried an X chromosome, then a female zygote results and if the sperm carried a Y chromosome, it results in a male zygote. The zygote is a diploid cell, carrying the genes of both parents.

The meiosis process (figure 2.15) creates the gamete haploid cells (sperm or egg cell) in an adult individual who, in turn, has the genes of his or her mother and father. In meiosis, in addition to the interphase of mitosis when the diploid cell creates duplicate chromosomes and the first stage of mitosis in which the parent cell divides into two cells, there is a second round of meiosis, in which the homologous paired chromosomes divide further into two cells (totaling 4 cells) each containing a daughter haploid cell. Therefore the process of meiosis ends in 4 daughter haploid cells made from the parent cell. Now, it may be noted that the 4 haploid cells are not identical and the portions of genes may cross over and provide different variations. Therefore even within the individual sex cells, diversity is already built in. This is the reason siblings tend to be different.

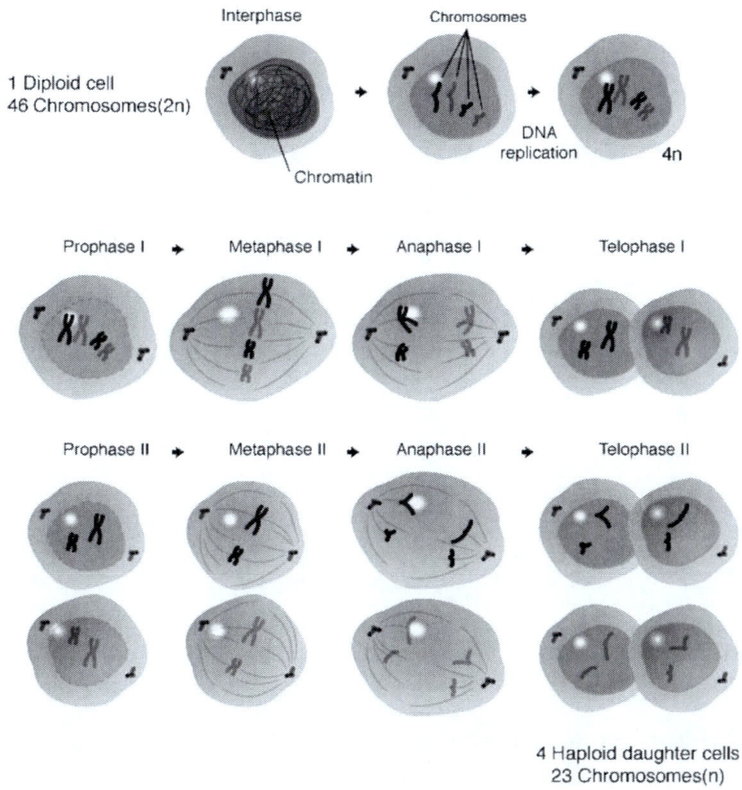


Figure 2.15. Steps in Meiosis, (Source: National Library of Medicine, National Institute of Health, US Government. (GeneED, Cell Biology, Meiosis). Free Use).

## Dominant and Recessive Alleles

As we saw before, sexual reproduction of an off spring involves the pairing and reconstituting of chromosomes from the male and female parents. The chromosomes and therefore the DNA of the offspring have a mixture of these inherited genes. Of these, the sections (or a set) of the DNA (genes) that give an identifiable characteristic, are called alleles. A dominant allele is one, which (from male or female) codes for a specific characteristic, irrespective of what the other allele (from female or male) codes for.

The figure 2.16 shows the difference. Here we designate alleles as N or r for two alleles with r being the recessive allele. In the top figure, the recessive gene rr has to be present for the recessive trait to be inherited, which gives (only) one in four chance for inheritance of the trait even when both parents carry the gene. (This was the case for Mendel's pea plant experiment). If one of the parents does not have the recessive gene, no rr group will be created and so, the trait will not be exhibited, though it may be carried. But, for the dominant gene D, (figure 2.17), it is sufficient to have one of the homologous pairs to have D, and so even with only one parent having one chromosome with dominant gene (ND), there is a 50% chance of inheriting the trait. If one parent has both pairs of chromosomes with the dominant gene, all the offsprings will carry the trait, independent of the other parent. The table gives the probability for other combinations of the recessive and dominant genes. Note that for the parent as well, although having one recessive gene is insufficient to cause him or her to be affected, he or she will nevertheless be a carrier. However, the parent is affected if he/she has even one dominant gene. (Also Table 2.1 and 2.2).

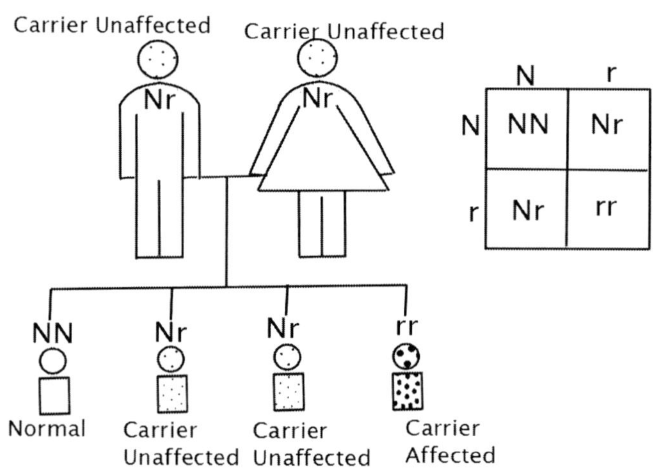


Figure 2.16. Recessive Gene and inheritance of trait. With both parents unaffected but carrying the recessive gene there is 1 in 4 chance of being affected. (This corresponds to Mendel's experiment).

**Table 2.1. Table of combinations of normal and recessive genes, giving chance of offspring to inherit the recessive trait.**

| Mother | Father | Carrier off spring | Affected off spring | % affected |
|---|---|---|---|---|
| NN, not a carrier | NN, not carrier | NN, NN, NN,NN | none | 0 |
| NN, not a carrier | Nr, carrier | NN, Nr, Nr,NN | none | 0 |
| NN, not a carrier | rr, affected | Nr, Nr, Nr,Nr | none | 0 |
| Nr, carrier | Nr, carrier | NN, rN, rr, Nr | 1 | 25.00% |
| rr, affected | Nr, carrier | rN, rr, rN,rr | 2 | 50.00% |
| rr, affected | rr, affected | rr,rr,rr,rr | 4 | 100.00% |

If NN, Nr, rN (same as Nr for our discussion) and rr were equally probable of 2 each (one male and one female), there would be 16NN, 16rr and 32Nr (or rN). So 16 would be affected out of 64 offsprings.

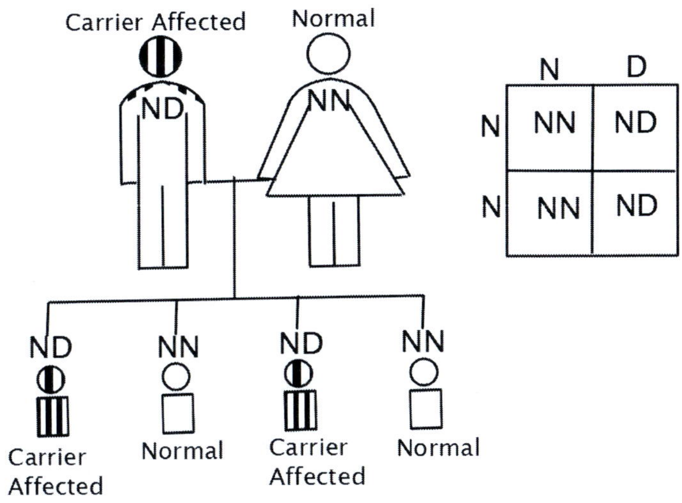


Figure 2.17. Dominant Trait and inheritance for one parent with dominant gene. Carriers are affected and even with only one parent as carrier, the chances of off spring being affected is 1 in 2.

**Table 2.2. Combination of Dominant Trait and normal traits giving offspring chance of inheritance of the dominant trait**

| Mother | Father | Carrier off spring | Affected off spring | % affected |
|---|---|---|---|---|
| DD, affected | DD, affected, | DD, DD, DD, DD | 4 | 100.00% |
| DD, affected | ND, affected | DD, ND, ND,DD | 4 | 100.00% |
| DD, affected | NN, not a carrier | ND, ND, ND,ND | 4 | 100.00% |
| ND, affected | ND, affected | DD, ND, ND,NN | 3 | 75.00% |
| NN, not a carrier | ND, affected | ND, NN,ND, NN | 2 | 50.00% |
| NN, not a carrier | NN, not a carrier | NN,NN,NN,NN | 0 | 0.00% |

For initial equal probability of DD, ND, DN (same as ND for our discussion), and NN of 1, there would be 16DD, 16NN and 32ND (or DN), so that 48 offsprings out of 64 are affected.

So getting back to Mendel's pea plant experiments, what he found was that that one in four pea plants had purebred recessive allele, two out of four were hybrid and one out of four were purebred normal. But the hybrid always expressed the normal trait. In the example of yellow and green peas, the yellow color is normal and green color is recessive and therefore, only one in four peas was green, while the purebred and the hybrids were yellow.

The recessive allele has a chance of 16 in 64, while the dominant allele has a chance of 48 out of 64. (In the above discussion, the designation of normal is only for illustration, the pairs are either recessive or dominant). These chances would remain the same in further generations. Both the dominant and recessive traits would keep increasing as the population grows, if the traits do not affect the fertility and survival of the offsprings. The importance of recessive and dominant genes is to have a joker-in-pack, so to say, a property that stacks inheritance in favor of certain genes in natural selection. So, one can see that dominant genes and traits will spread faster than recessive genes and traits. Curly hair is a dominant trait, while straight hair is recessive.

While the above is correct for autosome chromosomes, different rules apply for sex cells, because these have to be haploid cells when the egg fertilization takes place. An example of this is baldness. Even though it is a recessive gene, it is attached to the sex chromosomes. Since Y chromosome does not have a pair, the two Xs of the mother and the X and Y of the father act independently. This results in baldness in male off spring in 50% of the cases, even if only one of the Xs in the mother carries the gene. (For female (XX) off spring, the baldness gene is not active). These types of exceptions, give additional mechanisms for the persistence of diversity- as in the above example, baldness and hairy heads continue through generations.

## MECHANISM OF NATURAL SELECTION

So, one can see that, if all things are equal, genetic selection will result in the dominance of the yellow pod over the green pod, the dominant over the recessive. Now, if there are additional environmental factors, such as worms preferring green peas, then one could see that green peas could be selected out completely over time. The mechanism of natural selection describes the descent of species from one another and then success of one species over another in a given environment. This involves three mechanisms (1) generation of variations in the traits of an original species through mutation (an abrupt change in the gene due to radiation, chemical etc.), genetic drift and other processes, (2) heredity or passing on of these variations to offsprings and lineage with the factor of dominant and recessive nature of genes playing a role (3) the advantages and disadvantages of these variations in terms of survival and proliferation in the environment that the different species find themselves. In addition, there are traits, which help a species to co-evolve with another organism. All this is derived from, (a) fossil evidence – extensive and intensive collection and study of fossil from hundreds of millions of years to more recent. Darwin and Wallace and all the geologists, zoologists, paleontologists, archaeological and anthropological scientists have collected and studied the evidence, over the last centuries, (b) molecular evidence- evidence from the identification of genes. There is commonality of genes for all plants and animals showing a common origin and it is seen that more genes are common between related species. An example is that the difference in genes between humans of all races is only 0.1% and the genetic difference between chimpanzees and humans is less than 2%, showing common ancestry.

### Genetic Variation and Introduction of New Species in a Population

A species may continue without change or suffer a change in its traits due to a change in its genes (alleles). This change can come due to,

(i) Mutation –a modification of the DNA, a damage to the DNA or an error in copying a parent gene to the off spring, or alteration of gene due to a chemical agent or radiation. This can be substitution, insertion or deletion of nucleotide sequences in one or more codons (the nucleotides, which code for the manufacture of specific proteins). Though, this is rare and this phenomenon is under-emphasized in text books because of its rarity, *mutation is the process that can create brand new variations.* The mutations at the molecular level must be much greater than the occurrence of a particular trait, because the mutations have to be specific so as to make a nucleobase which is used in the DNA and a combination of mutations has to occur at different gene loci, so that a new allele is created. There are probably large numbers of mutations going on that accumulate junk DNA not useful for any trait or mutations that do not produce useful nucleobases. But a mutation can easily turn off a gene function.

(ii) Recombination- When a DNA string is broken and is reconstituted into different strands, a new variation can be created. (See figure 2.18 and 2.19). This process mixes up DNA sequences and the changed sequence can then code for a new variation in traits. Cross over recombination occurs when during meiosis, the four haploids chromosomes are placed close together. There is a finite probability that two of the homologous pairs of chromosomes can recombine and depending upon the location of cross over (where the centromere forms), the resulting zygote could be a new variety. When the sperms or the eggs are made, they are utilizing the chromosomes of the individual which, in turn, are derived from its parents. In making an egg, the female's chromosomes first find their matched partners and exchange parts of the DNA with each other. Because of this shuffling, genes from both the parents of the female wind up together on the same stretch of DNA. (The same thing happens in the father's sperm.). The cell undergoes meiosis and the eggs are made. When the egg and sperm combine, the zygote has alleles of all the four grandparents and any mutations that might have occurred in the parent cell.

(iii) Gene Flow, a transfer of a gene due to interbreeding because of migration or a visit by a new member in a population, or pollen carried by air to a different area. A gene flow between different species can also occur due to infection from a bacteria or virus.

A variation that occurred in a particular individual or individuals will not affect the population if the traits are not inherited by the offsprings. (See reproductive barriers in the later section). But if they are, then these traits would be propagated.

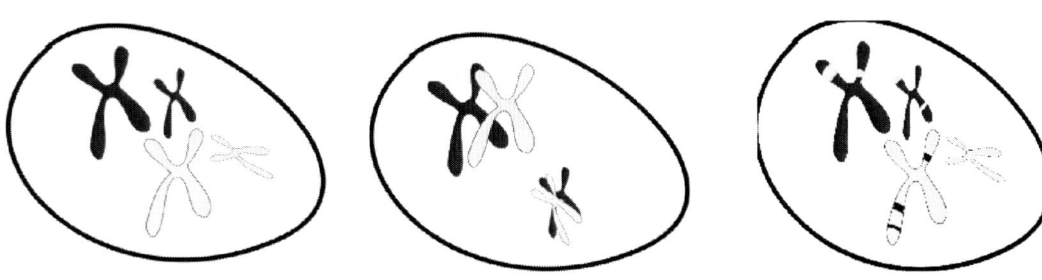

Figure 2.18. Recombination allows shuffling and mixing of genes. It can happen between homologous chromosome pairs during mitosis and also between non homologous chromosomes in meiosis.

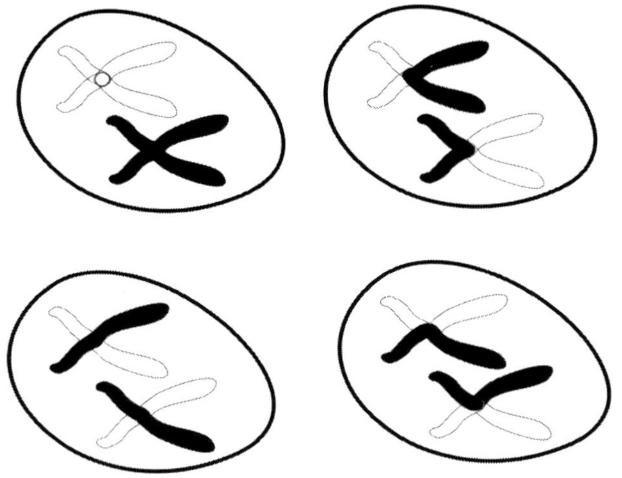


Figure 2.19. Meiosis with a cross over makes for additional configurations of the reconstituted chromosome, producing more diversity.

## Mechanism of Change

Let us take the reported case of moths. Let us say that there are two kinds of populations at the beginning- gray winged and dark winged. Now the population of, say, the dark winged moth is increasing over the gray ones. This can be caused by the following mechanisms.

***(a) Genetic Drift***

The parent has genes for both gray and dark wings and gives birth to an equal number of gray and dark winged moths, but in the next generation, this sampling has an error and fewer gray moths appear. Now if this difference is significant, it can lead to significant population difference after a few generations. But, generally, a drift in a particular direction has to take place for several generations for a significant change. Random chance, in which a disproportionate population with one type of variation is lost, say by forest fire, may cause a population imbalance and may result in a genetic drift.

A genetic drift is a particularly sensitive effect when the population is small. Random chance events can spur a significant population growth or a reduction, altering the chances of future survival of the species. Though one species may have equal chance of succeeding or failing as another, the chance works disproportionately when the population is small. This is like flipping coins. If the flipping is done a very large number of times, the chance of heads or tails is 50% each. But if the coin flipping is done only a few times, then these chances may not be equal. Specifically, since the probability of inheritance of the recessive allele (not the gene) is low, chance deviations to reduce the pool of recessive allele can have a greater effect. (Also see bottleneck effect, below).

Conversely, a variation introduced can propagate very fast within a few generations when the population is less. Say, a trait variation appears with the chance of one in six. Let us say, it is a dominant trait and the sexual reproduction with the individual with that varied trait guarantees offsprings with that trait. In addition, the one in six probability of appearance of the trait also continues to be inherited. This corresponds to an increase of population with the varied trait. At some point, the increase of the species expressing this trait may saturate or may overwhelm the original species, if the variation has an advantage. The net effect of random genetic drift on a small isolated population is therefore large and this is the reason why in the small isolated environments such as in Galapagos, there can be a lot more diversity than elsewhere. The example below shows the fast increase of such variations.

### *(b) Pressure of Natural Selection*

An environment which gives greater advantage of food or water availability or protection from a predator is a major driver in population changes. For example, let us say that a certain building in an area is the resting place of moths and this building is getting blackened by industrial pollution and soot. In this case, the black moths would be camouflaged and gray moths would be visible and spotted by birds as food. This would reduce the population of gray moths disproportionately. The picture (figure 2.20) below shows, what could happen: a gray moth has a variation in its genes that gives rise to off springs of both gray and dark moths. In the next generation, due to a natural selection, gray moths are preferentially lost. This leads to a significant increase in the dark moth population over that of the gray ones, due to both loss of the population by predation and also because of the smaller reproductive rate. If the natural selection continues to be unfavorable to gray moths, the population of gray moths would reduce and can even disappear.

In England, this particular instance of natural selection is likely to have happened to the peppered moths, where during the industrial revolution, when a lot of coal was burnt and the walls were covered with soot, the gray moths were visible to birds and dark moths were camouflaged. The population of the gray moths is now returning with the reduction of pollution and cleaning of the building walls.

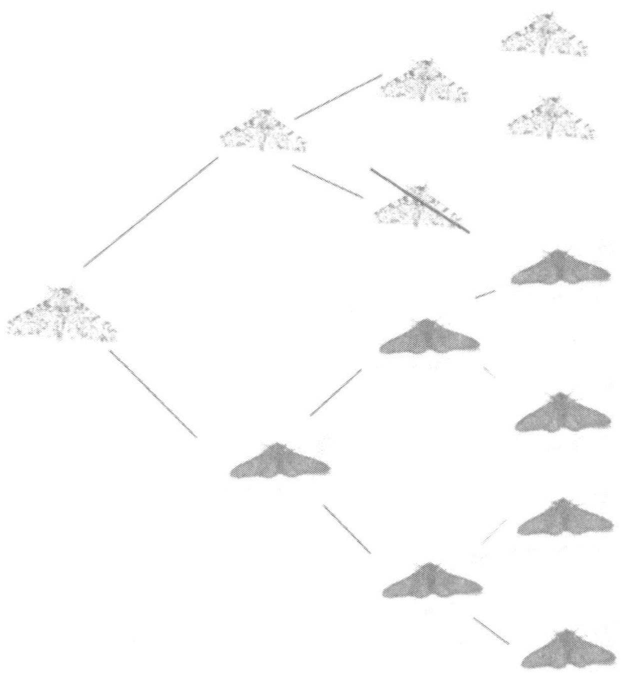

Figure 2.20. Selection of dark moth by environmental conditions.

In other cases, genetic drift and natural selection can be due to an advantage of one species over the other in a given environment of food resource. For example, in a dry area with few nuts, a finch with a strong and thick beak can crack nuts that have harder shells and can survive, whereas finches with longer and finer beaks can break and take the meat out of small nuts. This is the reason for the distribution of Darwin's finches. In the case of a wet environment changing to a dry environment, finches with bigger beaks would be able to eat better and have more and more surviving offsprings, than their cousins with smaller beaks. As a result, the developing dry area would see an increasing population of large beaked finches and a reducing population of small beaked finches. In isolated environments, the diversity is greater because the individual populations can form niche communities taking advantage of a specific resource without competition from another species. Such diversity breeds further diversity. Currently, this is happening in California in U.S., where some bird species are leaving habitats because of the decade long drought and this helps different (new to the area) birds to occupy the vacated environment and resource niche. (Murphy 2014).

One can readily see the effect of natural selection on the population from the dominant-recessive allele tables. We saw that in the case of Mendel's peas, yellow peas had 3 times the chance of being available from a crop than the green peas. Now, in a worm infestation, let us say that the yellow trait is susceptible to worm infestation. In a few generations and with the infestation continuing, the yellow pea would disappear and only the green peas with the recessive trait would exist. Since the remaining population is purely recessive, the species would uniformly have the recessive gene. The reverse scenario could also happen and the recessive green pea species would be wiped out, if the worms preferred them. But in this case, the recessive gene carrying yellow peas (Nr) would continue and green peas would keep appearing and not be wiped out. This way, one can see that the hiding of the recessive gene in a normal looking offspring may protect it. This offers support for the concept that gene looks out for itself rather than the animal body, that is it is the survival of the fittest gene that is occurring, rather than survival of the fittest species.

Therefore natural selection favors the success or adaptation of that species that has genes to confer traits such that random chance and environmental changes would give them the advantage. Similarly, natural selection will cause the reduction of that species whose genetic traits are a disadvantage. Though we call this adaptation, the species does not adapt. It is simply that when one species has favorable traits, it flourishes while others do not. When we look at the tableau of nature after considerable period of evolution, we find species that seem to be very adapted to the environment. We loosely say that a flower has nectar so as to attract bees and promote pollination and propagation of the plant, implying that the flower and the plant were making a choice to produce nectar. It is really that a variation with nectar is successful because it attracts bees which pollinate the flowers. Any other variation with a flower without nectar would not be able to attract bees and therefore could die out, if it does not have another means of propagation. The same is true of camouflages to hide from predators or for hunting preys. The Ophrys orchid bloom (see figure 2.21) that looks like a wasp, fools a wasp to "mate" with it causing the orchid to pollinate. It is just that this wasp look is a variation that developed over many generations, with increasing advantage and success in breeding with each generation and finally it looks almost exactly like a wasp. The fact that a stick insect is amazingly similar to a stick for camouflage is just the success of the particular trait of developing that look, such that it becomes practically a runaway process over many generations by natural selection, where the further improved trait is selected over the previous variation. This also means that the pressure that selects this trait remains powerful. Predation and mating competition are such powerful pressures and therefore, we see such runaway phenomena in camouflage, in features to attract mates, in bird nest construction or in propagation of seeds. But this is a deceptively static picture. A change in environment or a random event can upset this picture.

The biological fitness to find mates, get many off springs and ability to care for them can also be a process of natural selection. Members with stronger traits in this area would be the more successful species. Each trait provides a success or failure path and when carried over several generations, these traits end up in serious changes in population of that species. However, it is a path of progression not chosen by the species, but rather is due to the suitability or unsuitability of the combination of a particular trait in the particular environment. A white fox has great advantage in the polar region, but would be at a disadvantage in a poorly lit forest floor, where a dark brown fox would do much better. If the populations had been mixed at the beginning, then only one would eventually survive. Even though, Darwin used the term, "Survival of the Fittest", he did not mean the way it is interpreted sometimes- as the strongest or fastest or meanest, but as being the most fit to survive and reproduce successfully in the environment that the species lives in. Different species have different reasons for their success (sometimes, it is called strategy, but this is not to be confused as an animal planning its life, but a trait of behavior that it inherited over generations). For some, it is expending a lot of resources in finding a biologically fit mate, for others, it may be producing a lot of off springs to play the odds of survival of the young, in others spending a lot of resources in carefully nurturing a few off springs to adulthood. Another proof of adaptability is the case of invasive species, where plants or animals brought in to a new environment are already so adapted to the new environment that they spread wildly and use up the resources hence displacing indigenous populations.

Figure 2.21. Ophrys Orchid bloom, which imitates a wasp. Source: Wikimedia Commons Free Public Image.

### *(c) Descent of New Species and Tree of Life*

One of the most difficult and yet the most profound statement of the theory of evolution is that all life forms descended from a single form of life, a common progenitor. It is most likely that a particular gene was altered by mutation resulting in a minor or major change in an inheritable trait that finds itself in a favorable circumstance. The tree of life, the creation and propagation of species from that point, as we see today, is the result of slow changes in the traits retained by the process of natural selection. While the original progenitor may or may not continue, a new successful trait, say a small wing like feature on a dinosaur may provide an advantage for survival by the ability to hop higher than others and then this trait when inherited could benefit species with bigger and bigger wings, finally leading to a flying dinosaur- a bird. This tree of life, according to the fossil records providing anatomical and molecular evidence, is given in figure 2.22.

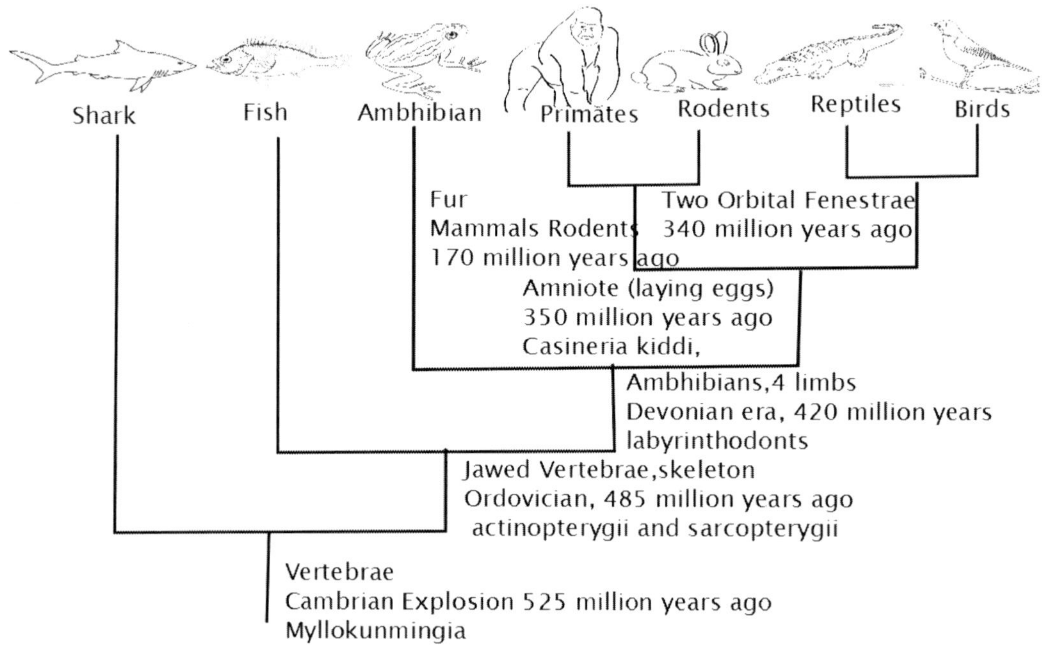


Figure 2.22. Descent of example species. (The vertebrae evolved from arthropods that, in turn, evolved, from tubes and sponges).

As shown in this figure, an original progenitor, a marine creature, without a vertebra, acquired a variation of a vertebra in the genes and developed this trait and this trait propagated, forming fish, which developed a variation of a skeleton and so on. (Figure 2.23). Today, we cannot see any visible commonality between the shark and the birds, but this is the amazing outcome of mutations, genetic drift and natural selection acting over hundreds of millions of years, over millions of generations. Actual observable example of such evolution that is easily demonstrated in a laboratory and seen in nature, is the case of bacteria, where large numbers of generations of fast multiplying bacteria evolve quickly depending upon environments, mutations etc. We see this in antibiotic resistance, appearance of fast changing HIV virus and new strains of flu virus.

Figure 2.23. Evolutionary steps.

***(d) Epigenetics***

One of the new findings is that the characteristics of a species with identical genes can be different because of environmental effects which activate or deactivate gene function. One is reminded of the *Ragas* in Indian Carnatic music where two ragas with identical ascending and descending solfa notes (*swaras*), can sound very different because the singing emphasizes different aspects of the raga or uses different strings of notes. For example, an African butterfly species when hatched in summer has completely different coloration than if it is hatched in the fall. The variations that occur due to epigenetics may also be inheritable. The epigenetics change because changes occur in the histones (the proteins that coat and package the genes inside the nucleus), or the change is due to methylation where a methane group is added to the cytosine or adenine base. So, though there is no change in the DNA sequence (same genotype), the organism exhibits different characteristics and is susceptible to different diseases (different phenotype). It is increasingly being realized that variations in humans and animals may simply be differences in epigenetics and as the environment changes (famine, climate change etc.) variations in a population would appear even without changes in genetics. The prevalence of the epigenetic process is restricted by the fact that most of the epigenetic changes get erased when the sperm and egg combine, by what is called "reprogramming". Epigenetics is becoming a very important field in modern biology.

## Macroevolution

When we see a change in species- not a simple variation, but a complete direction of change, say from fish to amphibians, this process is termed macroevolution. It is subject to the same laws of natural selection as the smaller changes in species. The same is also true of the more recent variations, where accented by environmental changes, humanoids evolved from ape like animals in just a matter of a few million years and the humanoids evolved into extremely intelligent beings in a matter of 100,000 years. The pressure of natural selection accelerates the process of evolution. Though, it might seem mind-boggling that a small mutation in fish could have brought out amphibians or a mutation in an ape-like being is the cause of the evolution into humans, this is what happens when multiple generations amplify this effect aided by environmental factors. When one looks at the fossil evidence, we see that the evolution is not directed, is chancy and is the result of many fortuitous (for that species) circumstances, such as mammals emerging as the dominant species because of lack of competition from the dying out dinosaurs. The evolution path is marked with many failures of a branch of a tree, is zigzagging and sometimes misdirected with random progress and sometimes, abrupt transitions. Natural selection is far from a sure success story because chance plays a major role.

## ADVANCES IN THE SCIENTIFIC THEORY OF EVOLUTION

Biology is a very complex discipline, involving empirical observations, quantitative and qualitative comparisons and assessments, filling in gaps and ascertaining the validity post facto and requiring several statistical analyses and some mathematical modeling. While no science linearly proceeds with theory, hypothesis, postulates and evidence in some specific order, this is particularly true of biology. But the recent technological revolutions and accuracy of models have stabilized biology into not only as a robust observational science but also has provided it with verifiability of theories. The spectacular advances in genomic research have yielded a cornucopia of proof for Darwin-Wallace theory of natural selection and evolution of species, Mendel's genetic principles and all associated theories, and have taken the field further into the realm of high precision with molecular genetics, where remote pasts and the distant future of species can be assessed quite accurately. While Darwin and Wallace could only investigate the species – the animal, the plant or the insect, biologists can study species at the gene level and know that genes are the evolutionary entities. Therefore, it is adequate to postulate that Darwin's theory of "Survival of the fittest" really applies to genes (see (Dawkins, 1976) . The Modern Evolutionary Synthesis or Synthetic view of evolution (produced in the 1940s and sometimes called Neo-Darwinism) synthesizes Darwin's theory, Mendelian Genetics and the later knowledge of Genetics. Before this theory, there was a conflict between Mendelian Genetics and the slow evolution. The understanding of macroevolution was also poor and it was not clear how broad global populations could be so radically altered by local population changes. The modern synthesis explains that it is the population genetics that causes speciation and macroevolution. Considerable impetus to the modern evolutionary theory was given by biologist-geographer-taxonomist Ernst Mayr, geneticists Theodosius Dobzhansky and Sewall Wright, botanist George Ledyard Stebbins and paleontologist George Simpson.

Darwin was severely handicapped in providing evidence for his theory of evolution. Though he could substantially prove the process of natural selection, he could not come up with how the diversity of species originates and is inherited. Mendel's findings explained some specific aspects of this, but the two never communicated. The modern evolutionary synthesis combines the Mendelian genetics with Natural Selection. (In fact, in early 20$^{th}$ century, Mendelian Genetics was considered to be arguing against natural selection, because Mendel dealt with only discrete -unit- traits, while Darwin focused on subtle differences in having great effects). The Modern Synthesis lays out the speciation and population genetics, so that genetics and natural selection come together to explain evolution.

## Speciation

Species in Latin means "type" or "appearance". In biology, species includes morphology (shape and structure), body functions, behavior, genome, etc. While we can understand that a new trait is developed in a population by mutation, gene flow or genetic drift, the fact that a variation develops into a unique species must also mean that there is no "dilution" of that trait by sexual reproduction with the older, non variant or other variant individuals of the population. Therefore, speciation greatly depends on the fact that the genetic divergence of members with the variation is isolated from the greater population by reproductive barriers. The diagram in figure 2.24 shows the reproductive barriers:

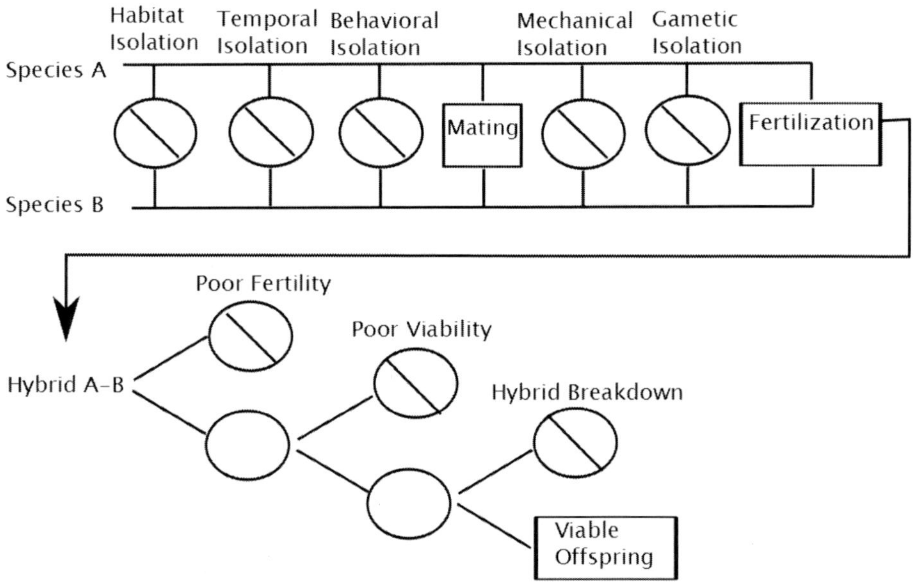


Figure 2.24. Stages of Isolation: Hurdles different species have to overcome in order to produce a viable offspring.

The habitats may not be suitable for both the species. They may not inhabit a given habitat at the same time of the season. The courting and mating behaviors may be different and not understood. (A good example of behavioral barrier is a courtship ritual of one species that does not interest a potential mate in another species. In Cichlid fish, the female chooses its mate by choosing its own coloration. But, it will mate with other colored individuals, if left in dark, producing fertile hybrids). The organs may not be suitable for mating. Gametic isolation is essential for many species such as fish, ferns, tree (pollen) etc., because these do not mate mechanically and spread their seeds in the general environment. The species stability is protected by this type of viability. This may happen because the gamete may not develop due to post-zygotic change in the female such as the swelling of vagina in fruit flies, hindrance in the growth of pollen tubes in plants etc.

In addition, developing geographic barriers help in origin of species. When a population gets isolated by a new geographical barrier, like a land mass rising in the middle of a lake, or a large tide pool becoming separated from the ocean by a rising rock, the reduced, isolated population becomes a crucible for origin of species, because there is a higher probability of a variation taking hold. This can explain a seeming abrupt appearance of a new species in a population, but this would be one that arrived from a different isolated population. (In contrast to this explanation for the sudden appearance of a species, Punctuated Equilibrium theories believe that in these cases, instead of a gradual change, there are stasis periods and a sudden speciation). Speciation can also occur among non-isolated population by accidental, low probability reproduction results such as generation of extra chromosomes. This can happen because of malfunctioning of, say a step in meiosis, resulting in polyploid species.

Speciation does not mean that a particular variation occurs only once and each species has a unique trait; it also does not mean that a species becomes a species only after say, full development of a specific organ. The human eye is an amazing piece of natural technology, unrivaled by human (artificial technology) endeavor. It did develop gradually from animals first getting photosensitive organs to just see light and dark and then evolving lenses and proper eyes to see shapes and colors. However, the same development has taken place in other clades, for example in a flatworm, which has the ability to see light or darkness. The idea of "exaptation" is the use of an existing organ for a different purpose, such as the light honeycombed bone frame of a dinosaur developing gradually into a wing frame.

## Population Genetics

Founded by Seawall Wright, J.B.S. Haldane and R.A. Fisher, population genetics states that the unit of evolution is a population (not individual), it emphasizes the central role of genetic variation in a quantitative way and studies the evolution of the population through natural selection, gene flow, gene drift and mutations. In this way, it integrates Mendelian type genetics with the Theory of Natural Selection providing a genetic basis for natural selection. This follows the general theme and some of the specific processes described above, but in a mathematical and statistical way. It examines how gradual and small changes accumulate in a population to result in a large change in the gene or allele frequencies, over a long period or many generations. Using population genetics, one can ask quantitative questions like, how long it would take for a new allele to be broadly fixed in a given population or how much selection force it would take for the allele to take hold. While this synthesis is not free of some questions, this theoretical synthesis is quite valid for what we understand today. Populations are concentrations of individuals belonging to the same species, so that these individuals are more likely to breed with the members of the same population. The gene pool of this population is the sum of all the various alleles (genes at different locations, working together to create a particular trait) of the individual members. In arriving at a given allele frequency in that gene pool, the alleles would be counted twice (twice the weight) for a homozygous individual with two identical alleles for a trait (say blue eye) as opposed to two single counting of the alleles for two different traits (blue and brown eye) for a heterozygous individual. Taking Mendel's pea plants as an example (see the table), let us say there are 240 plants (480 chromosomes because these are diploid organisms) and 70 have the recessive gene for white flowers (rr), 70 have NN and 100 have Nr (same as rN). So the total number of r alleles is 70x2 +100x1 = 240 and this is the same for N alleles. But if, for some reason 30 rr organisms were lost then the total number of r alleles would only be 40x2+100x1 = 180.

### *Hardy Weinberg Equilibrium*

This states that given equal mating between members of different traits (unaffected, carriers and affected), the gene frequency would be the same. This can be seen in the Mendelian diagram and it can easily be seen if all probabilities of generational mixing of N and r alleles are the same. If p is the fractional frequency of one type of allele and q is the fractional frequency of the other, then for all generations $(p+q)^2=p^2+2pq+q^2=1$ or 100%. That is, if initial gene frequency of N is 70% and r is 30%, then the NN occurrence ($p^2$) is 49%, Nr=42% (2pq) and rr =9% ($q^2$). Since rr frequency is observable for the appearance of the trait, the total number of r allele in the population can be back calculated. If a disease due to a recessive allele is present in 1 in 100 then the distribution must be rr=1%, rN+NN=99% or $q^2=0.01$ or q=0.1. Since, $(p+q)^2=1$, p is equal to 0.99 or NN fraction ($p^2$)=0.81, and Nr frequency (2pq)=0.18. The frequency of the allele r is then 1x2+18x1=20 in 200 (2x100) alleles or 10% of the alleles.

The Hardy Weinberg Equilibrium is unstable because of factors such as genetic drift, mutations, migrations, tendency for disease for one trait, and selection of mates (for example, hummingbirds will pollinate red flowers preferentially). In fact, the equilibrium may be disturbed by a change in one gene locus. Still, this equilibrium is a baseline against which one can compare results of natural selection. So, if the frequency of alleles or genes at a gene locus changes in the population from the equilibrium value, one knows some mechanism is active for the population's evolution.

Mutation, genetic drift and gene flow are mechanisms for disturbance from the equilibrium. As stated before too, small isolated populations tend to amplify the effect of these factors because of what is called sampling error. In random events like coin toss, the probability of error as a percentage is given by 100 /square root (N), where N is the number of coin tosses. Therefore if the sample size is 25, the sampling error is 20% while, if the sample size is 2500, the sampling error is 2%. Once a trend is set because of this error, the species balance can change in that direction rapidly- success breeds success, simply because of a head start. The so called "Bottleneck Effect", also a type of sampling error, is another perturbation that can change and imbalance the equilibrium. (See figure 2.25). Disasters like floods, drought, fire, earthquakes and climate change can reduce the size of the population, thereby creating a bottleneck where the original gene pool has diminished and the species goes into the genetic drift due to sampling errors. This results in a reduced genetic variability among surviving species. There is also an obvious effect called the "Founder Effect" where, one male- female pair or a single seed is the sole source of the population. The population will initially move along without diversity, but eventually genetic drift and mutation will increase diversity. Where one finds a high frequency of a single trait, such as a particular feature, behavior and/or disorder, then the population may have been "founded".

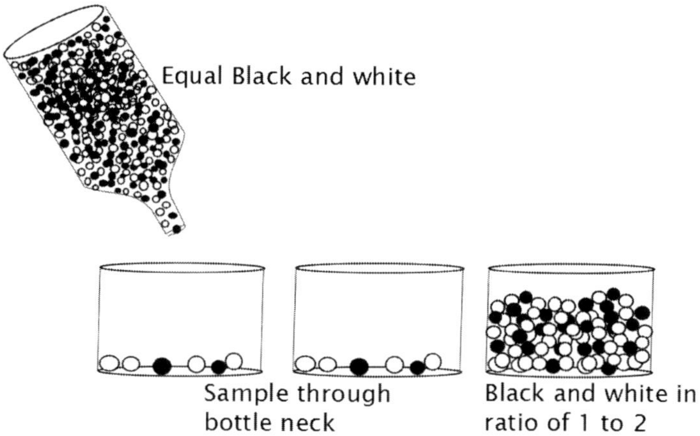


Figure 2.25. Bottleneck effect: The bottle neck may not allow a proper amount of sample to be taken out and when a small population breaks away into a new habitat, the diversity of variation in the new habitat does not reflect the relative population of variations of the old habitat.

Population genetics is like fluid mechanics. Instead of following every molecule of gas to determine the properties of the gas and the flow etc., one uses general laws of fluid dynamics, which describe the aggregate behavior. This is the same with population genetics. Instead of tracing individual variation, the population behavior is ascertained by factors affecting the aggregate population. The field is broader than what is described above and includes effects of recombination and other influences. Over the years, population genetics has become a well formed theoretical field with mathematical and statistical foundations.

Evolution is based on rules of change, but mediated by chance events. It is also the result of interaction of multiple systems, biology, geology, climate and social environment (competition, symbiosis etc.). In addition, there are statistically occurring errors in following the rules, even though good self- correcting mechanisms are built in. This is a perfect recipe for non-equilibrium processes and chaotic phenomena. As shown in a later chapter, at certain points (bifurcation), this makes for high sensitivity resulting in completely different trends, a confinement of changes not to go beyond a domain etc. This is an important aspect that can explain sudden changes in populations and sudden evolution of species. The phenomenon of chaos acts at every aspect of natural selection- at the level of resources, diseases, weather climate and catastrophic events. More often than not, populations are either growing unstably (human population in the past) or declining strongly (very fast rate of extinction that is happening now) or are in a dynamic equilibrium at the edge of criticality (see phenomena of sand hill in Chapter 4). It must be evident to the reader by now that flora and fauna we see today, those in the past and those in the future are, were and will be governed by no one time event and the reality is far more complex and far more wondrous than miraculous creation.

## Summary of Theory of Evolution

We define life forms as having 4 characteristic abilities (1) Reproduction, (2) Inheritance, (3) Catalysis of biochemical processes and (4) Consumption of energy (metabolism). The characteristic of inheritance intrinsically results in other 3 characteristics in generations. So, the evidence for the common descent in the Theory of Evolution are as below (all fully proven): (Despite various caveats and doubts of several people, the reader should note how astonishingly specific and therefore vulnerable to falsification the Theory of Evolution is and yet, the Theory comes out of these tests, intact).

1. All life forms are based on specific polymers of nucleotides (DNA and RNA).
2. No other basis for life has been found.
3. In order for an organism to maintain itself and reproduce, it must have catalyzed chemical reactions. The mechanism for this "enzymatic" catalysis is the manufacture of proteins in cells according to the code- sequence of nucleotides in the DNA.
4. The set of rules that govern the creation of a specific protein is universal among all life forms. A Universal Genetic Code is used by the cell in interpreting the code in a certain DNA molecule.
5. The small variations from this code, which actually were predicted by the theory of evolution, were found after 1979.
6. The metabolism is identical or nearly identical in all life forms, such as glycolysis to digest sugars, hydrolysis to absorb fats etc.
7. Similar molecules are used in all life forms for body organs or functions, such as lipids for membranes, keratins for hair and claws and use of carbohydrates for energy storage.

The GenBank which is a genetics databank owned by the U.S. National Institute of Health has over 300 billion bases in 200 million sequence records. One can probably extrapolate from this the total amount of data available in other countries. None of this mind boggling amount of data has given any indication that the above statements may be violated. No new genetic material or polymer or metabolic process has been found in any newly discovered species, such as the newly discovered species of a snub-nosed monkey in Myanmar or a nematode that lives in deep gold mines. This is a ringing endorsement of the Theory.

8. The hierarchy of the tree of life or the hierarchy of clades remains intact and no confusing species, intervening in the middle or belonging to two different branches in a discontinuous way, has been found. That is, there are no birds which give birth or there are no fish with bicuspid teeth and per J.B.S. Haldane, rabbits have not been found in the Precambrian era. The nesting of this "phylogeny" is rigorous and well founded in analysis and can be measured by such things as *consistency index* (CI).
9. If a new phylogenic branch (a new branch in the tree of life) is found, it should not remain separate from the main tree. This is like stating that if a new element is found, it should fit into the periodic table. If it did not, the whole periodic table paradigm would fall. So far, no disparate phylogeny has been found. The degree of match or congruency for new trees being found is stunningly close.
10. Fossils of transitional species to confirm the descent have been found, for example, transitional reptile-mammals, four legged seacows and ape to human hominid skulls. Fossils of Archaeopteryx, a dinosaur closely related to the origin of birds was found as early as the $19^{th}$ century. The pakicetids (Figure 2.26), which lived 53 million

years ago, are an extinct family of hoofed mammals that were the earliest whales. A crocodile like amphibian-dinosaur Ambulocetus Natans that lived about 49 million years ago was discovered in Pakistan in 1994.

Figure 2.26. An illustration of a pakicetidae. (Source: Wikimedia Commons Free Public Image).

Contrary to the naysayer's statements that there is no fossil evidence for macroevolution, the fossil evidence, in both plant and animal species, are too numerous to give an account here. Again as stated before, macroevolution is easily disprovable by finding one species of ape or fish in this transition that does not belong in the outline of the descent. So far, no such finding has taken place despite attempts by several people. (See for example, *Mammal-like reptiles and the origin of mammals* by J.S. Kemp. (Kemp1982)).

Much more can be said in support of the theory of Evolution and Theory of Natural Selection. The biology of today has matured and become too complex for easy description to a lay person and even so, the theory of natural selection and evolution remains inviolable. Even the above simple account must suffice for the reader to be convinced that the large number of scientists are not barking up a wrong tree and have arrived at this as the best theory to explain life as we see today. This conclusion is also not based on a single person's whim (such as this is Darwin's theory and others are just blindly following it). A mountain of evidence gathering, passionate disagreements, intensive collaborations have led to a truly amazing biological knowledge. The new scientific discoveries in molecular biology and botany, medicine, agriculture, horticulture, fisheries and behavioral studies are founded on this theory. 150 years of experience has only confirmed the usefulness of this theory, even if one may have questions about its Universal validity. In that, the Theory of Evolution is no different from the Special Theory of Relativity. They just work!

## THEORY OF EVOLUTION, SOCIETY AND RELIGION

Undoubtedly, the origin of species, particularly of humans, is of great interest to religions. The only other field with equal interest for religion would be cosmology. Also, the unique role of the people, religious or otherwise, as humans must dispose them to have a greater interest in knowing how humans came about. This latter interest is seen in popular culture, art and literature and entertainment industry. Even politics and rhetoric are not immune to the attraction of this field.

The theory of natural selection has received tremendous amount of attention in the society, because it has social implications or it can be reconstructed to advance one social theory or another. Fascism, Communism, Socialism and Capitalism have all interpreted the theory of natural selection to benefit specific themes or even the foundation. The letters exchanged between Karl Marx and Friedrich Engels, the founders of socialist theory, show that they considered the theory to be a natural premise common to communism. The famous Marxist revolutionary Leon Trotsky and Chinese revolutionary leader Mao Tse Tung held that Darwinism is the basis for the dialectic for examining social class struggle. The communists take to the theory because, to them, it stands for emerging victorious in conflict and struggle for existence. Free market capitalism also draws inspiration from the survival of the fittest in market economy. Specifically, the emergence, survival and fall of enterprises, the competition for the resources (customers) and the varying environments of demand and supply, governmental regulations and money supply mirror the processes in biological evolution. The symbiotic relations between companies, adaptations to environments and spin off of new variations in product lines seem to be similar to the biological evolutionary process. Darwin himself mentioned Adam Smith, father of Capitalism, in "Descent of man".

The Nazis and white supremacists hold that the white (Aryan) race is the superior of human species and interpret the theory of natural selection and the survival of the fittest (actually not a phrase belonging to the theory of natural selection, but to a writing 6 years earlier by Herbert Spencer and constantly misquoted) to mean that the whites should have a higher status and should carry out a struggle. In that effort, they tried to eliminate other races and committed crimes against them. Darwin's own cousin, Francis Galton, though a thoughtful scientist, coined the term Eugenics, and in 1901, proposed that high ranking members should marry early and produce children to improve the human species. Although Thomas Huxley pointed out serious arguments against such interpretation, the idea received much attention and was even practiced for many decades, "to improve the quality of human societies". Advancement in science to disprove such ideas, the Nazi atrocities in the name of eugenics and the general recognition that all races are of equal competence and "quality", has now led to the total rejection of eugenics. (This is another role science plays in society, namely it weeds out prejudicial ideas based on false presumptions and premises).

Today, most societies, except for sections with deeply and fundamentalist-religious populations, accept the theory of natural selection as a biological theory and are careful about its broad application to other fields. There is a considerable variation in the acceptance of it as THE explanation for the ascent of the human race and animals and there is insufficient public understanding of the power of natural selection in producing the species that we see around us. There is just a general feeling among individuals that it is reasonable and possible. To some extent, this is understandable because, indeed, to all appearances, the perfection of species, the exquisite functionality of organs and the beauty of many of the flora and fauna on land and ocean, seem like they have been engineered by someone. But as shown in Chapter 4, repeated steps with simple rules of change, criticality in balance of competing parameters, sensitivity to conditions, can create these exquisite entities. This, therefore, is not a matter of belief or realization in a religious sense, but is something to be observed and verified.

However, religion is square and center in the debate with science on this topic. Of all scientific topics, the Darwin-Wallace theory of biological evolution and the later evolution theories incite strong sentiments among the religious. As others have pointed out, there is an unmistakable spiritual theme in the modern science of evolution- nature is constantly exploring potentialities of life forms and this potential is expressed in evolving to the next step. This should certainly be palatable to all religions. Christianity struggles to explain biblical ideas on the creation of living things, and according to Christianity humans are special beings with souls and other life forms are just lower. The theory of evolution roundly refutes the narrative of the Bible. Hinduism, on the other hand, has no problem with the theory of evolution and has an inexplicable and astounding congruence. The Hindu creation narrative also does not have a special place for creation of humans, since it is the *Karma* of a soul to be born a plant, an insect, an animal or a human. Hindu mythology delineates the descent of species in broad terms, just as it has happened. The response from Islam is mixed. In early history, Muslims have proposed theories of evolution and have anticipated those of Darwin, but there are others who have the same problem as Christians. Modern Islamic theologians accept the overall theory, but maintain that humans are miraculously created while conceding that there may have been proto humans or human-like creatures in the past. In the author's opinion, in many non-western countries, anathema to the theory of evolution is just due to the distrust of the west and its approaches. Islamic fundamentalism also breeds this distrust. (On a related but different topic, despite a variety of scriptural notions in religions, no religion has ever posited or even speculated on the past existence of dinosaurs, which are proven to have existed. It is science that discovered this prehistory of animals. This fact alone is sufficient to convince us that science seeks and discovers truths).

## Theory of Natural Selection, Christianity and Western Society

The theory of evolution runs counter to the Bible. In the first chapter of Genesis, there are ten repeated statements to the effect that God created plants and animals to procreate more of their (own) kind. The New Testament completely accepts the Genesis account. The evolution theory and natural selection runs against God's design and if ascribed to God, God would appear to be meandering and aimless, prone to mistakes, which is against the biblical description of God. As for the case of most religions, theory of evolution robs the people of a morality and value based origins where God's love and God's omnipotence created the living organisms with God creating everything as equal.

Your eyes saw my unformed body;
All the days ordained for me were written in your book
Before one of them came to be.- Psalm 139.16

and,

"But from the beginning of the creation, God 'made them male and female." *(Mark 10:6)*

Above all, the Bible's place of humans as the prime members of the living world is denied by the theory, which states that humans are products of millions of generational process which originated with a single cell organism; that Man is not created by the hand of God and Woman is not created from the rib of the Man. Also that man is no different from other living organisms including single cell creatures. While wildly successful now, in the past, human population came close to extinction. Such a fact does not conform to human-centric or teleological notions. The theory of evolution also frees itself of judgment about which species is successful on the basis of intelligence or consciousness. So a cockroach would be considered by the theory of evolution to be as successful as humans. The theory of evolution disagrees with the Bible, in origins, the descent and time scale of evolution, (Bible counts the age of humans as 6000 years and same as other life forms). So, some Christians believe that teaching of evolution is teaching of atheism. Many faith based writings denounce the theory of evolution in circular arguments, such as, Jesus cannot err or that the theory is against the Gospel and therefore it is wrong. For the faithful, these arguments are fine, because faith in the scriptures is primary to them. For other faithful believers, it would be sufficient to point out that the theory has a lot of holes such as the inability to find some transitional species when a species evolves from one kind to another gradually or that genetic relationship does not prove that the species evolved into another by natural selection or simply that there could be alternative theories that might be equally valid. For many, the lack of adequate proof for macroevolution, the clades of the tree, like amoeba to fish to vertebrates to primate and birds is a major shortcoming of the theory and thereby proving that Creationism is as good a theory. Then, of course, the mother of all questions is raised - the theory of evolution does not explain the origin of first life or in the words of Francis Collins, former head of the "Genome Project"- *"But how did a self replicating organism arise in the first place? ... That is not to say that reasonable hypothesis have not been put forward, but their statistical probability of accounting for the development of life still seems remote."* (Collins, 2006)

To the scientific mind, the Bible's statements are not acceptable and this also troubles many rational Christians. A statement such as, "- *Then Jacob took fresh rods of poplar and almond and plane, and peeled white streaks in them, exposing the white of the rods. He set the rods that he had peeled in front of the flocks in the troughs, that is, the watering places, where the flocks came to drink. And since they bred when they came to drink, the flocks bred in front of the rods, and so the flocks produced young that were striped, speckled, and spotted"* (*Genesis 30:37-39*) would immediately cause a lot of problems to the scientific minded person who knows that a visual suggestion does not produce new species.

Darwin himself struggled with his basic religiosity and his finding of the amoral character of Nature. Darwin was christened in Anglican tradition and trained to be a parson. He married a woman who was a devout Christian and they were unified in their Christian faith. But, he rejected Christian faith (and became an agnostic) after his young daughter's death. This also coincided with his development of the theory of natural selection which could explain life forms without the need for divine creation and intervention. The theology that he had grown with was always a burden to him and his own personal travails assailed him resulting his losing his faith and seeing the misery of life in the midst of this wondrous science of diverse flourishing of life. In the movie "Creation" released in 2009 and Directed by Jon Amiel, Screenplay written by John Collee, Jon Amiel and Randal Keynes on Darwin's life, there is an exchange between his long time friend, a reverend, that fully captures his disenchantment with the concept of a benevolent God -

> "**Charles Darwin**: It's been finished for me, actually. A Mr. Alfred Russel Wallace has arrived independently at exactly the same opinion. Expressed in a... In a mere twenty pages! Now there's brevity for you. I had covered two hundred fifty so far and have come to a dead end, so whilst having wasted twenty years on the project, I have at least rid of it.
>
> **Reverend John Innes**: Well... Well. The Lord moves in mysterious ways.
>
> **Charles Darwin**: Hmmm, yes he does, doesn't he? You know, I was remarking only the other day, how he has endowed us in all of his blessed generosity with not one but nine-hundred species of intestinal worm, each with its own unique method of infiltrating the mucosa and burrowing through to the bloodstream. And on the love that he shows for butterflies by inventing a wasp that lays its eggs inside the living flesh of caterpillars.

Neil Gillespie points out that Darwin dearly needed to break away from the doctrine of Godly creation so as to put the origin of species on a scientific footing. He states –

*"Darwin realized that he had opened up the most serious problems at the interface of science and religion, but in the end he could not resolve them even to his own satisfaction. He ended his career in a state of total confusion about the one problem which his great book purports to solve; namely, how to explain the origin and evolution of life in scientific terms without an appeal to religion. As he confided in a letter to one of his closest colleagues, "I am in an utterly hopeless muddle."* (Gillespie 1979). (The current advances in biology have now carried out his mission). Neil Gillespie also characterized pre-Darwinian creationism as*: "not a research governing theory (since its power to explain was only verbal) but an antitheory, a void that had the function of knowledge but, as naturalists increasingly came to feel, conveyed none"*, and then concluded that Darwin rescued the subject from the creationists.

It would appear that Wallace was never in doubt about the validity and sufficiency of the science and the scientists. In an unequivocal support for science against faith, he wrote in his letter to his brother-in-law-

> "... whether there be a God and whatever be His nature; whether we have an immortal soul or not, or whatever may be our state after death, I can have no fear of having to suffer for the study of nature and the search for truth, or believe that those will be better off in a future state who have lived in the belief of doctrines inculcated from childhood, and which are to them rather a matter of blind faith than intelligent conviction." (Fichman 2010).

In the modern assessment of the theory of Evolution and Natural Selection, there are as many shades of opinion as are color shades available in Sherwin Williams paints. These range from those opposed, with -everything about theory of natural selection is wrong (as in (Morris 1989)), to dismissive, like Cornelius Hunter's *"Darwin so lowered the requirements that anyone with a pen and a vivid imagination can now claim to have solved the problem of complexity*. There are those like John Lennox, who essentially says that though there is evidence for the theory of evolution, it is probably not incontrovertible, and even if the evidence were true, it only proves all this was created. Some just profess dubiousness like Colin Patterson, whose conclusion is that despite all his attempts, he is unable to grasp the validity of the theory of evolution. There are some like Thomas Huxley who gleefully declared that evolution is finally the theory that antagonizes the Catholic Church and Richard Dawkins who believes that the Theory of Evolution is the only theory there is, and anyone who does not believe in it, is ignorant or insane. In between are also, scientists like Stephen Jay Gould, who believe that the theory of evolution and Creationist beliefs belong to different "magisterial domains". Unfortunately, there are also some authors who have critiqued the Theory of Evolution without doing their homework. The cleverly but incorrectly argued book *"Darwin's Black Box"* by Michael Behe uses a poorly applicable theme of irreducible complexity (body systems are just too complex) to counter gradual adaptation by natural selection. The thesis has been poorly received in the community. (See for example, (Perakh 2004)).

Mathematician–theologist John Lennox, in his book, *"God's Undertaker"* (Lennox 2007), compiles nearly all the shades of views of many scientists and philosophers dissenting from the theory of evolution and natural selection, either because of their belief in creationism or because they are uncomfortable with the gaps in the theory, in order to claim that science does not have the answers. Using mainly the statements and quotes of many scientists from history and contemporary ones, like Fred Hoyle, David Berlinsky, Robert Wesson and others and using very few of his own arguments, he attempts to discard practically all the main conclusions and claims of the theory of evolution. The book states its opposition to the science of natural selection as the final theory of species to the exclusion of creation or design.

This view is just a matter of semantics, because naturalism is the study of science and theism is present in religion. Also, the argument that all scientists would be atheists because they do science is false. Humans are fully capable of compartmentalizing their work and religion or lack of it. It is probably not infrequent that atheists and agnostics work in places of worship. John Lennox uses the confidence of the backers of the theory of evolution against the theory itself. In response to geneticist Richard Lewontin's statement that to deny the evolutionary facts is like denying that the earth revolves around the sun, Lennox argues that " *Of course, Lewontin's admitted a priori materialism ...we can now put his protest in the context: no other option is open to him. However, there is reason to suspect that at least part of the vehemence of this kind of protest comes from the ambiguity in the very definition of the term 'evolution'.*" But, he appears to have, for some reason, ignored the fact that the context is actually the requirement for overall scientific consistency and attitude which is common in all scientific theories, be it the denial of a flat earth, proposing a gravitational field, proposing that blood circulates in the body or that species evolved through natural selection.

The illustrious philosopher Karl Popper held a very deep respect for the theory of natural selection, calling it "his cup of tea" and yet had reservations in embracing it wholeheartedly. His statements, *"The Mendelian underpinning of modern Darwinism has been well tested and so has the theory of evolution which says that all terrestrial life has evolved from a few primitive unicellular organisms, possibly even from one single organism"* (Popper 1978) and *"Indeed, the recent vogue of historicism might be regarded as merely part of the vogue of evolutionism—a philosophy that owes its influence largely to the somewhat sensational clash between a brilliant scientific hypothesis concerning the history of the various species of animals and plants on earth, and an older metaphysical theory which, incidentally, happened to be part of an established religious belief"* show his trusting of the theory. But his reservation is shown in the prudent statement," *"There seem to be exceptions, as with so many biological theories; and considering the random character of the variations on which natural selection operates, the occurrence of exceptions is not surprising. Thus not all phenomena of evolution are explained by natural selection alone. Yet in every particular case it is a challenging research program to show how far natural selection can possibly be held responsible for the evolution of a particular organ or behavioral program"* (Quoted in (Radnitzky 1993)).

Climatologist Sir John Houghton's statement, *"Just because we understand some of the mechanisms of living systems does not preclude the existence of a designer..."* (quoted in (Zacharias 2008)) and writer Charles Kingsley's *"Evolution being a noble conception of Deity"* (Kingsley1859) are mild statements, not worthy of elevating them to the level of conflict, because these are definitely not made in the sense of militant opposition to the theory of natural selection. In fact, in his PBS interview with Bill Moyers on the Public Broadcasting Service in U.S., Sir John states that "*There is widespread suspicion of science, fostered by the feeling that science goes against the Bible. This is very unfortunate; it takes a very small view of God, and a very inadequate view of science.*" (Houghton 2007). These points of view show that scientists can form their own smaller worldviews not requiring God to be in one area and needing Him in another. To many, both disciplines are comforting and they do not let contradictions come in the way or they reconcile the contradiction with non-logical, non-rational arguments such as intuition, taking encouragement from the fact that 'science has not found out everything, at least, not yet'. So, for them, it is okay to take comfort in religious beliefs, particularly with respect to creation of life, where there is the big unanswered question of 'how did life arise?".

The well known biologist at the British Museum of Natural History and a leader in the field of "Transformed Cladistics" (the field of biological trees), Colin Patterson (who gave the famous answer to the question, "Can you tell me anything about evolution?", as "Any one thing that is true?"), exhibits a more nuanced reluctance to accept the theory of evolution. Contrary to the meaning that is derived from this statement by creationists and Intelligent Design advocates, Patterson admits only to agnosticism of the theory of evolution. He states that he was of the same view as Gillespie that pre-Darwin theories were anti-theories in that they were creationist theories which were no theories at all. He does trust the evidence for the theory of evolution. For instance, he argues: *"Convergence between molecular sequences is too improbable to occur, just as similarity between sequences is improbable to be explained except by common ancestry. Some might view this argument as viciously or vacuously circular, but the same argument is routinely advocated in morphology...This is the argument from complexity: if two structures are complex enough and similar in detail, probability dictates that they must be homologous rather than convergent."* (See (Patterson 1988)). Patterson believes the data and only questions the interpretation and language of the Theory of natural selection. He argues that molecular explanations explain the data, but do not prove it. He states that it is more correct to say. *"'I know that evolution explains hierarchy' rather than to say 'I know that evolution generates hierarchy'."* (Patterson 1994). Actually, Patterson is simply reflecting the fact that is actually a creed in science- one does not have to believe in theories. One accepts scientific theories as explanations for observed phenomena and uses them for forecasting other unobservable or as yet unobserved phenomena. Science does not generate natural order. Science explains natural order based on concrete and observable evidence.

There are many supporters of the Theory of Evolution. The biologists of today would wholeheartedly embrace the Theory, even if they reserve their right to believe in scripture for their spiritual reasons. To quote some examples: During his successful campaign for the presidency in 1912, Woodrow Wilson, Ph.D., the former president of Princeton University, was asked whether he believed in evolution. He replied, "*Of course like every other man of intelligence and education I do believe in organic evolution. It surprises me that at this late date such questions should be raised.*" (Quoted in (Mikicic 2010)). It is ironic that he should have used the phrase "at this late date", considering that a century later such questions are still asked. Theodore Roosevelt, his predecessor in the White House, wrote in "*My Life as a Naturalist*" about his childhood reading: "*Thank Heaven, I sat at the feet of Darwin and Huxley.*" (Roosevelt 1918). (Figure 2.27).

Figure 2.27. The edifice of the science of evolution has equaled the houses of worship in its appeal and importance.

Then we have the strong supporters of the Theory of Evolution, for whom it is a well proven scientific theory. By the nature of their affirmation of this theory, they tend to be against Christian version of Creation as the tyranny of its doctrine. Thomas Huxley was one of them. He stated, "... *In addition to the truth of the doctrine of evolution, indeed, one of its greatest merits in my eyes, is the fact that it occupies a position of complete and irreconcilable antagonism to that vigorous and consistent enemy of the highest intellectual, moral, and social life of mankind – the Catholic Church.*" (Huxley 1896). If Thomas Huxley was called Darwin's bulldog for his stout advocacy of the Theory of Evolution, then the geneticist and author of *"The Selfish Gene"*, Richard Dawkins and the physical chemist and author of the book, *"Creation"*, Peter Atkins, and Daniel Dennett, author of *"Darwin's Dangerous Idea"* should be called Darwin's wolf pack. These scientists fight on the side of Evolution at two fronts, one explaining the Theory of Evolution from a scientific view point, for example, *"The Blind Watchmaker"*, where Dawkins explains the slow and gradual evolution of the human eye, and the other, for example in *"The God Delusion"*, argues that religion or God based explanation for the Origin of Species is bad and unhealthy.. Dennett is a staunch Darwinist-adaptionist and believes that evolution is a type of algorithm and even sees religion as a naturalistic process happening in human social evolution**.** (Dennett 2006). Dennett's famous statement in his book *"Darwin's Dangerous Idea"* that Darwinism is a 'universal acid,' "*eating through every traditional concept and leaves in its wake a revolutionized world-view, with most of the old landmarks still recognizable, but transformed in fundamental ways.*" (Dennett 1995), incenses religious people and proves to them the Theory of Evolution is a treacherous scheme of the atheists. Peter Atkins is very direct and says that "*...It is important to distinguish fact from theory, observation from mechanism, phenomenon from explanation. Evolution is a fact; natural selection is a theory of how that evolution came about. I think it rather muddling to speak of the 'theory of evolution'. Although natural selection is the currently accepted theory of how evolution occurs, to refer to it as the 'theory of evolution' colours the term evolution to suggest that it, evolution, is a theory whereas it is a fact. This is perhaps a pedantic point, but the issue is of such sensitivity for some people that it is better to be precise."* Indeed, this is a pedantic point and going back to Colin Patterson's point, the definitions of explanation, theory and fact depend on who is using them for what purpose. For many, mere fact is not science. To this author, the theory of evolution by natural selection is a sound theory, wholly based on evidence much like the Special Theory of Relavity or the quantum mechanical theory of atom.

In a vigorous response to the book *"The Language of God"* by Francis Collins (Collins 2006) in the eminently readable book "*Decoding the Language of God*", George C. Cunningham counters Collins statement that even 150 years after the publication of Darwin's Origin of Species, there is a public controversy. He states, *"(Collins) needs to state that this is a public relations controversy, not a scientific controversy. The essentials are accepted by all but a marginal few in the scientific community. The only scientific controversy is in expanding and applying the theory and filling out the details of the evolutionary mechanisms."* (Cunningham 2009a). Indeed, the nature and methodology of science demands that modification of theories and expansion of theories into new regimes be rigorously debated before acceptance and if necessary, disallowed when found inappropriate. In that spirit, scientific controversies currently surrounding the theory of natural selection and the origin of species are very healthy and will only strengthen the theory. Cunningham goes on to say that the broad media, being in the business of making money, promote controversies. In that vein, the "Evolution Controversy", he states, is manufactured by *"pitting religious proposals and pseudoscience against scientific facts to stimulate the audience's emotional response."* (Cunningham 2009b).

All this debate has resulted in a polarization among the lay public in some countries, though majority of people are now in favor of the theory of evolution, though not necessarily because they are persuaded by the details of the theory, but more because it rings true. A Pew Survey of December 2013 shows that 33% of Americans (most of them Christians) believe that human beings have existed as they were since the beginning of time (and 60% believe in evolution). A very high percentage of Western Europeans believe in the theory of evolution. Among Western countries, Turkey, with a significant Muslim population, has the lowest percentage of people who believe in the theory of evolution.

## Hindu Views on Evolution

If the anthropomorphic God of the Christians is counter to the Theory of Evolution, the connection of Hinduism with the Theory of Evolution appears to be, well, miraculous. The evidence for human evolution, in broad terms, states that living beings came from the oceans to evolve into fish and then to amphibians and then to mammals, then to protohumans, then to a primitive man and then to the modern man.

There are great many similarities in the organization of Hindu thought with the Theory of Evolution. Like anything else in Hindu oral tradition, the dating of scriptures is not clearly traceable, but the *Puranaas* are older than 500 C.E. The final written script is found in the 13$^{th}$ century, any way well before any Theory of Evolution. One of the most revered texts of *Vishnu Puraana* is the *Bhaagavatam*. It describes the mythology of creation and interaction of *Vishnu* in protecting and preserving the world and its inhabitants. In this effort to preserve the world and people against the tyranny and cruelty of demons, *Vishnu* takes 10 incarnations (Sanskrit root word *Avataaraha-* abbreviated as *avataara*). The description of these incarnations is in different cantos and verses of the *Puraanaa*.

1. *Matsya-* or Fish: Water Borne life form
2. *Kurma-* or Turtle: Water/Land borne - Amphibians.
3. *Varaaha-* or Boar: Land borne- mammals.
4. *Narasimha- or* Human-lion: Animal-human
5. *Vaamana* -or Dwarf:- primitive human.
6. *Parashuraama* – Semi-cultured, semi barbaric, agricultural conscious human.
7. *Raama-* the ideal man - social, familial modern human

*Bhaagavatam* goes on to describe the incarnation of Krishna, a superhuman incarnation with abilities to perform miracles. In a twist which might give credence to the notion that this text might have originated before 500 B.C., the *Bhaagavatam* uses the future tense in describing the next incarnation of *Buddha*, a historical figure. (The previous incarnations are described in the past tense and the Canto 1 and Chapter 3, verse 5 describe only the first 7 incarnations),

> Then, in the beginning of Kali-yuga, the Lord will appear as Lord Buddha, the son of Añjanā, in the province of Gayā, just for the purpose of deluding those who are envious of the faithful theist.- Bhaagavatam 1.3.24

Here, *Kali Yuga* refers to the Hindu era which started in 3102 BC in the proleptic Gregorian calendar and has a duration of 432,000 years. (One has to be careful about interpreting this verse as a prophesy that came true. Gautama Siddhartha was actually a grandson of Anjana. It is likely that he was named Buddha because of this association. In fact, Buddha did not preach theism). (As stated before, Hindu scriptures also do not have any ideas about the past existence of dinosaurs, although there are adherents who would claim that the mythological serpents and crocodiles were the dinosaurs known to Hinduism).

Given this background, it is not a surprise that Hindus wholeheartedly embrace the Theory of Evolution, more so, because like evolution, the Hindu time scale of creation involves eons, unlike Christian and Old Testament scriptures that teach creation and appearance of life forms in a matter of thousands of years. Various studies by Pew Research Foundation and British Council show that 75% or more in India agree that the Theory of Evolution is a good description of the descent of the species. The believers would, of course, maintain that all that is part of God's design.

The author C. McKensey Brown differs with this assessment. The précis of the book states that "*...the book raises broad questions regarding the frequently alleged harmony of Hinduism, the eternal Dharma, with modern science, and with Darwinian evolution in particular.*" (Brown 2012). Similar theological dissension is found in more strict adherents and interpreters of Hindu religion such as the International Society for Krishna Consciousness, which disagrees with the Theory of Evolution, much along the lines of Christian denouncement, mainly pointing to evidence or lack of it and in general, because life does not, they believe, originate and evolve only with matter. (The ISKCON trust published a book reflecting these views (Jensen 2013)). It is probably certain that since the Theory of Evolution makes its point with non divine explanations, many devout Hindus would have problems, if pressed on this issue.

## Evolution and Islam

There is not much agreement among Muslims on their view of the Theory of Evolution. It ranges from acceptance to rejection. Many, like Hindus, would accept the mechanisms but still place them under the jurisdiction of God. Like Christians and Jews, the sticking point is still the human descent from Adam. But, Islamic views are moderated because in the hierarchy of scriptures, Quran is supreme and Quran does not have a descriptive narrative for Creation, other than God being the creator. The Ahmadiyya sect believes that the theory of evolution is in harmony with Quran's teaching. Among the liberal Muslim scholars there is a great deal of acceptance of the Evolutionary theory. A good example, representing such wide thinking, was Ibrahim Hakki Erzurumi, who was a Sufi philosopher of the 18$^{th}$ century, living in Turkey and a contemporary of Darwin. He was a very devout Muslim and yet he stated that there is sponge between plants and animals and monkey between animals and humans. (Ziaee 2010). Although this may not be more than an observation in evidence, it does exhibit the acceptance of evolution rather than creation. The 19$^{th}$ century activist-ideologist Jamal ad-Din al-Afghani agreed with Darwin's theory of Evolution, but maintained that God is the intermediary in this process. The ambiguous response of Muslims to the Theory of Evolution continues to this date, and theological attempts have been made by some to reconcile Quran with the theory of Evolution.

An example is, *The Bible, The Qur'an and Science*, by Maurice Bucaille, a physician who was the personal physician of King Faisal of Saudi Arabia. This immensely popular book was published when Bucaille had returned to France after an illuminating period of research into the preservation of mummies in Egypt and the study of Quran. He states in his précis of the book, "*Anyone who reads the Qu'ran for the first time sees a book which abounds with accurate and easy –to- understand scientific subjects... This prompts us to say that if the author of this Qu'ran was a human, then how did he write about facts which never belonged to his age!?"* The book is detailed with many comparisons of modern scientific ideas with statements in Quran, such as "*Do not the Unbelievers see that the heavens and the earth were joined together, then We clove them asunder and We got every living thing out of the water. Will they then not believe?"* - Sura 21, verse 30, showing that Quran states that life originated in the water as proposed in the Theory of Evolution. While Bucaille overreaches on the meaning of other verses in comparing with scientific facts, there are a few striking examples - *"There is no animal on earth, no bird which flies on wings, that (does not belong to) communities like you."* - Sura 6, verse 38, *"(God) fashioned you in (different) stages*." -Sura 71, verse 14. Whatever we take away from this book with respect to anticipation of evolution in Quran, it seems that Quran is probably not against the Theory of Evolution.

## The Origin Problem

While the arguments against the Theory of Evolution include a lot of fault finding with various parts of the Theory, the defense of Creationism or Design and the attribution of devilishness to ideas which turn people against the worship of God as the omniscient and omnipotent entity, focus mainly on two points: The lack of evidence in the theory for the primordial origin of life and the conclusive evidence for macroevolution showing majority of the transitional creatures in the branching of the major limbs of the phylogenic tree. It seems strange that the people of faith would use these arguments against the Theory, because they are the ones who are prone to believe that "Absence of evidence is not evidence of absence." But, to be fair, what they may be saying is that science is more like religion in this sense and therefore you cannot claim absolute objective evidential proof. Indeed, these are two difficult topics. (There is a second level of objection- even if these ideas work, there is no real proof or that there might be a better theory, such as Intelligent Design. The second objection belongs in the category of Huxley's "aimless rhetoric", because, there is adequate proof for the Theory of Evolution, not complete, but enough to be able to work with, without going astray and there is no other theory that will stand up to the methods of science).

The deniers of the Theory of Evolution have two main thrusts- First, the claim that no transitional species have been found. As was shown before, this is not true. In a robust dismissal of this claim, Steven Jay Gould, originator of the idea of punctuated equilibrium, states,

> "Since we proposed punctuated equilibria to explain trends, it is infuriating to be quoted again and again by creationists – whether through design or stupidity, I do not know – as admitting that the fossil record includes no transitional forms. The punctuations occur at the level of species; directional trends (on the staircase model) are rife at the higher level of transitions within major groups." (Gould 1981)

The second point is a stickier one: The Theory of Evolution does not show how life originated in the first place. This is a problem that, in the absence of a mass of evidence, occupies human thought, a lot. The current scientific notion is that life either originated on earth or came with a meteorite, but must have originated in a primordial soup of ingredients that over several billion years, chanced to come to form the nucleotide polymer of DNA. Darwin himself suggested this, stating

> "(The) original spark of life may have begun in a "warm little pond, with all sorts of ammonia and phosphoric salts, lights, heat, electricity, etc. present, so that a protein compound was chemically formed ready to undergo still more complex changes". He went on to give a beautiful insight, "...at the present day such matter would be instantly devoured or absorbed, which would not have been the case before living creatures were formed." – (Murray 1887).

This thesis, indeed, faces many hurdles. It requires almost perfect primordial conditions, such as the correct range of temperature, the correct range of oxygen and chemicals content, a certain amount of geographic, climatic and oceanographic stability, the correct distance from the moon to have the right amount of tides and so on, for life to form and survive. So the probabilities for all these "good" conditions to exist, when combined, are extremely small. On top of it, if one thinks of formation of life as a consequence of Universal laws, then the probability of fine tuning of Universal constants for life to come into existence, becomes extraordinarily minuscule. Then, principles like Anthropic Principle- "We observe and therefore it is", and multi-universe or multiverse theories have to be invoked. The chance based origin of life is stated to be akin to the oft quoted classical problem of a large number of monkeys typing randomly on a typewriter and by chance coming up with a Shakespearean classic like Hamlet.

The possible process of life forming out of inorganic chemicals is termed Abiogenesis. In 1952, at the University of Chicago, Stanley Miller and Harold Urey reported their experiments which were successful in creating amino acids, the building blocks of life, (Miller 1953). (See figure 2.28). At the Iowa State College, Sidney Fox and Kaoru Harada were able to create the amino acids in a soup of inorganic chemicals and dried them out to produce strings of proteinoids- cross linked amino acids. Fox, when working with material of lava cones, saw that these proteinoids clumped to form microspheres, which were cell like specimens with cell like membranes, though these did not exhibit metabolic functioning. In 2009, Matthew Powner, Shao Liang Sheng and Jack Szostak at the University College of London were able to synthesize a nucleotide – a sugar linked to a molecule called AICA (Powner 2012) and synthesis of a DNA type nucleotide is now only a matter of time. However, progress in generating life-like molecules from inorganic molecules has been slow, although introduction of artificial forms of life, starting from existing bacteria is moving rapidly. Let us therefore say that the theory that life originated in a primordial soup is far from proven.

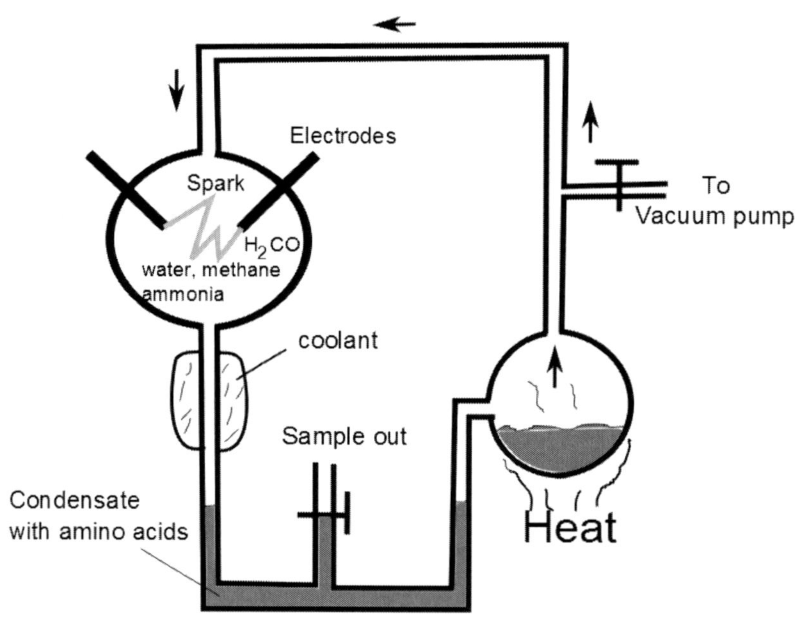


Figure 2.28. Miller and Urey's experiment to create the building blocks of life.

The difficulty with this problem is that it is poorly posed. As is true in physics and theory of complexity, a specific property or randomness is the simplest to model in a simulation. If one were writing a code, it is a one line, x=blue or x= random(101), (that is, x is a random number between 0 and 100). Here, in the case of origin of life with millions of chemical pathways and catalysis possibilities, physical and organic chemistry, geological and atmospheric conditions generating extraordinarily non -linear processes, a theoretical model is not likely to emerge. Whatever the probabilities are, the fact is that life had to be created only once or may be just a few times. For this unique event, the rather naïve estimates based on random chance are as ad hoc as positing that God created life. Biological origins, reconstructions and evolution are a long chain of events connected by a large range of values of probabilities, high sensitivity to prevailing conditions and are also highly iterative and progressive. These are the characteristics of non equilibrium and chaotic processes and are unlikely to be predictable or even retraceable. It is more than likely that this topic will require an enormous amount of experimental work or a deeper understanding and reckoning of the mechanism with no guarantee that the actual process can be pin-pointed. In this sense, the problem of finding of life, a few billion years back is of the same scale as finding alien life on other planets. But, in the meantime, it would be foolish to conclude that creationism is the only answer.

The Theory of Evolution and Natural Selection explain a lot about what we cherish most in our experience, our own body and mind, our partners, our parents, our children and our relatives, our friends and above all the diverse, eye-filling, terrifying, awe-inspiring flora and fauna. It gives us an understanding of how complexity can come from baby steps and it gives us hope that in our own lives, sociological evolution can take place along these baby steps. It tells us that we and all life forms are made of the same fiber and there is much to protect, preserve and sustain in life as we do our own. It promotes harmony among living forms by showing that we are all the same sharing a common ancestry and it allows us humans to feel that we are part of something larger, wholesome and one that promises further evolution to greater heights. The fact is that the Theory of Evolution is a theory of science, which continues to be verified and can be falsified easily. Yet, it has withstood the test for 150 years and most scientists do place their trust in it. A brilliant mind like that of Karl Popper can keep searching for the truth, in the middle of many paths with an open mind, while accepting the theory as a practical theory of evolution with no additional assumptions that controvert it.

However, the mind of the general public is different. Their view of the Theory is not only informed by science but by other influences, sometimes disproportionately. Very often, general understanding of a "Theory" is that it is just a theory and not necessarily true, as in opinions and rumors. The perception that the Theory of Evolution is only a theory, is a failure to understand the methods of science. To quote Karl Popper,

> "The fact that the theory of natural selection is difficult to test has led some people, anti-Darwinists and even some great Darwinists, to claim that it is a tautology... . I mention this problem because I too belong among the culprits. Influenced by what these authorities say, I have in the past described the theory as "almost tautological," and I have tried to explain how the theory of natural selection could be untestable (as is a tautology) and yet of great scientific interest. My solution was that the doctrine of natural selection is a most successful metaphysical research programme...." (Popper 1978)

Daniel Nettle points out that the reason people find it easy to reject Darwin and Wallace's theory is that the Theory is a mixture of deceptively simple ideas and deep consequences. (Nettle 2009). Therefore, one may run with an impression that one understands it and later be influenced by other arguments, for which there are good answers in the Theory. While religious people, with their strong faith in God's role in the origin of life and species, struggle to reconcile or oppose the Theory of Evolution, the Theory is here to stay and has already had a deep and indelible influence on our culture, science and technology. At the same time, people who do not understand the power of order, disorder and chaotic phenomena and the fact that complexity and sharp variations can arise out of simple but profound laws (law of natural selection in the case of evolution), attribute the complexity to the power of a Creating force. In simple words, Nature is the creator of this diversity and complexity, it is not a directed and mindful creation and it is –Evolution.

The public would do well to learn this field and, in general, the field of biology. Biology has become the science of the 21st century with deep consequences for the health of individuals, society, the environment and the earth. The issues that the society will focus on in the near future, such as the effect of climate change on flora and fauna, the rise of new diseases, the technology of maintaining human health and improving quality and length of life and conservation of resources while maintaining quality of life will require a public understanding of the theory of evolution and biology that is much deeper than what it is today. Hopefully, this chapter enables readers to get a good start on this task.

## REFERENCES

Archibald, D.J.; *'Edward Hitchcock's Pre-Darwinian (1840) 'Tree of Life'.', J. History of Biology,* Vol. 42:No. 3,(2009), pp.561–592.

Atkins, P.; *On Being: A scientist's exploration of the great questions of existence"*, Oxford University Press, Oxford, U.K. (2011)p. 203.

Behe, M.*; Darwin's Black Box*, The Free Press, New York, NY (1996).

Brown, C.M., *Hindu Perspectives on Evolution -Darwin, Dharma, and Design*, Routledge Hindu Study Series, Routledge, New York, NY(2012),.

Bucaille, M.; *Bible, the Quran and Science*, ABC International Group Inc., Anaheim, CA (2003) pp.81-143. (Reprint). Ebook: https://archive.org/details/TheBibletheQuran ScienceByDr.mauriceBucaille.

Campbell, N., and Reece, J., *Biology*, 6th Edition, Benjamin Cummings, San Francisco, CA (2002), p.292.

Chambers, R.; *Vestiges of the Natural History of Creation*, W. & R. Chambers, London, U.K. (1884), p.152. Ebook: https://archive.org/details/vestigesofnatura00unse.

Collins, F.; *The Language of God: A Scientist Presents Evidence for Belief,* Chapter Four, Free Press, New York, NY (2006) p.90.

Cunningham, G.C.[a], *Decoding the Language of God*, Prometheus Books, Amherst, NY (2009)p.153.

Cunningham, G.C[b]., *Decoding the Language of God*, Prometheus Books, Amherst, NY (2009) p.191.

Darwin, C.; *On the Origin of Species*, 4th edition, John Murray, London, U.K. (1866), p. xiii.

Darwin, C.; *The Origin of Species by Means of Natural Selection, or the Preservation of Favoured Races in the Struggle for Life,* 6th edition, John Murray, London, U.K. (1872), pp.104-105. Ebook: On the Origin of Species, 1st Edition, Project Gutenberg, http://www.gutenberg.org/files/1228/1228-h/1228-h.htm.

Darwin, C[a].; *Journal and Remarks, 1832—1835 in The Narrative of the Surveying Voyages of His Majesty's Ships Adventure and Beagle,* Vol III, Henry Colburn, London, U.K. (1839), p. 464. Etext: von Wyhe, J.; Darwin online, http://darwin-online.org.uk/converted/ published/1839_Voyage_F10.3/1839_Voyage_F10.3.html.

Darwin, C[b].; ibid, p. 462.

Darwin,C.; *"Darwin to Leonard Horner, 29 August (1844)"*, Darwin Correspondence Project, Cambridge, U.K. Etext: https://www.darwinproject.ac.(2014) uk/entry-771.

Dawkins, R.; *The Selfish Gene*, Oxford University Press, Oxford, (1976).

Dawkins,R., *The Greatest Show on Earth: The Evidence for Evolution,* Simon and Schuster, New York, NY ( 2009), p.147. Also, see: Newsweek for excerpts, Etext: *Excerpt: Richard Dawkins's New Book on Evolution*, Newsweek, New York, N.Y. (2014), http://www.newsweek.com/excerpt-richard-dawkinss-new-book-evolution-79345.

Dennett, D.; *Breaking the Spell: Religion as a Natural Phenomenon*, Viking Press (Penguin), (2006).

Dennett, D.; *Darwin's Dangerous Idea*, Simon and Schuster, New York, NY (1995) p.63.

Fichman, M., *An Elusive Victorian: The Evolution of Alfred Russel Wallace,* University of Chicago Press, Chicago, IL(2010), p.34 quoting from *Alfred Russel Wallace: Letters and Reminiscences*, Editor: James Marchant, Harper, New York, NY (1916), p.65.

Gillespie, N.C.; *Charles Darwin and the Problem of Creation,* University of Chicago Press, Chicago, IL (1979), p.87.

Gould, S. J.; *Evolution as Fact and Theory*, Discover Magazine (May1981) pp. 34-37.

Houghton, J.; *Interview with Bill Moyers, Faith and Reason show,* Public Broadcasting Service (July 16th 2007).http://www.pbs.org/moyers/faithandreason/watch_ houghton. html.

Hunter, C.; *Darwin's God: Evolution and the Problem of Evil*, Brazos Press, Ada, MI (2001) p.75.

Huxley, T.H.; *Mr.Darwins Critics* in *Darwiniana,* D. Appleton and Co., New York, NY (1896) p.147.EText: https://archive.org/stream/darwinianaessay00huxlgoog#page/n6/mode/2up.

Jensen, L.A.; *Rethinking Darwin, A Vedic Study of Darwinism and Intelligent Design*, Bhaktivedanta Book Trust, Grödinge, Sweden (2013).

Judd, J.W.[a]; *The coming of Evolution*, London, U.K.: Cambridge University Press (1910) p.12. Ebook #31316, http://www.gutenberg.org/files/31316/31316-h/31316-h.htm.

Judd, J.W.[b]; *The coming of Evolution*, London, U.K.: Cambridge University Press (1910) p.84. Ebook #31316, http://www.gutenberg.org/files/31316/31316-h/31316-h.htm.

Kemp, J.S.; *Mammal-like reptiles and the origin of mammals*, Academic Press, New York (1982) ).

Kingsley, C.; In a *Letter to Darwin*, 18 Nov 1859, Darwin Correspondence Project, Letter 2534. http://www.darwinproject.ac.uk/letter/entry-2534.

Lennox, J., *God's Undertaker*, Lion Hudson plc, oxford, U.K. (2007).

Levine, G.; *Joy of Secularism*, Princeton University Press, Princeton, NJ (2011) p.191. Etext: Notebook 135e, Darwin Online, http://darwin-online.org.uk/content/ frameset?viewtype=side&itemID=CUL-DAR123.-&pageseq=112.

Lyell, K.; - "*Life and Letters of Sir Charles Lyell*", Vol. II Cambridge University Press, London. U.K. (1881), p.5.

Mayr, E.;*The Growth of Biological Thought,* Belknap Press, (Harvard University Press), Cambridge, MA (1985) pp. 479-480.

Mayr, E.; *The Growth of Biological Thought*. Harvard University Press, Cambridge, MA (1982) p. 330.

Mikicic, A.; *God is redundant*, Dog Ear Publishing, Indianapolis, IN (2010) p.19.

Miller, S. L.; *Production of Amino Acids Under Possible Primitive Earth Conditions*, Science, Vol. 117 (May 1953), pp.528–9.

Morris, H.M.; *"The Long War Against God"*, Baker Book House, Grand Rapids, MI (1989).

Murphy, S; *Drought Could Take Toll On San Diego Bird Populations*, KPBS online evening edition, February 11, 2014, 10:25 AM. EText: KPBS online Evening Edition February 10, 2014, http://www.kpbs.org/news/2014/feb/10/drought-could-take-toll-san-diego-bird-populations/.

Murray, J.; *"Letter to Joseph Hooker",* in: *The Life and Letters of Charles Darwin, Including an Autobiographical Chapter*, Vol. 3,(1887) p.18, Etext: Darwin Online, http://darwin-online.org.uk/content/frameset?itemID=F1452.3&viewtype=text& pageseq=1.

Nettle, D.; *Evolution and Genetics for Psychology*, Oxford University Press, Oxford, U.K. (2009).

Patterson, C.; *Models in Phylogeny Reconstruction*, Editors: R.W. Scotland, D.J. Siebert, and D.M. Williams, *Systematics Association Special*, Vol. 52, Clarendon Press, Oxford, U.K. (1994).

Patterson, C.; *"Homology in Classical and Molecular Biology", Molecular Biology and Evolution,* Vol. 5, Issue 5 (1988): 603-625.

Perakh, M., *Unintelligent Design*, Prometheus Books, Amherst, NY (2004).

Popper, K.R.; *Natural Selection and the Emergence of Mind*, *Dialectica*, Vol. 32, Issue 3-4, (1978), pp. 339-355. Quoted in Isaac, M., *The Talk Origins Archive*, http://www.talkorigins.org/indexcc/CA/CA211_1.html.

Powner, M., Sheng, S.L., and Szostak, J.; *J. Amer. Chem. Soc.*, Vol.134, Issue 33, (2012) pp. 13889–13895.

Radnitzky, G.; *Evolutionary epistemology, rationality, and the sociology of knowledge* by Radnitzky, Bartley and Popper, Open Court Publishing Company (1993) pp. 144–145.

Roosevelt, T.; *"My Life as a Naturalist", American Museum Journal* (1918). Etext: http://www.naturalhistorymag.com/picks-from-the-past/12449/my-life-as-a-naturalist.

Wallace, A.R.; *My life; a record of events and opinions,* Dodd, Mead & company, New York, NY (1905) pp.360-363. Ebook: Hathitrust Digital Library, Ann Arbor, MI http://catalog.hathitrust.org/Record/001490070.

Zacharias, R.; *Beyond Opinion: Living the Faith We Defend*, Thomas Nelson, Nashville, TN (2008) p.124.

Ziaee, A.A.*; Islamic Cosmology and Astronomy*, Translation of *Marifetname – Book of Knowledge* by Ibrahim Hakki Erzurumi, Lap Lambert Academic Publishing, Saarbrücken, Germany ( 2010)).

*Chapter 3*

# MEDICINE AND HEALING: OUR HEALTH, ILLNESSES AND TREATMENTS

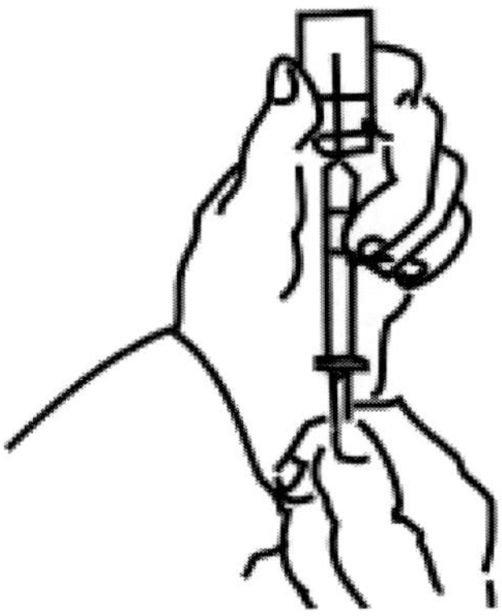

"Life is short, and Art long; the crisis fleeting; experience perilous, and decision difficult. The physician must not only be prepared to do what is right himself, but also to make the patient, the attendants, and externals cooperate." (Hippocrates, 2004)

"...nothing is more estimable than a physician who, having studied nature from his youth, knows the properties of the human body, the diseases which assail it, the remedies which will benefit it, exercises his art with caution, and pays equal attention to the rich and the poor." (Voltaire 1764)

The field of medicine is indeed an art that is based in science. Medical science is more empirical than any other and it is the most complex of all fields of science. There are numerous and clear reasons for it. The primary one is the intimacy of the science to the patient and because of this, the patient feels the products of science more than any other application of science. So, it becomes personal to the patients and engages them at every emotional level, in addition to the curiosity and learning. The emotional response is both an advantage and a detriment to the science and to the practitioner. The expectations from the physician can be likened to that from a court astrologer of early times, about to predict the outcome of the proposed war a king is planning to wage. The patient, at once, reveres the doctor as a god and reviles his or her failure as a failure of the whole field of science. The doctor, in turn, takes the practice as very personal and is impacted by the outcomes emotionally and professionally. Such intimacy with subjects and emotional involvement on multiple sides makes for a science that is not completely objective. Unlike other fields of science (excepting deliberate malevolent applications), medicine can invoke distrust like no other field can. It is a field in which a patient dearly hopes she can understand a lot, but is also a field where the practitioner himself or herself does not know adequately. Often, the doctors have to pretend that they know more than they do and may not adequately explain the details that may confuse and instill fear, in order to give confidence to the patient. These dichotomies breed an atmosphere of confusion, fear and distrust and can block the progress of the science of healing. Again, like astrology, it confers powers to the practitioner well beyond the powers of the science itself and some have misused this power, while others have shunned it to make progress. The history of medicine is replete with examples of this.

There is another compelling reason which makes medicine a very difficult scientific field. The complexity of medical science is due to the complexity of the human body and its processes. A human or animal body is the result of the evolution of life over 3 billion years, through processes of mutations, gene flows, natural selections etc. It involves physics principles, chemical and biological processes and every aspect of a human body is complete with all these sciences incorporated in it. On top of it, because of the simultaneous involvement of various chemical, electrical and biological processes in precise ways, the body functions in the chaotic regime, much like the weather. But unlike the weather, the feedback controls are so exquisite that one does not see this chaos. One gets the illusion that a human body is a highly stable entity. The processes themselves are incessantly responding to diet, exercise, temperature, humidity, composition of air, chemical and other agents in the environment, emotional and physical inputs and responding also to other organisms, all the time. The processes also radically change with the age of the individual. The response of the body to medicines and other therapies is varied depending on the specific nature and habits of the individual. Under these conditions, concrete conclusions can be drawn from the example of outcomes only in rare cases. In this complex interconnected system, even if a cure for an ailment is obtained, it may have a detrimental effect in some other function. To isolate and administer a treatment to the affected area is a mean challenge.

The third difficulty is that medicine and healing are sprawling fields, hard to integrate, at least at this point. The only analogy is our society itself with various field of activities, emotional, intellectual and spiritual components and the integration and disintegration of its various parts. Science, which simultaneously reaches for a high level understanding while focusing on the microscopic level, finds it hard to pursue such an approach. But, so far, a human body has been much too complex to understand using a systematic method. Therefore the field employs different investigative branches to deal with this complex system. The study and practice are broken into different subjects and specializations namely, knowledge of the architecture of the body (anatomy), the development of observational (diagnostic) methods and tools (pathology) and the research and development of drugs (Pharmacology), the repair of organs by physical means (surgery), understanding of change of the character of the body and the bodily processes (physiology) with age and gender (pediatric, adult and geriatric, gynecological divisions), detailed knowledge of the different types of ailments (infectious diseases, malfunction of organs, cancer etc.). In addition to this, the function of the enigmatic brain and the aura of non-organic nature of mental processes, create uncommon difficulties for medical researchers and doctors in the field of neurology and mental diseases. But these are artificial divisions and the body is indeed a single, though complex system and eventually as medical science matures, one can expect more integration of disciplines. This is the reason there is also a great push for holistic treatments, which is more of a goal.

Because of the combination of some or all of the above major difficulties, the progress of medicine was very slow until the 20$^{th}$ century and until then quacks were more common than real medical scientists and doctors. Even celebrated and well intentioned historical medical personalities would appear to be fools today. At the same time, ancient medicine systems that had the benefit of the knowledge of centuries with well demonstrated outcomes in some cases, compete with modern scientific medicine and expose the vulnerabilities of modern medicine, even when traditional medicine is poorly understood. This makes for modern day competition between systems, although modern medicine outperforms traditional medicine in most areas of cures of communicable and non- communicable diseases and ailments. In effect, this competition becomes the third issue for modern medicine: how to include successful traditional therapies in the context of modern science without invalidating scientific methodology and, in the meantime, how to educate the public about the scientific methodology and how to encourage them to be objective on the issue of traditional versus modern medicine. This is a typical problem of an application driven science.

Medicine has to fight superstition, magical perspectives and religious doctrines more than other fields because the first obstacle to a treatment is the personal belief system of the individual, while in the other fields (excluding evolution), the personal attitudes are not immediate.

The last but not the least difficulty, the obvious fact that medicine involves the living, complicates the science exceptionally. Unlike other sciences where one can experiment with some abandon on non- living subjects, medical researchers have to be conscious of pain, privacy and dignity of the patient and ethics of the profession. The uncoerced and informed consent of the patient is necessary for any procedure or treatment. The physician has to consent to the wishes of the patient and the patient's family. Other ethical codes of the profession are additional constraints. Therefore the freedom of experimentation, trial and error approaches and free exchange of information about the patients are not available to the modern physician. An interesting dichotomy stymies the whole process of healing. Human physiology and diseases are subject to statistical and phenomenological processes. Yet, a patient thinks of herself as an individual not to be treated as a statistic with probability of outcomes and needs assurance that the science, in her case, is truly applicable to her. No other science has this problem.

Despite these difficulties, the celebrated and unsung heroes of the science of medicine have done a phenomenal job of bringing the field of science to where it is today. The 20$^{th}$ century has clarified the sources of illness, the molecular nature of the physiology, the physical and chemical processes of organs, developed medical technology to diagnose and prognosticate and has found cures for a large range of diseases and bodily ailments. The merit of this advancement is often measured in the increase in life expectancy, reduction in child mortality, reduction in the mortality of mothers in child birth etc. For example world life expectancy which remained stubbornly below 30 years even in early 20$^{th}$ century, has now risen to 67 years, coincident with development of medical science and availability of cures. But, the truth is much bigger than that. While, admittedly, medicine remains more behind in its understanding and application than where the society would like it to be, the quality and the relative certainty of it, has increased a hundred fold in this century. Nature is very reluctant to reveal its secrets particularly in this field and we owe a great debt to the said recognized and unrecognized heroes of medicine, both researchers and healers.

## Before Modern Medicine

This 'before' was not that long ago, since modern medicine, like modern cosmology, is a recent success and is yet to receive the recognition and accolades it deserves. The history of medicine reflects the confusion and the difficulties mentioned in the above introductory section. The quest for health and cure is, obviously, as old as animals themselves. The instinctive attempt of wild animals to cure their own ailments is well known, though sometimes exaggerated. Deer eat the toxic skunk cabbage to improve their nutrition balance and bears eat the root of the cabbage to purge themselves when they have stomach problems. A cat or a dog eats grass to do the same. Capuchin monkeys rub themselves with millipedes because the benzoquinones in the millipede ward off insects. Chimpanzees eat the pith of the bitter leaf plant fruit (*vernonia amygdalina*) to cure themselves of intestinal worms. Kenyan elephants make a pilgrimage to Mount Elgon to grind up and eat a mouthful of volcanic rocks in order to supplement their diet with minerals. Perhaps, human interest in curing illnesses by doctors is partly due to loss of instincts. In any case, evidence of the cures and efficacy of some of those cures is present from the early days of civilization. Prehistoric humans with tool making abilities knew enough to stabilize broken bones with clay, use herbs for sicknesses and even used trepanning (drilling a hole in the skull) presumably to cure epilepsy.

The hunter-gatherer societies were sparsely populated, had a diverse diet and lived in coherence with nature. Though their life expectancy was not very high, they were less affected by infectious diseases. With agriculture came a host of medical problems: denser populations for transmission of contagions, relatively poorer nutrition of monocrops in advanced civilizations like the Sumerians, contaminated water, poorer hygiene and the incessant wars between kingdoms and tribes, increased illnesses and injuries. The increasing populations and high fertility rates demanded greater maternal and child health care. Above all, close proximity to animals brought out new horrible diseases that were harmless to animals but after jumping to humans, caused devastations, until the contagions evolved to coexist or disappeared. Examples are influenza from pigs and ducks, and rhinoviruses from horses. Festering animal faeces gave rise to diphtheria, hepatitis, polio, typhoid, cholera and whooping cough. Rodents propagated mites and ticks. Typhus jumped from rodents to humans. The ruts and little water ponds created by traffic and waterholes and troughs for animals created opportunities for parasite bearing mosquitoes bringing filarial and malarial diseases. The Ascaris worm evolved in human bodies to foot long lengths. Humans were infested with other worms- ringworms, hookworms, tapeworms and so on. Conversion of forest to agricultural lands brought harmful species close to human populations. An example of this is malaria, which, originating from sub Saharan Africa, raged all over the agricultural worlds of Babylon, the Indus valley, China, Mesopotamia and the Mediterranean. Over millennia, natural climate changes modulated human illnesses and infectious diseases through animal populations, susceptibility to weather and nutritional availability. Cities attracted high levels of immigration and fostered hosts of diseases in the densified and unhygienic conditions. The Old Testament narrates the story of ten plagues and pestilences as God's punishment of Egyptians. Some of these might have been based on real plagues brought on by global warming. Thucydides states that the Great age of Athens was terminated by a devastating infectious disease which affected the Athenian troops killing a third of them and then moved on to the mainland killing a similar proportion. Similar stories are told about the Roman Empire having plagues of small pox and measles killing tens of thousands of people.

A scholarly and encyclopedic history of medicine is given in *Greatest Benefit to Mankind*, by Roy Porter (Porter 1997) and a highly readable text book on this subject is *The History of Medicine* by Jacalyn Duffin (Duffin 2000). Readers interested in more details of the history are encouraged to read these and other books. What follows is a brief introduction to the history of medicine.

Humans in early civilization did make some scientifically valid connections between illnesses and causes and medicines and cures, but attributed much of the problems to divine causes, magic and destiny. In some cases, they made false connections and pursued ineffective treatments. The Mesopotamian civilization, existing at around 3000 B.C. along the rivers of Tigris and Euphrates (near present day Iraq) consisted of Sumerians and Assyrians. The physicians of these times combined magic, religion, astrology and empirical observations to diagnose and treat illnesses. Illnesses were said to be caused by breaking taboos, sorcery etc. The diagnosis and prognosis were omen based by reading entrails of animals and consulting astrological charts. A seer divined the influences, a priest conducted incantations and exorcisms, a physician administered drugs and carried out procedures and a head healer presided. Attending a royal or noble family was dangerous because failure could mean the dire punishment of having the hands cut off. Despite these primitive methods, the Mesopotamians could set fractures, and diagnose and cure a number of eye and ear diseases, distended bellies and stomach ailments. They used several drugs, bandages and ointments.

The next millennial Egyptian civilization employed a greater range of drugs and treatments and the physicians were deemed by the Greeks as very proficient. The Ebers papyrus, the oldest known medical text of 110 scrolls of 20 meters length, lists diseases and contains 700 remedies and magical incantations. It describes the heart as the center of blood supply. The vessels were supposed to be carrying all the fluids (blood, urine, semen etc.) and solid wastes around the body to keep it in balance. The Egyptian doctors regarded mental disorders in an enlightened way as a physical illness and thus treated them similarly. The 800 dietary prescriptions and formulations, such as ox liver for night blindness (which worked because of Vitamin A), extract of the stem of juniper for eye disorders, extract of lizard vulva and penis for baldness, treatment for worm invasion and cancer, were all included in the papyrus. Minerals like antimony, silver and copper were known to the Egyptians as medicines. Since dissection was not permitted, anatomical knowledge was limited among Egyptian doctors and was mostly related to surface anatomy and bones and joints. Healthy living was considered essential for avoiding sickness. Diet and digestion were considered to be the root of good health and intestines were the seat of illnesses. So, a conscientious Egyptian would spend 3 days a month evacuating his bowel completely.

## Hippocrates

As Greek medicine started with a good knowledge of Egyptian medicine (the Minoans were in close touch with Egypt) and Egyptian medicine was known in Crete, it expanded somewhat along Egyptian lines importing Egyptian pharmacopeia. In 700 BC, the first medical facility was opened in Greece and the idea of observing the patients before treating, started here. Early Greeks revered Asclepius (thought to be the legendary son of Apollo) as the great healer and physicians were called Asclepiads. Like Egyptians and Mesopotamians, they also went in for magical cures. For example, epilepsy was considered to be a sacred disease brought on by a god's curse. Around 400 B.C., the situation changed with the arrival of the legendary physician Hippocrates of Kos, who is considered to be the father of medicine. All accounts of his achievements and his aphorisms are second hand statements narrated by his followers. These Hippocratic texts were assembled into a Corpus in 250 B.C. and remain a guiding document even to this day, the shining example of which is the Hippocratic Oath taken by all medical practitioners.

In the intellectual atmosphere prevailing in these times, the advancement of physics and astronomy and the philosophical presence of Socrates, Plato and Aristotle helped medicine to be separated from religion, but divinity always played a role in the overall scheme of illnesses and cures. The Corpus showed that the Hippocratic practitioners believed that diseases and illnesses could be cured by appeal to reason, by analysis and search for laws governing the human body, much like laws governing the Cosmos. This, then, was the birth of a scientific medicine. Hippocratic physicians scoffed at sacred diseases and made the patients the center of their investigations. They used analogies and theories for their human models: Hen eggs model pregnancy, digestion is similar to cooking, heat is the source of living, breathing is for cooling the body, liver is the source of blood etc. Dry/wet and hot/cold were states of organs. Since dissection was an indignity, this was not allowed. So, while surface anatomy was well developed, the inner organs were revealed only by wounds and many incorrect assumptions existed. The Koan school of Hippocrates, humble and passive, made general diagnosis admitting poor knowledge of diseases and applied passive treatments. Patient care was most important and allowing nature to cure the body was the key aspect. This gentle method has continued to find favor among physicians of all times. The humble nature of their treatment, using drugs sparingly and making the patient comfortable but immobilized, was primarily therapeutic. Surface treatments of giving rubs or fomentation etc. were administered. Yet, the observations of patients in care gave considerable insights into various ailments. Hippocrates himself was the first to describe clubbed fingers as the sign of lung disease. He described the symptoms for the infection of the lining of the lungs and his teachings remain relevant even today for pulmonary medicine. Hippocrates was the first to develop a system of symptomology, which is taken seriously in fields like homoeopathy even today. Hippocratic methods described diseases and illnesses as acute or chronic and endemic or epidemic-terminologies that are critical today.

The most important concept developed by Hippocratic medicine and one that persisted for close to two millennia was the idea of humors (derived from the Greek word for sap), fluids that circulate in the body and give its characteristics of health and illness. The four humors were blood, phlegm, yellow bile and black bile. The roots of this may have come from the older Indian *Ayurvedic* three- fluid system – air, phlegm and yellow bile, along with hot and cold conditions of the body. (This concept is deeply ingrained among Indians even today and they are liable to describe their smaller illnesses in these terms and treat them accordingly. For example, boils are caused by heat and a cold is because the body is cold. Vomiting and nausea are caused by excess of bile). The discredited but highly influential idea of previous millennia supposed that the balance of these fluids was essential for health and illnesses were caused by an imbalance. (The four humors may have been posited on the basis of observing a glass tube of blood that separates into three fluids, the dark purple bottom layer, the serum being yellow bile and the top translucent liquid, the phlegm). The above noted states of wet, dry, hot and cold, juxtaposed together with 4 humors, provided ample permutations and combinations for describing most illnesses. This also was congruent with Aristotle's four elements of nature, providing a comprehensive schema. Excess of blood was often cited for excessive passion among the young and lack of it, passivity of the old. There were blood-rich and blood-poor foods. Pooling of blood in the affected areas was the symptom for disease or the reason for it. So Phlebotomy or blood-letting became a practice pursued for the next 2000 years as treatment. (It must be noted that while the broad generalization was incorrect, there are specific instances where this diagnosis is correct and therefore, using leeches to suck out blood is sometimes included in modern treatments). Surgery was a last resort in Hippocratic medicine. Injuries were treated appropriately, but suppuration (formation of pus) indicated that the wound was healing. The most prized talent of a physician was prognosis. Admittedly, this has been so for ages, but its value lay in sounder reasons than just the prestige of forecasting. It was more in the nature of estimating chances of cure and probability of efficacy of the treatment. This would also cultivate the trust of the patient resulting in the patient being able to distinguish between real physicians and charlatans. The humble, first Hippocratic principle of doing no harm is a brilliant concept inculcating in all future medical practitioners the idea that a patient is not an experimental subject and also, the philanthropic notion that made this profession a service. This latter idea was fully integrated into the Christian concept of charity and compassion so that many priests and nuns took to healing through medicine, even while believing in divine cures. Despite the inaccuracies, lack of knowledge of anatomy and wrong theories of illness, the theories of Hippocrates ushered in the era of medical science. Until this period, medicine was not amenable to observational research and was only an acquired skill developed through tradition.

## The Roman Medicine and Galen

This access of medicine to non- traditional approaches and permission for anyone to enter into the medical field was expanded in the Roman era. Unlike in later centuries, physicians, particularly ones using Greek methods, were not held in esteem in the early Roman era. This might have been due to xenophobia and the arrogance of the Roman Empire. As a result, the early Roman physicians resorted to atomist, corpuscular theories of medicine where every disease was caused by tightness, narrowness and free versus constricted flows. So the treatment proposed was simple- apply opposites. Consistent and extensive bleeding to the point of the patient collapsing was needed. Between 200 B.C. and 50 C.E. these Asclepiadeans held sway. The Sicilian physician Scribonius changed the trend to return to Greek and Hippocratic medicine. The first to use the Hippocratic Oath, Scribonius was deemed influential because he was respected by Emperor Claudius. The first survey of medicine in Latin was the work of a wealthy land owner-physician Celsus, whose 21 volume Artes (8 of them survive) made him a highly respected physician and won him the title of Cicero (after the famous philosopher) of Medicine. Celsus was a great proponent of reasoning and held that experience, observation and divining were insufficient. He wrote broadly on all aspects- diet, exercise, surgery, diseases, inflammation etc.

### *Galen*

If Hippocrates is held in high esteem for his ideals in medicine, the Roman physician Galen is the Aristotle of the medical field, ruling over it for more than a millennium. Deriving much of his medical theories from Hippocrates, the man swaggered and professed comprehensively on all aspects of medicine. He was born in what is now Turkey to a wealthy family and was educated in the best liberal schools. Following a dream his father had of Asclepius, Galen was put in medical school by his family. He learned and travelled widely and then became the physician for the gladiators, where he learned about anatomy, injuries and treatments intensively. Like Hippocrates, he emphasized patient care and development of patient's trust and showed that some illnesses were psychosomatic and therefore could be treated by treating the patient's attitudes. Galen studied animal anatomy and made medical theories on the assumption that human anatomy was the same as animal ones. (Human dissections were not permitted during this period too). He stated (as the Egyptians thought) that, food was absorbed from the stomach and intestines and sent into the liver. Blood was produced in the liver and this passed through the lung, pumped by the heart (by suction during diastolic phase), to aerate it and imbue it with life and the life-blood was absorbed at various organs and tissues. (See figure 3.1). Blood pumped from the right to the left chambers of the heart through pores in the heart although no (animal) dissection agreed with this hypothesis. When the life-blood reached the brain, animal motions and sensory information were generated. Galen knew of the ideas of blood circulation very well and yet chose to maintain his theory to be in line with his other philosophical beliefs.

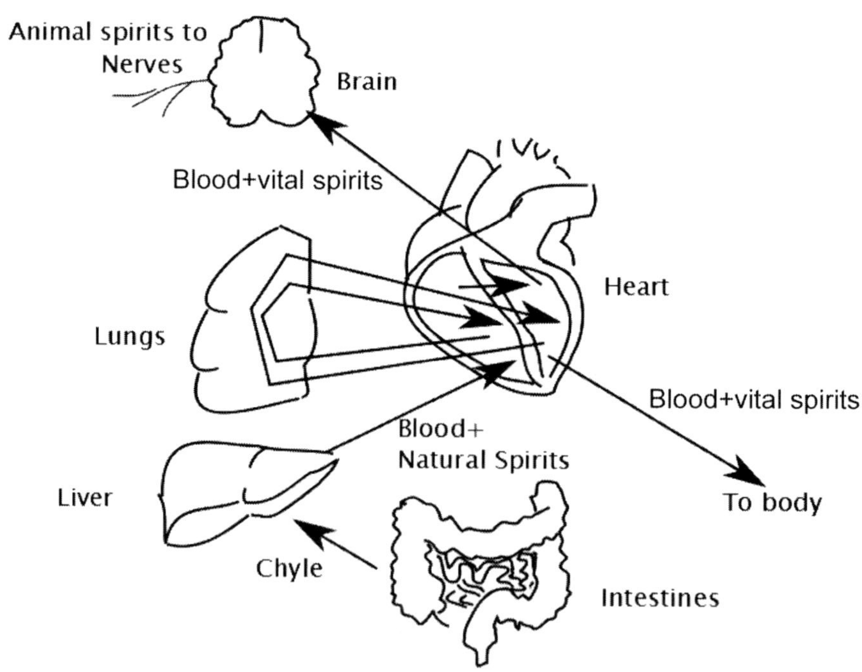


Figure 3.1. Galen's Theory of Blood Circulation.

Hippocratic humors were still the main stay of physiology and pathology for Galen. The balance of humors and qualities were related to proper functioning of the organ. Inspection of urine and faeces revealed the balance of the humors. He too emphasized extensive blood-letting when diagnosis of surplus accumulation of humors was made. Galen followed Platonic ideas of Cosmos and drew analogies to posit anatomical structures. He studied pulse extensively and taught methods and interpretations. (This is also consistent with the methods of *"nadi"* employed in Indian *Ayurveda*). Galenic medicine combined Hippocrates' ideas of humors with theories on anatomy and also of pulse. Drugs were the preferred method and surgery was a last resort, to be performed by practitioners of lower rank. Galen described nearly 500 drugs of animal, vegetable and mineral origin and prescribed methods for compounding. Undoubtedly a genius, Galen was prone to loudly proclaim his own methods and equally loud to criticize others whose methods were different from his. Galen was a showman as well as a scholar and impressed his audience with public displays of animal dissections. Like well-oiled salespersons, his belief in his own importance was concealed beneath a veneer of pedantic language stressing the dignity and honor of the profession. He married his ideas of teleological basis (divine purpose of nature) with the nature of the human body and diseases. Ultimately, he claimed that he had done as much for the field of medicine as the Trojans had done for the Romans in building the infrastructure of roads and bridges. He declared that while Hippocrates had shown the way, it was he, Galen, who had made it passable. Galen's doctrine was received by physicians without protest for a long time.

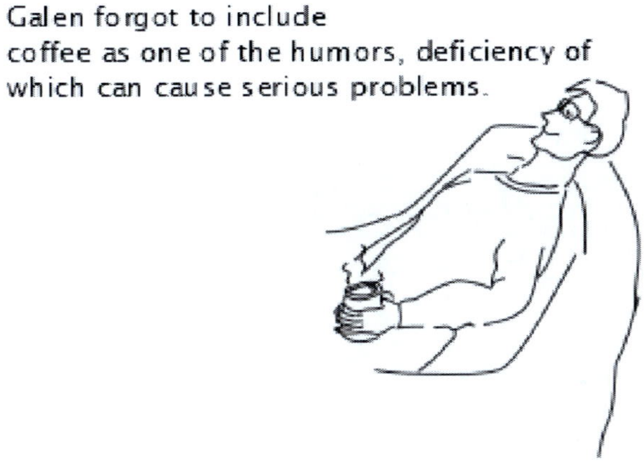


Figure 3.2. Coffee humor.

Despite the advancement of science and its methodologies, medical science remained a controversial field, with reputations of physicians always being suspect, even well into the 19$^{th}$ century. One has only to see the quotes about medicine in the popular literature, such as in Shakespeareans plays, to know that medicine was a field filled with scholars as well as quacks, and exploiters as well as those who practiced to be of service. During the medieval and Renaissance period, when Christianity played a major role in Europe, Galenic ideas thrived because these were consistent with Christian theology. In this period too, there were no ideas on contagion despite occurrence of serious plagues. One outcome of the plagues and epidemics was that the state had started taking a larger role in public health and construction of hospitals started becoming common. However, the medical profession prospered and expanded due to the availability of new texts from the Islamic world (such as the Qanun or canon of medicine by Avicenna – Abd Allah Ibn Sina, commentaries of Haly Radoan- Ali Ibn Ridwan), through Spain. Universities in Padua, Naples, Montpellier, Cambridge and Oxford became centers of medical study, still teaching the Greek and Galenic versions of medicine and using texts of Avicenna. These public engagements expanded the field of science into a scholarly pursuit, even as the progress in practice of healing remained slow. Human dissections were still prohibited, but with permission, executed criminals' bodies could be autopsied. This limited permission itself was responsible for considerable progress in medicine. Epilepsy and Leprosy remained in the domain of supernatural punishments and civic rules regarding people who suffered these conditions showed this bias. In the second millennium, influence of different methods waxed and waned, with Greek (Hippocratic) medicine coming to the fore since all things Greek had a great reputation. One significant contribution to the methods of medicine was the adoption of scientific methodology. A proponent of this was the Padua physician Gian Battista da Monte, who is credited with the formal teaching of clinical medicine. Such methodologies demanded new explanations for the epidemics and plagues. As a precursor to the germ theory, the scientist- mathematician – physician, Giroloamo Fracastaro proposed that disease was carried and created by action at a distance by "disease-seeds". In this period, several treatises on surgery were written such as the famous ones by Lanfranc of Milan detailing anatomy, embryology, ulcers, fistulas etc. The most comprehensive work of this from the medieval period was the 13$^{th}$ century work of Guy de Chauliac which had 3300 references to surgical treatments. However, despite the progress in scholarship, Galenic tradition and Church doctrines still loomed large in medical practice. (See figure 3.3)

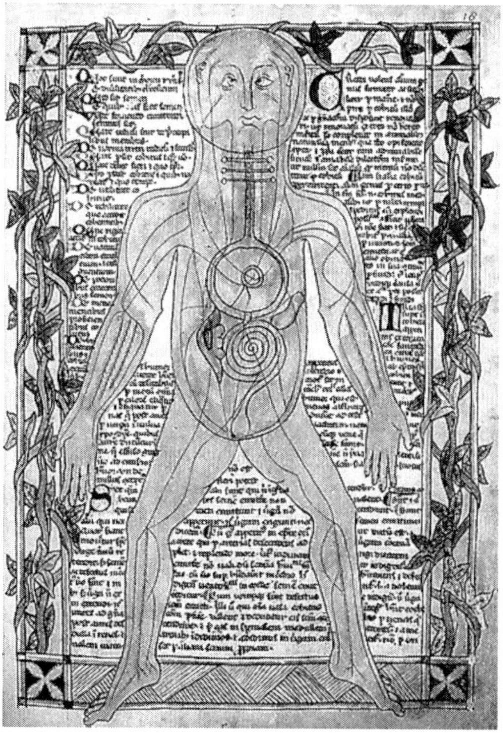

Figure 3.3. 13th century illustration of blood flow (Source: Wikimedia Commons Free Public Image).

## Andrea Vesalius and Anatomy

The 16th century physician Andreas Vesalius is deemed to be the founder of modern anatomy. The son of a pharmacist to the Emperor, Vesalius achieved his anatomical fame in the medically renowned city of Padua. His technically accurate anatomical illustrations created in 1543 with the assistance of Jan Stephan van Calcar were the first for students and practitioners. The first three sheets, drawn by himself, were for liver, blood vessels and male and female reproductive organs, the venous and arterial systems. While starting out with Galenic ideas of anatomy he realized that Galen had made some large mistakes and went on boldly to correct them thus creating one of the most important documents of the Renaissance period in the history of medicine. The documents showed precise arrangement of skeletons, muscles, vascular system, nervous system, abdominal and reproductive organs and the thorax-heart.

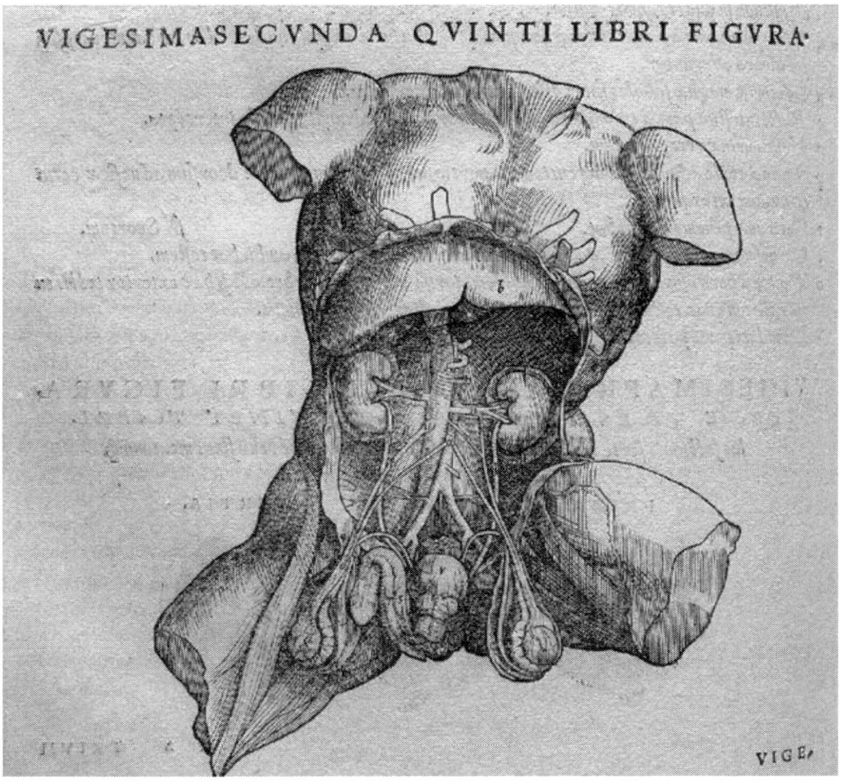


Figure 3.4. Illustrations from Vesalius' De Humani Corporis Fabrica (Source: Wikipedia Free Public Image).

Vesalius made new anatomical observation with examples like the discovery of the canal ductus venosus which passes in the fetus between the umbilical vein and the vena cava. His 6 volume treatise showed that blood flows from the right to left ventricle through a passage that escapes eyes. This would later guide anatomists like Realdo Colombo and William Harvey to conceive of pulmonary transit and circulation. He was the first who dared to show that Galen was wrong in his anatomical statements and assumptions. Vesalius was instrumental in the invention of the mechanical ventilator to assist in patient's breathing. Though the anatomical descriptions of Vesalius were corrected later, history has judged him to be the father of modern anatomy and the inspiration to include anatomy as one of the pillars of medicine. After the publication of the Fabrica, Vesalius was invited to the imperial court of Charles V, but endured the humiliations of a profession which looked down on anatomy and surgery as the work of barbers. His life had its ups and downs until finally he died in an unknown island where he had gone during a pilgrimage. The research work and illustrations by Vesalius are amazing in their detail and medicine owes a big debt to this great scientist. (See figure 3.4).

A view, counter to Greek medicine, also started to appear so that it was no longer regarded as sacred and there was a quest to start from scratch. In the early 16$^{th}$ century, Paracelsus (an assumed name, meaning beyond Celsus), a Swiss physician, learned medicine on the streets from the common people and challenged the then current state of medicine. He had no respect for the academics and believed that one learnt medicine at the bedside of the patient and while living with everyday people. Despite his wrong- headed ideas, his emphasis on medicine being a natural science, the use of chemical analysis rather than visual inspection and the importance he gave to mineral drugs such as sulfur and mercury, set him out to be different. He was followed by Marc Aurelio Severino, Theodore Tourquet de Meyerne and others, who established the use of chemical formulations as the basis for drugs.

## Mechanistic View of the Body

Anatomists of the Renaissance period questioned Galen's anatomy, once Vesalius's observations in Fabrica were seen. The mechanics of the heart was also seen to have been described wrongly by Galen and in general, Galen's cardiovascular description did not correspond to the observations of Realdo Colombo who saw the so called pulmonary transit. Michael Servetus, (born in 1506 C.E.), considered to be the symbol of the beginning of religious tolerance and the first Unitarian to question the orthodox trinitarian Christian theology, was also the first to question Galen on his description of the heart and was the first to conceive of blood circulation and proposed that the movement of the blood from the right ventricle to left, does not happen through pores but does so through the lungs where the blood is enriched with life giving air and then it returns to the heart. He derived these ideas, which were then considered heretic, from his scientific approach as well as his deep faith that the human spirit is bestowed on man from the hand of God. He stated (Cattermole 1997) (O'Malley 1989),

> "In all these there resides the energy of the one spirit and of the light of God. The formation of man from the uterus teaches that the vital spirit is communicated from the heart to the liver. For an artery joined to a vein is transmitted through the umbilicus of the foetus, and in like manner afterward the artery and vein are always joined in us. The divine spirit of Adam was inspired from God into the heart before (it was communicated into) the liver, and from there it was communicated to the liver..... The notable size of the pulmonary artery confirms this; that is, it was not made of such sort or of such size, nor does it emit so great a force of pure blood from the heart itself into the lungs merely for their nourishment; nor would the heart be of such service to the lungs, since at an earlier stage, in the embryo, the lungs, as Galen teaches, are nourished from elsewhere because those little membranes or valvules of the heart are not opened until the time of birth."

He then goes on to cite the anatomy of veins, arteries and walls and declares that the middle wall, lacking in blood vessels, could not send copious volumes of blood through, as Galen claimed. He also states that the coloration of the blood comes from the lungs where it is enriched with divine life. Unfortunately, Servetus, a vigorously intellectual and original thinker, was burnt at the stake for his heretic ideas. But the awakening continued and increased pace in early $17^{th}$ century.

At this time, the lack of any understanding about plagues was particularly irritating to logical professionals and in the increasingly free thinking world, more complete understanding was warranted. Servetus and Colombo had made an impression. Girolamo Fabrizi d' Acquapendente or Hieronymus Fabricius, the influential teacher in Padua and one who had established large theatres for public show of dissections, discovered the membranous folds that prevented blood from backing up, described all the valves of the heart and showed how the blood flowed in the heart. Andrea Cesalpino showed how the valves of the heart would work in a pulmonary transit and used the word "blood circulation" confidently. In the light of the increasing influence of natural science such as astronomy and chemistry, Aristotelian concepts were reinterpreted by Cesalpino in favor of natural explanations as far as one could go.

## Dawn of Modern Medicine and Dusk of Galenic Medicine

William Harvey (born in 1578 in Cambridge) was a student of Fabricius of Padua. He got an extensive education in anatomy and with his prestigious and advanced learning, keen intellect and energy, quickly rose in the profession to become a royal physician by 1618. His essay du mortu cordis, ("Anatomical Essay Concerning the Movement of Heart and the Blood in Animals"), is a major contribution to medical science and became the inspiration for mechanical explanations of the human body (mechanisms). Harvey pointed out, more accurately than Servetus and without resorting to supernatural allusions, that among other things, Galenic anatomy could not explain how the air and the particulates breathed in were kept distinct in the veins in the lungs. Using vivisections (dissection of living animals and tissues), Harvey actually demonstrated blood flow and one visually saw the circulation. He arranged to do this in frogs, which have a slow heart process and thereby one could actually see the blood leaving different compartments. He was the first to announce that circulation was real. He demonstrated this by calculations on how much volume of blood was being pumped by heart and showed that the amount of blood going through the heart in an hour was far larger than the total amount of blood in an animal and therefore, the blood had to be circulating all the time. He demonstrated the pathways and connections for blood flow in detail. His demonstrations showed that by tightly tying the forearm the arterial blood flow would be stopped, but as the strap is relaxed a little, the arterial blood would flow, but not through the veins and veins become swollen. This meant that blood had flowed through the arteries and then to the veins, where it backed up. This demonstration showed one way transit of the circulation and then he completed his theory with arguments such as lack of pooling of blood in lower portions that the blood circulated. The circulation and the intricate branching and pathways then explained the fast absorption of medicines and poisons. Harvey also showed that the pulse was pulsating due to the alternating pressure from the heart and was not an intrinsic nature of blood vessels. This, for his time, was extraordinarily scientific method applied to medicine and would set the standard for all mechanisms in the body. However, Harvey still believed in vitalism (body was constantly energized by vital life energies unique to living things) and continued to draw the Aristotelian analogy of Cosmos and the body such as heart is the Sun of our body.

Harvey inspired a new line of thinkers, including the famous philosopher-mathematician Rene Descartes, who stripped the non mechanistic wordings from Harvey's concepts and described sensations and feelings by mechanistic processes, as small particles travelling through nerves. In order to be consistent with his truly machine-like body functioning, he placed the soul as a separate entity in the mind and thereby subscribed to the mind-body dualism. Instead of life energy, he posited heat as the source of life.

The mechanistic view was to change not only the way of thinking on how the body worked, but also treatment strategies. The discoveries of new elements contributed to a greater number of explanations and mineral drug treatments. In the 1770s, oxygen was discovered by Joseph Priestley and Carl Wilhelm Scheele, by heating oxides such as that of mercury and was demonstrated to be the key element required for combustion (burning). Priestley breathed it by itself and wrote "The feeling of it to my lungs was not sensibly different from that of common air, but I fancied that my breast felt peculiarly light and easy for some time afterwards." (Emsley 2001). Antoine Lavoisier, the famous chemist, before his untimely death under the guillotine during the French revolution, studied respiration extensively. He had the benefit of the discovery of carbon dioxide by Joseph Black and therefore he demonstrated that oxygen was the life giving gas and respiration was the breathing in of oxygen and breathing out of carbon dioxide. This was a remarkable improvement in understanding of body through chemistry. With such discoveries, quantitative analysis also became prevalent, such as calculation of heat absorption, retention and loss, volume of air flow etc.

Medical science started being influenced by great minds like Francis Bacon, who found great flaws in Aristotelian ideas because they lacked any method to arrive at the structure of nature. He believed that human knowledge could be unlimited if the key to this structure could be found. He exhorted scientists to seek empirical proofs without resorting to religion or giving in to idols. In particular, he did not separate medicine from natural sciences. Today that is the constant concept that animates the field of medicine.

## Stethoscopes and Microscopes

In his Inventum novum, published in 1761, Leopold Auenbrugger of Vienna described a method to deduce the health of the chest by the percussion technique- thumping the chest, much like thumping melons and water barrels to know what is inside. A muffled sound or high pitch when rapped with two fingers would indicate congestion and disease in the chest. His method became very useful for future diagnosis and later physicians listened directly into the chest to observe breathing. In 1816, the French physician, Theophile Hyacinthe Laennec, a prodigy who started publishing in the New Journal of Medicine at the age of 16 and became a doctor at the age of 18, practiced this auscultation method. But in the particular case of a stout woman, he found that it was difficult to use the percussion with hand and fingers. Because of her age, sex and her modesty, he could not place his ears on her chest. So he made an acoustic device using a tightly rolled up sheet of paper as an acoustic tube and was rewarded suitably with the sounds of breathing and heartbeats, with a greater clarity than he was used to when listening directly on the chest. This significant discovery led him to his first stethoscope, a one ear instrument with a 9" long, 1.5" diameter wooden tube with detachable ear and chest pieces. (See figure 3.5 showing early stethoscopes).

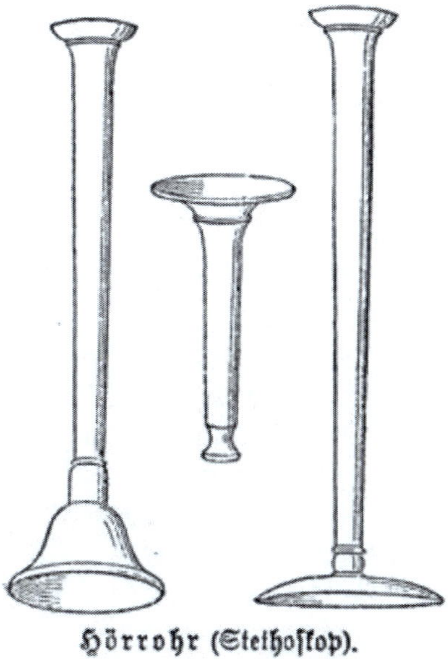


Figure 3.5. Early versions of stethoscopes (source: Meyers Konversationslexikon (1885–90), 4th Edition, Wikimedia).

Laennec developed detailed knowledge of and the ability to detect various chest diseases such as tuberculosis, bronchitis, pneumonia, normal cough and cold etc., by listening to the sounds from the chest. This added to the new methods of diagnosis of the health of the patient in great detail such as comparison of the left side with the right, signs of improvement in disease and correspondence of breathing and heartbeat with other symptoms etc. He followed up his diagnosis with prompt autopsies in case of mortality and perfected his art. Anatomic acoustics became a principle part of diagnosis and prognosis. The stethoscope (meaning chest explorer) was born and its construction kept improving until, in 1852, the modern version of the stethoscope was made. Laennec became very popular and he, along with Gaspard Laurent Bayle, studied tuberculosis at great lengths and established the amazing fact that the disease was present well before it showed up in the system. Laennec distinguished different forms of TB, acute and chronic. With the arrival of the stethoscope, anatomy was integrated into clinical medicine.

Like the telescope, the microscope was to change the view of the world. Once the telescope had been invented and used by Galileo with earth shaking results, microscopy was soon to follow. Galileo himself built one of the first microscopes in 1610. The Dutch built sophisticated microscopes, as they had built telescopes before Galileo and the first compound microscope was available in 1620. The first report of the view of biological tissue was published in 1644. In a couple of decades, the use of microscopy became prevalent for seeing tissues and anatomical structures in the lungs in high resolution. Marcello Malpighi was first to describe the detailed lung anatomy with its membranes (alveoli), the fine texture of bronchioles and their terminal structures. In 1674, Jan Swammerdam and Antoine van Leeuwenhoek discovered red blood cells through microscopy. Though some such thing was expected, the importance of this discovery took more than a century to be understood. In 1851, Hemoglobin with iron component was identified as the red pigment in the cell by the German scientist Otto Funke, and another German scientist Felix Hoppe-Seyler showed that this pigment took up oxygen. This was to finally explain the age old question of life through blood. Life metabolism is driven by combustion with oxygen, producing heat. This had been proposed by Lavoisier in 1777, but had been ignored after he was guillotined during the French Revolution. (Rene' Descartes had also suggested this). In 1876, van Leeuwenhoek saw micro-organisms through the microscope. Rudolf Virchow studied tissues under a microscope and declared that the cell was the fundamental unit of life and its reproduction was life's activity. (Virchow was the first to recognize the work of Robert Remak who showed that the new cells are born out of the division of old cells, a brand new idea. However, he published his own version of it in 1855 without crediting Remak.) Francisco Redi had coined the phrase "Omne vivum ex ovo (Every living thing comes from an egg)" and Virchow generalized it to state that living cells come only from older living cells and dismissed spontaneous generation. Virchow established cellular pathology and studied diseases at the cell level. He constantly urged his students to think "microscopically". After this, the discovery of the germs was only a matter of time. In 1876, after his wife presented him with a gift of a microscope on his birthday, the Nobel laureate Robert Koch began his studies on infectious diseases. Louis Pasteur's study of crystals under a microscope became a boon to humanity. Microscopy was to shake up the world of medicine.

## The Germ Theory

The history of medicine is not a story of clean and steady progress and yet, nor is it devoid of amazing and steadfast progress in a specific area, once a critical point is reached. One such area of medicine is infectious diseases and cures. The germ theory is a revolution in medicine, which is logical and supported emphatically by science. Far from being an idle theory, it is a theory that pointed to powerful cures and preventive measures for diseases that had plagued humanity forever.

For centuries, it was known that, contagions like small pox, syphilis and typhus were spread from one infected person or animal to another by contact. As noted before, Girolamo Frascataro also suggested that "disease seeds" spread disease through air and water. After the invention of the microscope, these wriggling seeds or "animalcules" were readily seen. The seeds were not spontaneously generated as demonstrated by Francesco Redi when meat was protected against flies (in this case maggots did not appear). When Redi's experiments were repeated in broth etc., the results could not be reproduced (because the other experiments contaminated the meat in other ways) and as a result, there was no agreement on where the animalcules came from. The production of animalcules generated religious controversies as to their origin, because after all only God can create organisms. Strangely too, Darwinists, were divided on the issue, though Thomas Huxley, the staunch Darwinist, wrote a brilliant essay on biogenesis (generation from living matter) and abiogenesis (spontaneous generation) and opined that spontaneous generation, though theoretically possible, has not been seen. This was a heartfelt debate encompassing religion and different theories in science. To illustrate, Joseph Lister, one of the founders of the Germ Theory, was not a Darwinist.

With the discovery of disease agents, such as fungus on silkworms and ringworms, the notion that disease could be caused by living organisms, was born. By the middle of the $19^{th}$ century Agostino Bassi and Jacob Henle declared that infectious diseases were indeed caused by living agents. Henle suggested experimentation by infecting animals using isolated parasites and looking for parasites in patients. Everyone understood that the knowledge of the source of infectious diseases would lead to cures. The breakthrough would come from 3 individuals, the French chemist Louis Pasteur, the German physician Robert Koch and the British surgeon Joseph Lister.

## Louis Pasteur

Pasteur, in search of the difference between living and the non-living at microscopic level and thereby understanding what lay behind "vitalism", engaged himself in studies of crystals of compounds, their symmetries and asymmetries in structure and the correlation with their effect on the polarization of light. Because of his proximity to the food and wine industry, the study of fermentation of beer and wine, formation of the acids in vinegar and souring of milk etc., were a natural subject to study and it was monetarily rewarding too. In 1860, Pasteur showed that fermentation was caused by microorganisms which were anaerobic (do not need oxygen). On the matter of souring, Pasteur devised elegant experiments to show that the agents came from the air and were not self- generated. He first passed air through a tube stuffed with cotton and open to the atmosphere. The soluble cotton was dissolved and under the microscope, the solution was shown to contain the same organisms as in the fermentation flasks. When the air was heated, such organisms were not detected showing that the heat killed the organisms. He showed that the organisms were affected by gravity since a flask did not get the organism if the neck was pointing the wrong way. (See figure 3.6). He identified the microbe Mycoderma aceti that soured wine and found that heating the wine to 55 deg C eliminated the bacteria. This gave birth to the process of "pasteurization", where liquids such as juices, milk and beer are heated well above the temperature where the organism could survive. In the case of milk, it is heated to 145 deg (about 65 deg C) for 30 minutes or to somewhat higher temperatures for shorter times. The first disease organism Pasteur identified was a protozoan (a unicellular Eukaryote) that was devastating the silkworm industry. Pasteur convinced the academic community about the nature of the spread of microorganisms. In his speech at the Sorbonne Scientific Soirée, in 1864, Pasteur described and demonstrated his experiments and stated –

> "No, there is not a single known circumstance in which microscopic beings may be asserted to have entered the world without germs, without parents resembling them. Those who think otherwise have been deluded by their poorly conducted experiments, full of errors they neither knew how to perceive, nor how to avoid" (Pasteur 1864),

and then finally on the negative experiment showing that a sterile water droplet does not develop organisms,

"Never will the doctrine of spontaneous generation recover from the mortal blow struck by this simple experiment".

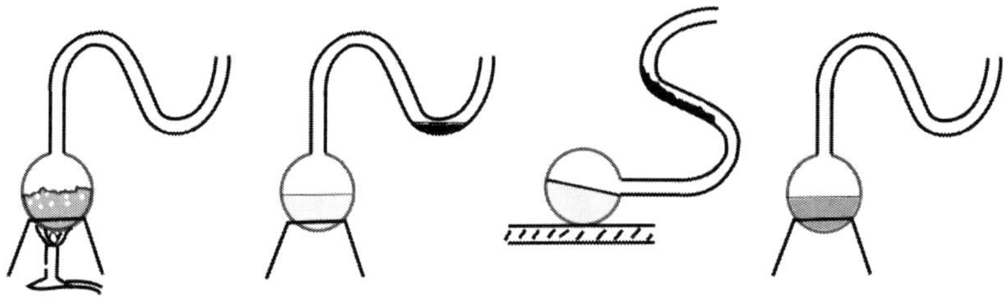

Figure 3.6. Pasteur's swan neck flask experiment: (left to right): 1. A flask with infusion is boiled to kill all microorganism; (2) Microorganisms are added at the swan neck but kept separate, no microorganisms develops in flask; (3) the microorganisms allowed to come into contact with flask infusion; (4) Microorganisms grow in the flask infusion.

Pasteur was already well recognized in the scientific community and had been awarded several prizes including one for physiology at the French Academy of Sciences where, on 19th February 1878, along with his assistants Jules Jourbert and Charles Chamberland, he presented his "germ theory" together with the evidence for it. Later he published the results as a paper. His summary is as follows,

"It appears from the preceding facts that it is possible to produce at will, purulent infections with no elements of putrescence, putrescent purulent infections, anthracoid purulent infections, and finally combinations of these types of lesions varying according to the proportions of the mixtures of the specific organisms made to act on the living tissues." ("The Germ Theory and Its Applications to Medicine and Surgery". (Pasteur 1878)).

Being a man with deep interest in practical applications and medicine, Pasteur ventured into etiology (causes or manner of causation of diseases). In his study of anthrax bacilli, he had developed methods for harvesting infectious germs by infecting the animals with them. His first experiment with disease was with chicken cholera. Two farm researchers, Perroncito of Turin and Toussaint of Toulouse, approached Pasteur with the idea that their chickens were infected with cholera by an organism they had discovered, but they did not know what the organism was. He was ready for this type of experiment. First he collected the infected tissue of the chicken and grew the organism in vitro (in a lab and nutrient environment). When he returned from an absence of 2 weeks, the microbes had not grown much and had gone "stale". Since his in-vitro method was not very effective, he injected this culture of germs into healthy hosts of chicken thinking that the chicken would develop the disease and he could harvest a greater amount of the culture from the chicken. He was surprised that the chickens did not contract the disease. Confused, he got a fresh batch of cultures from infected fowls and injected these into the already injected chicken and also some new ones. This way, he had two batches of chicken, one which had been injected with the stale bacteria before and re-injected with fresh bacteria and another batch which had only injections of the fresh bacteria. To his utter amazement, he found that while the freshly injected chicken developed the disease, the ones previously inoculated with the stale bacteria did not develop the disease. Pasteur now thought deeply about Edward Jenner's work on the development of immunity and recalled that sheep exposed to anthrax for a length of time, had developed immunity. Thus was born Pasteur's idea of vaccination. In 1881, he made a thrilling public demonstration using Robert Koch's anthrax bacillus and showed how 24 sheep, 6 cows and one goat remained immune to the disease when they were inoculated in three stages with serum of increasing strengths of the germ concentration, while an equal number which were not inoculated, developed the disease. This was the boost needed to solidify the germ theory. Pasteur successfully demonstrated the vaccination procedure in rabbits and then dogs to protect them against rabies. The moment of truth came when, in 1885, a 9 year old boy, bitten by a rabid dog, was brought to him as a last resort. Pasteur took the risk and gave the boy one vaccinating injection per day for 14 days with increasing virulence of the rabies germ. The boy remained healthy and grew up to work in Pasteur's household. 3 months later a shepherd boy was saved from rabies by Pasteur. The public knowledge, of children being saved from the jaws of death, inspired confidence in the procedure. Within a year, vaccinations had been carried out for 2000 people and in a decade 20,000 people were vaccinated against rabies. This enabled the establishment of Institute Pasteur, now a celebrated tribute to Pasteur and a powerful house for medical research.

## Robert Koch

In contrast with Louise Pasteur's non-specific methods, Robert Koch worked with well developed scientific methods. In 1866, Koch, who was in his fifth semester of medical school, was asked by the renowned Jacob Henle to join in his research on contagions. This then became his occupation, and his meticulous methods evolved into what are now the standard methods of medical research. Robert Koch did much of his work with microscopes and in vitro, unlike Pasteur. His insistence on developing pure cultures and microscopic understanding of the shapes and movements of organisms under the microscope established the germ theory as a scientifically sound theory in which the shapes of microbes would be very important. As a result, he is considered to be the father of microbiology. His was the contribution that provided the basis for bacteriology with ability to discriminate between different bacteria and provided firm causations between a given bacteria and a particular disease. This is a daunting work, since it is not an easy connection to make, considering how many small and large diseases exist in the body and how many thousands of microbes there are. In addition, as viruses cannot be seen under a microscope and may still cause serious diseases, it is easy to be led astray. Koch formulated 4 postulates that remained the cornerstone of today's research and provide the basis for identifying and ascertaining, without error, that a specific organism causes a specific disease. 1. The infected body must show the presence of a particular microorganism in abundance but should not be found in healthy bodies; 2. The microorganism must be isolated, grown as pure culture and if possible identified under a microscope; 3. If the culture is injected into a healthy animal, the microorganism must infect the animal with the associated disease; and 4. A new set of microorganisms should be extracted from the body and shown to be identical to the original microorganism.

Koch was the first to identify the bacillus Mycobacterium Tuberculosis in 1882. When invited to Egypt and India to identify the cholera pandemic, his laboratory method was superior to that of Pasteur's colleagues who were also invited, because this version of cholera infects only humans and Pasteur's method of culturing the bacteria in animals would not work. Koch identified the cholera bacterium using his in vitro cultures and microscopes. The identification of the microbes causing these two major diseases was a cause of jubilation among the scientific community as well as the public.

In 1890, Pierre Emille Roux, Alexander Yersin in France, Karl Fraenkel, Emil Behring and a visiting Japanese researcher Kitasato Shibasaburo in Koch Institute in Germany showed that the leathery throat causing virulent disease diphtheria is caused by bacteria that emit a poison into the blood stream leading to blocked arteries and airways. The same year, in a major breakthrough, Fraenkel and Kitasato showed that attenuated (diluted) blood or serum from infected pigs can be used to immunize other animals and the first patient to receive this immunization was a child in Berlin. Large scale production of such serum was carried out from pigs and horses all over the world and in a matter of a few years the diphtheria rate dropped by 50% and by 1940, diphtheria became a less significant disease. The serum based treatment was revolutionary and was a major boon because such vaccines can be mass manufactured and do not need culture based vaccines that have to be grown and isolated in a laborious process.

Unfortunately, tuberculosis, which was a major scourge, was not susceptible to such easy progress. First of all, it was known that while many people were carriers, only some (10%) actually get the disease. So there was considerable disagreement about whether it was a communicable disease. Many thought that all tuberculosis diseases were not the same. Once, Koch identified the tuberculosis (tubular) bacillus (TB) as the cause of the disease, these disagreements disappeared. But, contrary to scientific wisdom, Koch, in his eagerness to outdo his rival Pasteur, developed the so called "tuberculin" as a treatment in secret and did not reveal its make up even as he released it for use. Its initial success was received with spectacular enthusiasm and he was feted by celebrities and he was constantly besieged with requests for the tuberculin. But as the long term ineffectiveness of this vaccination became clear, he was in serious trouble. He then released the content of the tuberculin- it was just the bacillus in glycerin. Koch went off to Egypt with his new bride, and his associates had to deal with the aftermath of the scandal. One serious mistake Koch made was to deny another researcher's claim that the cattle bacillus M. bovis and human tuberculosis were similar and for a long time, his influence prevented people from realizing that one could get the TB from drinking milk. M. bovis can be transmitted to humans. (In early 20$^{th}$ century, there were enormous losses of farm animals due to M.bovis). The development of effective vaccine has been ongoing ever since this attempt. The Bacilli Calmette Gue'rin (BCG) inoculation, developed at the Pasteur Institute in 1906, has been used in children in limited ways and success, but its use has been mostly discontinued, because it is effective only against tuberculosis meningitis and its efficacy in pulmonary disease is questionable. In countries like India, one of the first countries to introduce BCG, there was an outcry against it even for use on infants when one of the leaders went against it. The largest trial of BCG was carried out in Chingleput, South India, results of which were reported in the Indian Journal of Medicine, August 1999. In this study, about 367 thousand individuals were registered and 281 thousand were administered the vaccine. 109 thousand of the group had negative skin test and showed normal chest X-ray. After the treatment, no differences were seen between the three groups and the authors concluded that BCG conferred no protection for adults and only a low level of protection (27%) in children.

## Joseph Lister

A British surgeon, son of a renowned doctor and one who developed compound microscopes with high magnification, Joseph Lister is a key figure who studied the germ theory and understood its implications to surgery. His methods saved and continue to save countless lives. Lister attended University College of London and entered the Royal College of Surgeons at an early age of 26 and became a first assistant to James Syme in Edinburgh. 6 years later, he was appointed to the Regius chair of surgery in Glasgow. In those times, surgery was just a procedure requiring no preparation other than tools. Surgical aprons were worn only for protection of surgeons' clothes and hands were washed after the surgery. Lister and other surgeons were keenly aware that nearly half of the heroic efforts in surgery were rendered useless because of sepsis (a whole body inflammation that is caused by the infection of blood leading to organ failure and death). Surgeries were threatened with erysipelas (a superficial infection now known to be caused by streptococcus bacteria) and pyaemia (a type of septicimia now known to be caused by staphylococcus bacteria). Abdominal surgery was considered extremely rarely because of the high probability of infection. It was known that if the wounds were kept covered and infection-free, they mostly healed. But even so, some developed gangrene. In his days, infection was supposed to be caused by the chemical reaction of tissues getting oxidized. Based on this, different surgeons preferred their own methods of dressing wounds, dry, wet etc. The other theory, the old "miasma" in the air, was also always there. Lister knew that air could never really be excluded from the wounds, even if scabs formed. During his work Lister made the following observations: When conducting some experiments using frogs, he found that gangrene was associated with the process of decomposition or rotting. He also concluded that suppuration (formation of pus) was a form of decomposition. Lister used to perform several wrist surgeries and found that carbolic acid helped the patients in healing.

In 1865, Lister heard about Louis Pasteur's work, from his colleague and chemist, Thomas Anderson. Pasteur had, in his "Mémoire sur Ies corpuscles organisés qui existent das l'atmosphère. Examen de la doctrine des générations spontanées" (1861) and "*Recherchés sur la putréfaction*" (1864), claimed that putrefaction (rotting) was caused by fermentation of the tissue by microorganisms carried by air and if the air were made free of these agents and kept under sealed conditions, urine or blood could last for a long time. He also stated that if certain (anaerobic) bacteria were present, air itself is not necessary for the fermentation. The concept clicked in Lister's mind and he came to the conclusion that the decomposition of wounds was the same process. Thus was born the idea of keeping wounds sterile, adding to the scope of the germ theory.

Lister developed and prescribed a ritual for surgeries. During any operation, the room and the personnel would be sprayed with carbolic acid. Remove the clotted blood and clean the wound, bathe it in carbolic acid, apply lint soaked in carbolic acid, cover with tinfoil to prevent evaporation and pack the wound with wool and apply bandage. The bandage would be refreshed periodically. This way, aseptic conditions were maintained and antiseptics were applied. The results, applied in many places, yielded very gratifying results. In Lister's own hospital, mortality fell from 46% to 15% with the introduction of these measures. However, some surgeons, who had been operating already in relatively clean conditions and therefore did not see a major change, and few others, who did not believe in little beasts of disease, ridiculed his claim. But Lister's report in 1867, with zero mortality among 11 fracture cases, convinced the broad community. The 76% mortality outcome, among the wounded in the Franco-Prussian wars, convinced the medical community that something needed to be done and Lister's prescription was embraced. In a Munich clinic, where mortality after surgery had risen to 80%, the introduction of Lister's methods brought the rate down to about 20% and the head of the clinic declared that the clinic would gladly submit to all the trouble this treatment involved.

It took another three or so decades before the actual import of the germ theory in surgery was understood. In the early periods, no one, even Lister, scrubbed though they washed their hands in carbolics and people operated in street clothes. Surgeons used to make a fetish of wearing coats in which they or their mentor had performed successful surgeries. They must have been overly reliant on the efficacy of carbolic acid. Other clinics opposed the procedures because they hated the fumes of carbolic acid. The British surgeon Berkely Moynihan instituted stricter aseptic procedures and adopted the use of rubber gloves which had been developed by William Halsted in the U.S. for dissections. The Polish surgeon Johannes von Mikulicz-radecki introduced the mask to be worn by personnel to prevent "droplet" infections. Robert Koch proved that sterilizing or heating of instruments was more effective than carbolic acid. The use of sterilization, face masks, gloves, caps, scrubbing and surgical gowns rendered the use of sprays unnecessary.

Lister's work advanced medicine directly and in a practical way. In some ways, this is the more understood aspect of germ theory because it is so clearly demonstrable in day to day experience. Lister's contribution to the health and safety of medical procedure is highly celebrated. Some quotes are: *"he [Lister] saved more lives by the introduction of this system than all the wars of the 19th century together had sacrificed"* (Nester 2007). *"Attention to 'detail' and striving for 'improvement' were crucial to Listerian practice, and he sought to convey his credo in three main ways: first, his publications aimed at 'bringing the subject out in the same sort of way as it had been worked out by himself'; second, he set out strict protocols and information on materials and methods, yet also encouraged surgeons to improvise; and third, he made himself an exemplar of a new form of professionalism, which made constancy and vigilance in practice a moral duty for surgeons."* (Worboys 2013).

In the following period of 25 years, almost one disease causing bacterium a year was discovered. However, the history of control over other diseases through vaccination is a story of mixed success. The new genus bacteria *Rickettsia* was identified by and named after Howard Taylor Ricketts and in 1909, it was found to be transmitted through lice, ticks and fleas (zoonotic agent) by Charles Nicolle, who received a Nobel Prize for it. It was not until 1934 that bacteriologist Hans Zissner developed a vaccine for rickets, after a long research. In 1837, typhoid was recognized to be a separate disease from typhus. George Gaffky at the Koch institute identified the bacteria and immunization was introduced by Almroth Wright in 1897. But controversies raged about it and it was not until 1913 that immunization was adopted. By World War I, the incidence of typhoid was down to 2%. Tetanus is another story of success. *Clostridium tetani* is a ubiquitous bacterium causing tetanus and is found in soil. So its transmission is through any cuts and wounds received. It emits a neurotoxin tetanospasmin which travels through the spinal cord and causes spasms and lockjaw. Arthur Nicolaier produced the bacterium in mice and Kitasato grew it in pure culture that led to the production of the vaccine. The tetanus toxoid antitoxin is a deactivated toxin that confers immunity only for a period of about 10 years. The dreaded disease plague, on the other hand, is a difficult disease. Kitasato and Yersin isolated the bacillus *Yersinia pestis* and found that it is transmitted through rats. Because the plague acts by spreading widely through the body quickly, a small culture of the bacillus does not confer immunity. Vaccines, developed from killed cultures of the bacillus, are also not very effective.

The germ theory is the basis for a paradigm shift from the stifling, non-scientific approaches of the past and changed the field of medicine to the extent that the minds of doctors are never away from germs. The germ theory also explains how, on earth, the smallest can overwhelm the mightiest. Today, the anti-germ approach is an emphatic part of public health, be it every one washing hands after going to the restroom or sneezing into the sleeve or wearing a mask when going out or not having open sewers or preventing vectors like mosquito to proliferate etc. It is also a profitable industry in producing antiseptic agents, anti-bacterial soaps and gels. The germ theory is a crucial step not only in identifying microorganisms, the etiology of the disease and developing vaccines, but also was a crucial step that helped in advancing the science of immunity. The molecular basis of immunity and drugs to cure diseases had much to do with the understanding of the germ theory.

Modern medicine is light years advanced over the previous knowledge because of the molecular biological understanding of the germ theory. The advancements in electron microscopy, the increased understanding of molecular and biochemical basis of cellular functions and correlation with DNA codes have enabled deeper understanding of infection mechanisms and drug discoveries, understanding of which has gone well beyond the abilities of lay people.

## IMMUNITY, DRUGS AND ANTIBIOTICS

Elie Metchnikoff, a Russian pathologist and senior member of the Pasteur Institute, had questions on the science of immunity conferred by vaccines. How indeed does the body acquire the defense just from the administration of a small amount of the killed bacteria? Metchnikoff came up with a theory of cellular immunity, when he saw that when fungi were placed on fleas, amoeba like flea cells "ate" the fungi. He wondered if pus was a similar agent that came to the wound spot and ingested the offending bacteria. In the middle of the 19$^{th}$ century, red and white blood cells had been identified and it was known that both changed in disease. Metchnikoff observed the flea analogy in white blood cells "fighting" and eating the disease germs. Metchnikoff called these the 'phagocytes' since Greek *phagein* means eating. The usual Franco German rivalry prevented Koch and other Germans from joining in further research and instead propagated their own ideas, such as white cells might be disease carriers. With due diligence, Metchnikoff isolated the large germ eating white cell (macrophage) granulocyte, and showed that the body's defense mechanism produced a lot of these cells to fight infection. So, the abundance of the white cell itself became a diagnosis of a malaise. Koch identified a third, smaller white blood cell, also one of the leukocytes. In 1888, in a beautiful level of experimentation, Emil von Behring and Kitasato Shibasaburo at the Koch Institute showed that an infectious disease need not be caused only by the microorganism. A filtrate with the toxins excreted by the microorganism, without the microorganism itself being present, caused the disease symptoms. However, the serum of immunized animals, also without the microorganism, prevented infection. This meant that the serum of the immunized animals had "learned" to produce the antitoxins and this humoral immunity was conferred on the vaccinated animal. This was the serum theory of immunity. The idea of these antibodies was set into firm footing by Hans Buchner and Paul Ehrlich and in 1890 the scientists had come to the conclusion that both immune actions, the immune system due to white cells and the immunity conferred externally by a vaccine serum, were active.

The well known British pathologist Almroth Wright, portrayed in George Bernard Shaw's play, theorized that the two types of immunities work together. The serum conferred additional strength to the leukocytes which develop a "taste" for the disease germ from the serum. This, so called "opsonin" theory, married the two immunity systems. This marriage also explained how some individuals could carry the disease germs, while not being affected by it. Sometimes, the vaccines gave adverse side effects. Clemens von Pirquet believed that these were not side effects but were themselves additional immune response and he called them allergies.

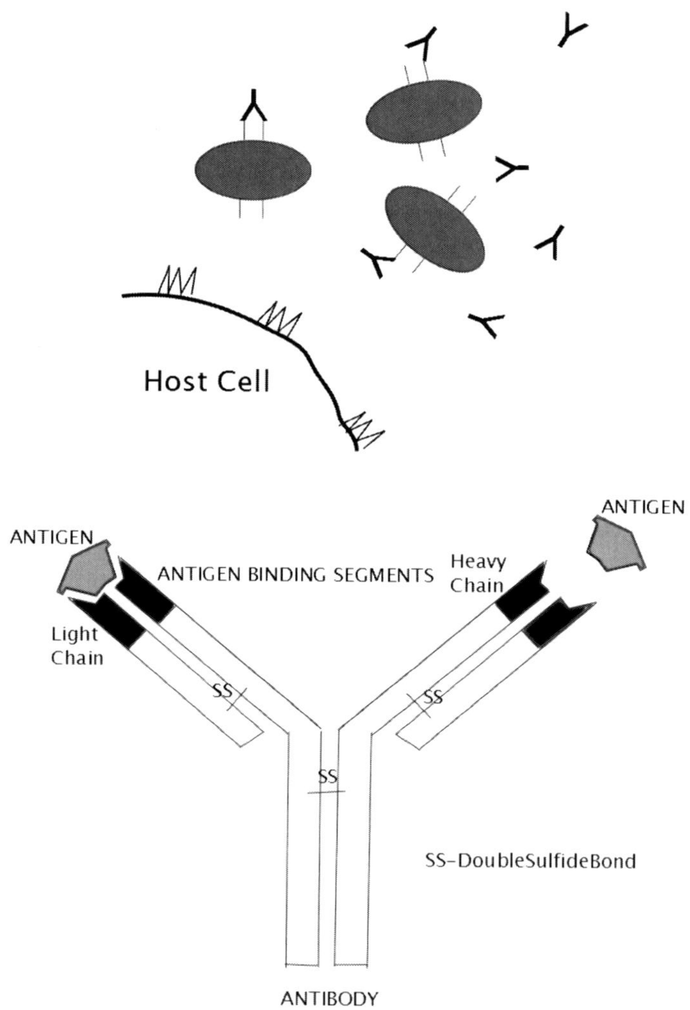


Figure 3.7. A conceptual figure of antibody matching antigen structure. The actual keys are finely detailed molecular structures, folded to generate the shape of the protein.

As stated in the previous section, the understanding of immune system at molecular level is a rapidly progressing field of medical science and is aiding in the discovery of new drugs to overcome immune deficiency and autoimmune diseases. Today, we understand the amazing immune system of animals in more detail. The humoral immunity is now known to be only a part of our immunity. The T cells, the macrophage and other cells (see below) form the primary immune system of the body. Typically, antibodies bind to a feature on an invading organism and render it inert. The antibody is a Y shaped molecule, with light and heavy protein chains on each of the free arms. (Figure 3.7). The ends of these free arms have three protrusions each. The three dimensional nature of these protrusions are such that these fit a particular protein on the surface of a particular bacterium and therefore, each type of antibody has a key to open the lock of the particular bacterium. The amazing fact is that since the body does not know what bacteria may invade and what key would be needed, it makes every possible combination of keys and keeps them ready. Once a particular key is found to be needed, lots of copies of those are quickly made and the invading army of bacteria is countered by an increasing number of antibodies. The killed bacterial carcasses are scavenged of by macrophages.

The immune system is a complex cooperative of different types of infection fighting cells, each like a specialized division of a defending army. Each of these cells "recognizes" the organism by molecular features on the envelope. (The actual process of recognition is chemical affinity and forming of molecular bonds). In case where there is innate immunity (for example, passed on from the mother), these cells recognize invading organisms automatically and when it is conferred immunity, a previous experience from the vaccination "educates" the cells. Initially the macrophages eat up the invading organism and go to the lymph nodes and regurgitate the antigens displaying the organism's structure, so that the lymphocytes at the nodes make the army of antibodies that attach to the features on the invading organisms to render them ineffective. The macrophages then come in sweep them out.

## Magic Bullets

A staining technique had been developed to selectively view or even separate particular cells of tissue (blood is also a type of tissue) or organisms in a tissue, by dyeing them. An example of this is the staining method developed by Hans Christian Gram in 1884. A dye would stain a particular bacterial cell, depending on the characteristics of the cell wall. One could select a particular dye which could color a particular organism or cell. Alternatively, one could color all tissues with India ink. If a certain type of cell is averse to being dyed by the ink, it would stand out under light. Paul Ehrlich, a Polish-German researcher, became the Director of the Royal Prussian Institute for Therapy, Frankfurt in 1899. As a teenager, Ehrlich watched his cousin demonstrate this dyeing technique and had been fascinated by it. When his turn in research came, he adopted this technique for studying immune response. Ehrlich, working with the pharmaceutical companies, Hoechst and Farbwerke Cassella, had a vision of developing synthetic drugs to imitate antibodies. In a new version of Pasteur's vision, Ehrlich believed that an understanding of the chemical structures (molecular arrangement) of cells would lead to proper drugs. (He was quite correct, because now drugs are "designed" from arranging structures of molecules). This impression came from his understanding that the tetanus toxin attached itself to the cells of the spinal nervous system, in particular to "side chains" on the cell protoplasm. With this attachment, the cell, he believed, stops functioning or malfunctions, causing disease. He speculated that the cell, in an attempt to overcome this malfunction, might develop additional side chains which attach to the attacking microbial part and protect the cell. The side chains would be the antibodies. Such antibodies would sponge off the attacking germs but would themselves be harmless to the host (human cells). Ehrlich's quest was to make such side chains synthetically and this would be called chemotherapy with magic bullets. Since dyes attach to the cells, he tried different dyes to create this effect and was successful with methylene blue aniline dye for malaria. For treating syphilis, a vexing disease, he tested chemical cures, still based on his original idea of magic bullets and decided to use milder derivatives of arsanilic acid salt atoxyl, which was known to cure syphilis but was intolerably toxic. After several years of Ehrlich's work, a reexamination by the Japanese bacteriologist Sahachiro Hata yielded a drug in 1914, then called Salvarsan, now known as arsphenamine. But this too was quite toxic and many injections were required and so the treatment was arduous for the patient. After that no rapid development of the magic bullet drugs were made for over a decade.

In 1927, in a new breakthrough, Gerhard Dogmagk, a research director at I.G. Farbenindustrie, developed a red dye named Prontosil that cured streptococcal infections in mice. In a bold application, Dogmagk treated his own daughter with it. After confirmation of its efficacy by the Pasteur Institute, the active part of Prontosil, sulfanilamide proved to be a boon and was recognized as a miracle drug. In 1938, A.J. Ewins, at May and Baker in England developed the now well known drug sulfadiazine which cured many diseases including pneumonia and saved the life of Winston Churchill. By early 1940s, sulfa drugs were the drug of choice all over the world, though overuse of it gave rise to resistant strains of the infections. Dogmagk was awarded the Nobel Prize in 1938, but Hitler prevented him from receiving the prize and he received it only in 1947, but did not receive the cash that went with it.

The most famous magic bullet that killed offending bacteria without harming the body is, of course, penicillin. Alexander Fleming, a Scottish biologist working at St. Mary's Hospital in London, was skeptical about chemotherapy cures and was of the French camp that the body's defenses were responsible for the containment of the disease. He had been following up on reports on lysis, the process of (selective) breakdown of cells and was looking for such an agent to kill germ cells. Researchers before him had, indeed, reported that when they had seen bacterial colonies disappeared with no apparent cause, the "juice" extracted from the bacterial colony had the potential to kill bacteria in a different culture. So it was suspected that, in nature, there are such lytic agents. Fleming's first accidental discovery came when in 1921 he was working on some of his bacterial colonies in a dish and his nose leaked and a drop fell on a dish of certain unidentified bacteria. The bacteria dissolved quickly. With diligent research, Fleming isolated this agent in tears and mucus and named it lysozyme. (In 1909, P. Laschtchenko had already discovered this substance. It is now identified as an antibacterial agent in hen eggs). This was clearly a lytic agent, harmless to the body and produced by the body. His follow through on this did not yield much exciting results but did confirm his theories. The second accidental discovery happened in 1928. Fleming had been sharing an experimental table with a colleague, had set up a number of dishes and started to grow the staphylococcus colonies before he went on a holiday for a few weeks. When he returned, he found his set up had been disturbed by his colleague and while grumbling about this and inspecting his colonies, he found a mold growing on one of the dishes and in the same dish the colony of bacteria was dead. Under a microscope he confirmed that the mold prevented growth of the staph bacterium. In 1929, he made his assistants work hard on the arduous task of isolating the agent. He finally identified it as the mold Penicillum and published the results without mentioning much about any potential therapeutic application. He found that the Penicillum had deleterious effects on staphylococci, streptococci, gonococci, meningococci, the diphtheria bacillus and pneumococci, without impeding the function of leukocytes. But the mold had no effect on cholera or plague bacilli (Gram-negative bacteria that have a specific type of cell envelopes). The penicillum was hard to isolate and harder to cultivate and Fleming did nothing beyond this. In that period a team, led by the Australian Howard Froley and a refugee biochemist from Germany, Ernst Chain, was researching on microbial antagonisms at the Dunn school of pathology at Oxford University. Since normally scientists worked on a topic all by themselves, this was an unusual arrangement for research, with a member working on a specific aspect of the penicillin. Chain discovered Fleming's papers and realizing its importance, the team, with the help of Norman Heatley, devised new production techniques to produce penicillin. They tested it in May 1940 on mice infected with streptococcus, with total success. The first human trial was a dying policeman who recovered from his wounds, almost miraculously. (But he died later, because the team did not have enough penicillin to complete his treatment). Florey seeing the great importance of this discovery, but knowing that a laboratory did not have enough resources to make large quantities of the drug, approached British pharmaceutical companies. But these companies were unwilling to manufacture the drugs. The reluctance of the companies to produce penicillin was due to the fact that they were too busy with war time needs, and sulfa drugs were already available. They were also hesitant to get on board with yet another miracle drug. Howard Florey and his colleagues traveled to the U.S. and, with the help of U.S. Office for Scientific Research and Development, they launched a production effort. Three American companies went into production and the almost miraculous efficacy of penicillin was proved against most pus forming diseases. Rates of death from pneumonia fell from 30% to 6 % after the introduction of penicillin and prolonged the life of older people. Penicillin would save a lot of soldiers wounded in WWII.

## Mechanism of Action of Sulfa Drugs and Penicillin

The sulfa drug works by being antimetabolite, disrupting the formation of a key vitamin-folic acid that is required for cell growth. In normal production of folic acid, an enzyme pteridine incorporates itself into a p-aminobenzoic acid (PABA) and the new compound reacts with another enzyme glutamic acid to form folic acid. However, the sulfa drug molecule has a very similar structure to PABA but has two additional radicals. The drug molecule replaces the PABA in the reaction and creates the intermediate compound of pteridine-sulfa. Now this compound when reacting with the glutamic acid does not produce the folic acid and the bacteria are unable to grow. (See figure 3.8).

Penicillin works by disrupting the process of cell wall production by bacteria. Peptidoglycon (or murein) is a mesh of sugars and amino acids that forms the cell wall of the bacteria. The bacteria constantly remodel the wall as they divide and multiply. The cell wall construction requires a specific enzyme to cross link the fibers of the peptidoglycon. The penicillin binds to these enzymes and prevents the catalysis of the process that creates these cross links. So, as the cell walls build in the new bacterial cells, these do not have any strength and break, causing the death of the cells. In the so called, Gram-negative bacteria, the walls are not fully broken, but the bacteria are degraded and therefore, penicillin does not work so well with such bacteria. Penicillin also inhibits the cell lysis prevention mechanism and promotes cell deaths.

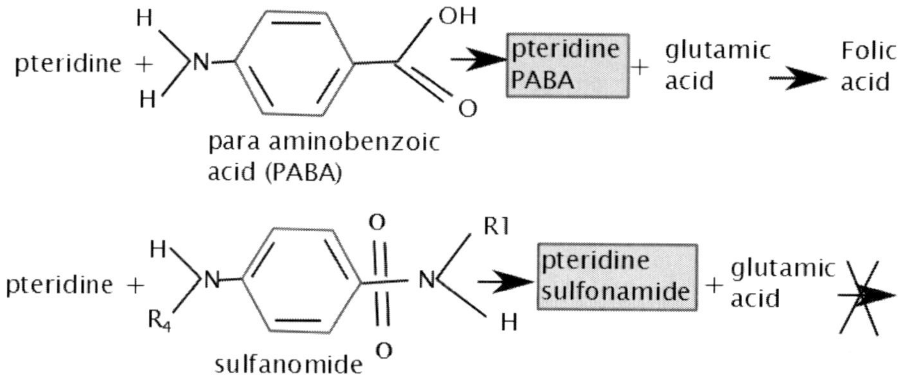


Figure 3.8. The sulfa drug replaces PABA and stops the folic acid production cycle of the bacteria.

Research along the lines of Fleming had been going on and, in 1944, streptomycin, a new class of antibiotic, was developed and showed efficacy against tuberculosis. Further series of discoveries gave powerful antibiotics against tuberculosis and, by the year 1953, TB became mostly curable with six month long doses of the medicine. However, there still remains TB of bone marrow or blood, that is difficult to cure. Also a serious problem has risen with the appearance of strains resistant to these drugs. Jonas Salk, at the University of Pittsburgh made a detailed study of the polio virus and its variations and when he came up with a vaccine, he set up the most elaborate field trial involving over 200,000 subjects. In April 1955, when the polio vaccine was released, he was hailed as a miracle worker and today, polio virus is close to being eliminated.

The rise and fall of discoveries in drugs and healing methods is a natural order of medical science. The public sees only the outcome at their end. The progress and the negation of progress is the scientific process. Progress, particularly in medicine, is chequered because of the empirical nature of the research and the extraordinary time and amount of work needed to verify any cure. So, as history points, there appear to be periods of dormancy and periods of course reversals. But progress is constantly being made, if not by the original inventor, by someone else who takes the baton and continues the race. $20^{th}$ century discoveries proved that diseases, that have destroyed untold populations, could be brought down by the power of science. Antibiotic resistance (both because of its overuse and due to the natural evolutionary process of bacteria) has now become a serious issue and development of new drugs to overcome the resistance is the current path. But future decades will see a better understanding of antibiotic resistance and development of counter strategies.

## ANAESTHESIA - ANOTHER GIFT FROM SCIENCE

The grim story of an amputation is told in the New York Herald of June 21, 1841,

> "The case was an interesting one of a white swelling, for which the thigh was to be amputated. The patient was a youth of about fifteen, pale, thin but calm and firm. One Professor felt for the femoral artery, had the leg held up for a few moment to ensure the saving of blood, the compress part of the tourniquet was placed upon the artery and the leg held up by an assistant. The white swelling was fearful, frightful. A little wine was given to the lad; he was pale but resolute; his Father supported his head and left hand. A second Professor took the long, glittering knife, felt for the bone, thrust in the knife carefully but rapidly. The boy screamed terribly; the tears went down the Father's cheeks. The first cut from the inside was completed, and the bloody blade of the knife issued from the quivering wound, the blood flowed by the pint, the sight was sickening; the screams terrific; the operator calm." (Tarlow 2010).

In the days prior to the invention of anesthesia, surgery was a horrible affair. Because of the grim nature of pain and agony accompanying surgery, often it was deemed permissible to let the patient die rather than submit him to the severe pain. The patient lived through every millisecond of the operation, experiencing every turn of the knife and every touch of the surgeon. The patient must have felt like a warrior going to battle. Surgery was permitted only for leg and hand amputations and cutting of bladder stones (lithotomy). Surface surgeries were allowed and only in some cases specialized surgeries (hysterectomy, tongue operation to cure stammering, removal of ovarian cyst etc.) were performed. Any opening of chest or stomach or the face etc., was forbidden by medical ethics, except in extreme circumstances. Even in permissible cases, alcohol or opium were given to sedate the patient, but more often than not, the patient fainted mercifully and it was not infrequently that the patient died on the table, partly because of the speed with which the surgery had to be performed. Robert Liston, a lion in surgery in the 1810s to 1840s, declared that surgeries like lithotomy should not take more than 3 minutes. Of course, surgeons were keen on reducing patient suffering and extending the time of the surgery to do a proper job. Surgeons knew that if the unbearable pain of the patient could be reduced there could be a wider range of surgeries to save many more lives.

Before 1800, in addition to alcohol and opium, decoction made by soaking the root of the mandrake plant, containing hallucinogenic alkaloids, in wine was also given. There were serious medical side effects to patients from these analgesics. Despite the long years of herbal and chemical experience, no culture seemed to have come up with a real solution to put a patient under. But this changed on a fateful day. British chemist Humphry Davy was the leader of the reform chemistry movement and pioneer of electrochemistry. The British physician Thomas Beddoes was developing a breathing apparatus and in 1795, while assisting Beddoes, Davy inhaled nitrous oxide to check its smell and found that it produced a giggle and also a tendency to dizziness. In his book "*Researches, Chemical and Philosophical, chiefly concerning Nitrous Oxide ... and its Respiration*" written in 1799, he suggested that while pure nitrous oxide was lethal, the gas diluted with oxygen could be used for destroying pain and may be useful in surgery. Across the Atlantic, Horace Wells had set up a dentistry practice Hartford, Connecticut in U.S. along with an associate William T.G. Morton. He volunteered to experience the 'Laughing' nitrous oxide gas when a showman Gardner Colton was demonstrating it. He felt the dizziness described by Davy and suggested to the dentist John Riggs to use it for extracting his own teeth. When Riggs performed it on him and the tooth extraction was painless, he realized its potential. But, when he tried to demonstrate it at the Massachusetts General Hospital, something went wrong and it did not work. Wells was driven out, humiliated with cries of "humbug", never to return to his practice and died a disreputable death. But nitrous oxide was used by many dentists after that and continues to be used in general anesthesia as a component.

The associate William Morton, however, continued to work on the anesthetics, discovered ether and tested it on his dog. On October 16, 1846, he asked dentist John C. Warren, the same person who arranged Wells' unsuccessful demonstration, to check the anesthetic use of ether. The experiment was an unqualified success and Warren declared that 'it was no humbug'. The use of ether became common and within months, it was popular in Europe and newspapers proclaimed victory over pain. Oct 16$^{th}$ is now celebrated as Ether Day. Anesthesia had come to surgery as a tool and saved many lives, but ether with its severe side effects needed a substitute. One was soon found in Chloroform by James Young Simpson in 1847, through another accidental discovery when a bottle of it tipped over during a party and all guests fainted. In 1853, none other than the British queen Victoria went under with Chloroform for childbirth and its use as an anesthesia agent became quite accepted.

In addition to this, topical anesthetics continued to be discovered so that the potential serious consequences of making the patient unconscious could be avoided. Cocaine and its derivatives are now common anesthetics. Modern anesthesia is a very highly sophisticated field where the practitioner has to specialize for several years, learning anesthesia and its consequences as well as aspects of monitoring physiological functions while under anesthesia. So an anesthetist is a so called perioperative physician monitoring and regulating the blood pressure, pH value, and temperature etc. throughout the operation. The rate of administration of anesthetic agents is also exquisitely monitored to allow the patient to go to sleep without adverse effects.

Though the detailed understanding of the biochemical mechanism of anesthetics is still under study, the pathways by which they act in the central nervous system are known. The purpose of the anesthesia is now more than just overcoming the pain. It now includes components for analgesia- removal of pain, amnesia- loss of memory so that the patient does not suffer any post surgical mental trauma, paralytics to negate any motor reflex to keep the body still and relax the muscle so that the surgeon can move and manipulate the body parts and for loss of consciousness such as with chloroform. There is now a long list of general inhalable, injectable or local anesthetics. The potency of each is compared by the so called "minimum alveolar concentration" which is the minimum amount in lungs that is required to immobilize 50% of the patients in response to a sharp pain stimulus. An anesthetist monitors the aspects of airway regulation, physiological status, motor reflex etc and continually adjusts these agents to maintain stasis. Additional premedication is given to patients to prepare them for anesthesia, to make the anesthetics more effective and stable, to reduce postoperative complications such as hypertension and to reduce patient anxiety. Anyone who has gone through a modern general anesthetic process during a surgery cannot but be grateful for the miracle of anesthesia.

# ADVANCES IN BIOLOGY AND MEDICAL SCIENCE

The pre 20th century advancements, as spectacular as they were (figure 3.9), were followed by a great revolution in science, namely the hypothesis and the subsequent discovery of the source of heredity, the DNA. It is difficult to comprehend the full impact of the discovery of DNA on the understanding of all biological processes. After 1953, when the American James Watson and Englishman Francis Crick described the model of DNA and observations revealed further details, the biological field exploded. Scientists everywhere looked at biology at the level of sub cellular activity, using the new DNA paradigm. The biological processes, that are organized by DNA codes and mediated by RNAs and proteins, can be seen in the light of molecules of hydrogen, carbon, nitrogen, phosphorus etc. and radicals formed with these elements, forming and breaking bonds. Therefore, this new "molecular biology" describes the biochemistry that drives all that happens to organisms, including us. A human cell consists of more than 20,000 types of molecules, 50,000 of each on average and their function and response are what decide the etiology of the disease. So, it is not surprising that molecular theories of diseases are being proposed in place of cell and germ theories. The replication of DNA, given in the chapter on evolution, and variables in that process has brought the diversity we see today and that diversity includes the capacity of organisms to invade animal bodies to benefit them or to cause disease. The investigation of the diseases at the molecular level and corresponding advances in biochemistry and microbiology have revealed the nature of old diseases that have eluded the scientists. A public announcement on the discovery of Cholera mechanism by a University of California research group in September 2013, illustrates the level of details to which diseases are now (required to be) understood. (McDonald 2013)

> "We uncovered a mechanism by which cholera toxin disrupts junctions that normally zip intestinal epithelial cells together into a tight sheet, which acts as a barrier between the body and intestinal content," said Ethan Bier, a professor of biology. "A consequence of these weakened cell junctions is that sodium ions and water can escape between cells and empty into the gut.... High levels of cAMP (a chemical messenger molecule) activate a protein channel called CFTR that allows the negatively-charged chloride ions to rush out of intestinal epithelial cells into the contents of the gut," said Victor Nizet, MD, a professor of pediatrics and pharmacy... Through basic physiological principals known as electroneutrality and osmotic balance, these secreted chloride ions must be accompanied by positively-charged sodium ions and water, altogether leading to a profuse loss of salt and water in the diarrheal stools.... The UC San Diego researchers found that cholera toxin acts by two entirely distinct, but cooperating mechanisms to produce diarrhea. In addition to increasing the efflux of chloride ions through the CFTR channel, it weakens cell junctions to allow a rapid outflow of counterbalancing sodium ions and water between the cells...."

Bacteria are mostly prokaryotes in that they do not have a membrane bounding the nucleus. While, most prokaryotes are ecologically beneficial, the reasons why prokaryotic cells are also the ones that cause most of the diseases are that they are small and prolific, have the ability to move (motility) and can be symbiotic (derive benefit from the host while they confer benefits to the host). The fact that the DNA is naked in their cell means that it is easier for the bacterium to transfer its DNA to the host in the several possible mechanisms.

Figure 3.9. Major historical figures in medicine: left to right and top to bottom, Hippocrates, Galen, Andre Vesalius, William Harvey, Louis Pasteur, Robert Koch, Joseph Lister, Kitasato, Laennec, Metchnikoff, Paul Ehrlich, Alexander Fleming and Horace Wells. (Source: Wikipedia, Free Public Images).

One mechanism is the attachment of the disease organism to the envelope or a structure on the envelope, gaining entry to alter its function. (This is similar to the lock and key arrangement, described in the immunity description, except it is the bacterium which has the key to the receptor on the cell envelope). In the example of salmonellosis, which can cause severe gastrointestinal disease, the salmonella organism attacks the intestinal cells. The intestinal cells have fingers of the "microvilli" which increase the cell surface area to absorb nutrients and digest them. The salmonella swims up to the cell envelop and inserts its syringe like organ into the cell and injects certain proteins into the cell cytoplasm. The cell responds by rearranging the cytoskeleton to expand the membrane, thereby engulfing the bacteria. The bacterium is then pulled in and enters the cell, cocooned by a vacuole made of the host cell's own actin filaments that make up the cytoskeleton of the cell, in the cell's defense mechanism of phagocytosis. Such a vacuole is exactly what the cells mean to do to destroy an invading foreign cell. Normally, a lysosome would attach to the vacuole and hold it while the vacuole is showered with digestive enzymes to dissolve the invading cell. But the salmonella, ahead of the game, injects a second set of proteins, which alters the vacuole shape so that the lysosome cannot attach to it anymore and once this process is impeded, the salmonella remains safe and uses the cell resources to multiply. The vacuole itself expands as more cells are made. It is not understood how the new pathogenic cells escape the host cell and infect other host cells. In the case of the bacteria that cause listeriosis, the bacteria, after the vacuole is formed, secretes enzymes that dissolve the phagosomal membrane, It then uses the cell's actin filaments as "comet tails" to swim to the cell's margin pushing out and infecting other cells. When the bacteria enter the epithelial cells lining the intestines, they damage the microvilli and cause severe malabsorption and diarrhea.

In another common mechanism, the invading bacterium cloaks itself in polysaccharide sugar coatings - exopolysaccharide capsule, by using its 23 different genes to manufacture the capsule. Once the bacterium is hidden in the capsule, the opsonizing (antibodies making the bacteria "tasty" for the white blood cell microphages and neutrophils) is disrupted, the macrophages have nowhere to go, are frustrated and cause increased inflammation of the tissue. The inflammation sites are vulnerable to the bacterial cell wall component "Lipopolysaccharides" (LPS) that can cause the host to produce mediators that, in turn, lead to sepsis and septic shock. This response is triggered in presence of inflammation by toxins that act from outside the cell. Exotoxins act by injecting toxins as described above or in the general ambience. Exotoxins are actual poisons from the bacteria and therefore can cause disease even without the bacteria being present. Some exotoxins, like botulism toxin, are some of the most potent poisons. The actual action of toxin varies from bacteria to bacteria, some damage the membrane and some prevent cellular function and so on. The toxins released by various disease agents are being identified slowly and drugs are being discovered to counter the activity. Yet another mechanism important in infection is the adhesion of the bacteria to the skin or other parts to aid in transmission and absorption. Capsules also act as adhesins.

## Viruses

Viruses are too small (less than 20 nanometers in size) to have been detected by optical microscopes and reproduce only in a host. Therefore, filtrates of disease organisms would have had the viruses pass through the filter and cause disease. This could easily be mistaken for an exotoxin due to a bacterium. But no bacterium could be found in many cases, such as the tobacco mosaic disease. The Russian scientist Dmitri Ivanowsky suggested that these disease organisms were very small bacteria. It was later shown that the filtrate contained some pathogen that could not be killed by alcohol that would kill all bacteria. In 1935, Wendell Stanley crystallized the infectious particle and identified the mosaic virus. Now viruses can be viewed through electron microscopes. They are found to be DNA with an envelope of protein called capsid and occasionally an outer membrane , like in the HIV. Each virus can only infect certain cells and this is called its host range. Swine flu virus can infect both pigs and humans and rabies can only infect raccoons, skunks, coyotes, dogs and humans. Again, this is because of the previously mentioned side-chain lock and key mechanism, where a particular radical on the envelope of the virus fits into the slot of a molecule of the surface of a cell, such as lipopolysaccharides, acids, proteins and flagellum (a whip or tail like feature on the bacterium used for sensing and locomotion).

Interestingly, viruses infect bacteria, and viral infections can be studied by observing infected bacteria. One such well studied process is the infection of E-coli bacteria by a bacteriophage virus. The phage virus T4 has a head and a tail piece with long legs to stand and short legs to steady it. (See figure 3.10). Like a moon lander, it lands on the surface of a cell and after a protein needle penetrates its tip layer, a set of additional enzymes break up the second layer over the membrane and then the lock and key of the receptor on the cell membrane opens up and contraction of the sheath of the virus leg, ejects the viral DNA into the cell protoplasm. Once inside, the cell mechanism obeys the additional new DNA code from the virus and makes the phage proteins from the bits and pieces of the broken surfaces as well as from additional resources of the damaged cell. Three different proteins are commandeered to make the head, tail and the legs. Once the pieces are made, these come together to form new viruses automatically because of their affinity, and the virus multiplies.

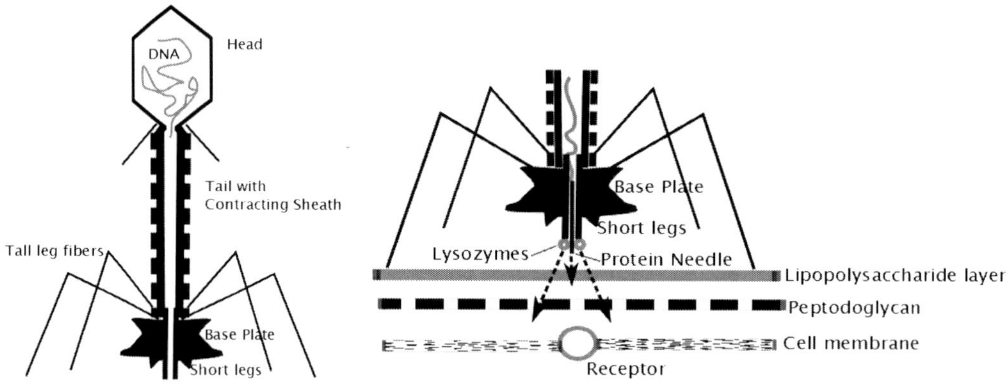


Figure 3.10. (Left) A sketch of the T4 bacteriophage, the tail piece contains a protein needle and lysozyme proteins to penetrate the layers of the cell membrane. (Right- Head not shown)The sheath contracts to set on the host membrane, the protein needle penetrates the top surface of lipopolysaccharides of the host membrane, the lysozyme dissolves the next peptodoglycan layer and the third layer of the host has a receptor that is a lock and key and opens the cell to inject the phage DNA.

At this point there are two possibilities. If the virus is a lytic virus like T4, the phage produces an enzyme to digest the bacterial cell wall and weakens it so that the cell swells and bursts. Several hundred viruses may be released in one burst. If the phage is lysogenic like the $\lambda$ phage, the DNA molecule causes recombination with the host bacterial DNA after its entry into the cell and incorporates itself into the bacterial DNA. This recombination is enabled by a repressor protein that suppresses the lytic tendencies of the bacteria. This way, the bacteria is changed but the further process of cell burst does not happen. Now when the cell divides, it recreates the viral DNA as well as its own. This is called a prophage of the virus. If the change in DNA is benign, both the virus and the bacteria benefit from the alteration and coexist. However, the incorporated phage can also change the phenotype (character) of the bacteria and the bacteria may produce toxins. Diphtheria and scarlet fever are cases of bacterial infection due to lysogenic invasion of harmless bacteria by viruses. But occasionally the $\lambda$ phage can break from the bacterial DNA in the cell due to radiation or chemicals and if it does, then it can act as a lytic phage. This shows how mutations produce diversity and how a disease can change character.

The viruses that infect animals and humans have a great diversity of infection and replication mechanisms. Since, the animal cells are more complex and are eukaryotic with a variety of substances, the virus can and does employ additional strategies. For example, viruses with glycoprotein layers as an envelope (influenza virus is of this type) fuse with host cells surface at specific receptor sites (lock and key) and enter the cell with the capsid intact. This is possible because the virus glycoprotein layer itself is derived from the host cell membrane. The cellular enzymes strip the capsid. The cell mechanism copies the new DNA and manufactures RNA strands. These RNAs act as mRNA. The cell organelles of ribosomes, the endoplasmic reticulum (ER) and the Golgi apparatus etc., then do their usual work and produce the new DNA covered with the capsid. The ER vesicles transport glycoprotein to the surface and when the new virus assembles and buds, it receives the viral envelope as well. The HIV and Hepatitis virus have different structures and therefore use different mechanisms for replication.

One can see that in addition to causing diseases directly and through bacterial alterations and also by lysogenic interaction with animal bodies, viruses have been part of the history of evolution by natural selection. There is a much deeper philosophical meaning to this. The first realization is that humans, however egotistical they may be, have to admit that they are a collection of multiple life forms coexisting together and are really a colony of life forms. In addition to this, we should be humbled by the fact that tiny little viruses and bacteria have shaped us and made us who we are.

Today, modern biology rules medical science in physiological, pathological and therapeutic domains. The understanding of the molecular activity of the toxins and the process of their release by bacteria, the molecular mechanism of infection, such as the makeup of the receptor site and the enzymatic and protein weapons that the bacteria and the virus carry and the immune response of the host cells to these attacks, allow the scientists to establish in detail the cause of disease symptoms, progress and outcomes. These details range from the description of physical characteristics such as pain, heartbeat, hormonal secretions, temperatures, edema etc., to the diagnosis by chemical analyses of blood, tissue and secretions at molecular levels. The cure too aligns with this understanding where the countermeasures, both preventive and curative, of drugs and interventions are devised and designed at molecular levels. Once again the theme that structure defines function is applied together with centuries of medical wisdom.

## Diseases that Are not Caused Directly by Infections

Every organ of the body, including nerves, muscles and bones, is susceptible to disease, from mild to severe, from chronic to acute and from malfunction to actual damage. The non infectious diseases may involve blockages, improper secretions, aneurisms in blood vessels, swelling, tumors and cancers and damage caused by injuries that could also lead to infections within the body. Each organ has a different cause for the trauma and responds differently, creating different symptoms. Therefore, the symptoms provide the first diagnosis of the disease. The second line of diagnosis is response to different drugs, such as a headache responding or not responding to aspirin. The third diagnosis would be made by testing the blood and checking the balance of various constituents, such as hemoglobin count, white cell count, creatinine concentration, liver enzymes, blood sugar etc. Further diagnosis might require imaging the organ with X-ray or magnetic resonance imaging (MRI) or positron tomography (PET), biopsy (taking a sample of the tissue for histological and chemical analysis), taking spinal fluid sample etc. to further narrow down the identification of the disease. Finally, exploratory surgery may be necessary for some diseases. In the case of heart disease, the primary diagnosis is made by taking the record of the electrical activity through electrocardiogram (EKG) and additionally through stress tests. In all these cases, the pathology and diagnosis are arrived at by years of study by scores of medical scientists, technologists and practitioners. The diagnosis and cures, as in infectious diseases, are built on basic scientific discoveries as well as by epidemiological evaluations, trials and accumulated knowledge of case studies. In some cases, where no scientific basis is available, the good old traditional ways are practiced using what is demonstrable by practical methods. In this sense, medical science is different from other sciences where usually a concept is not put into practice until cause and effect are understood.

The non-infectious diseases of the animal body are many and are exacerbated by unhealthy habits such as smoking, drinking, undue stress, unhealthy food or lack of nutrition, lack of exercise and lack of sleep. In addition, environmental causes such as pollution of air, drinking water and food, radiation, accidental poisonings and injuries take a toll on the body. Medical science is working at all these ends, studying the effect of each of these, at the organ or molecular levels and conveys to civil society administrators, the controls necessary in matters of public health, nutrition and safety. The variety and the detail with which the medical aspects are being recognized and addressed and the time it takes to understand and come up with the solutions and discoveries, require that medicine be specialized in every which way- by organ, by the type of technology, by the type of disease, by age and gender group etc. Sometimes, this comes in the way of understanding a disease in its integrated form and causes some frustration to patients. This is unavoidable in all sciences since the amount of knowledge being created is too fast, vast and deep for individual scientists and doctors to learn and practice. This gives a separate impetus to "holistic" practices, both scientific and non scientific, where the details of biology and anatomy are ignored and the body is treated as one single entity with characteristic symptoms and responses, but based on traditional theory and practice and in some cases, by individuals who claim success with their methods. Patients would be well advised to accept such treatments knowing fully whether these may have a scientific basis or not. (See Chapter 8).

While heart, liver, kidney, lung, intestinal and skin diseases are reasonably understood and, in many cases, can be treated with successful outcomes, except in critical stages, cancer, brain diseases, immune and endocrinal diseases are still developing fields of medicine. Ageing is another area of medical research where the factors are being identified. Where, drugs were previously administered for various dysfunctions of the organs, by empirical experience, increasingly, the understanding of the disease directly suggests the direction of drug development.

As stated before, medicine is a sprawling field with poorly defined boundaries. In the past, as in traditional medicines, it relied on expressed symptoms and treatments using herbs, chemicals, physical therapy and surgeries. But now medicine includes radiation treatments, body imaging by a variety of modalities, procedures involving minimum invasion, such as stints, laparoscopic surgery etc. and pathological examinations of blood, urine and other fluids to great accuracy and detail. These have been made possible by the interaction of the broad field of physics and chemistry and the resulting development of technology, sometimes surprisingly so. For example, the magnetic resonance imaging came from the field of crystallography in physics and only the steadfast dream of the inventors, made it one of the most powerful imaging technologies, today. The technology of imaging and the interpretation and correlation of the images with anomalies and lesions with disease is a thriving field of successful developments. Since this chapter is more about the science of medicine, technology topics are not covered, but the impact of physics, chemistry, biology and technology on medicine is immense and readily observable in any good hospital. As stated before, medicine has changed since the molecular nature of disease has been revealed and therefore, fields like cancer and infectious diseases have been transformed into studying the chemical and electrical pathways, viewing organs and pathogens with great magnifications. Subsequently, the interactions with biophysics, biochemistry and electron microscopy have increased manifold.

## PLACEBO AND EXPECTATION EFFECTS

One cannot leave the topic of general medicine without acknowledging the occurrence of placebo and the expectation effects. Basically, the placebo effect is when a patient gets relief from a symptom even when an ineffective cure is given or suggested, in the form of inactive (say sugar) pills or remedies. Patient's response may also be in the opposite direction, say a stomach pain or dizziness or immune response. When this is included, the effect is called the Expectation effect. This is essentially what the patient expects to happen and has a psychosomatic response to the placebo or even the medicine. (Therefore, good medicines can also confer additional placebo effects or unaccounted for side effects). The placebo effect can also occur just after a visit to the doctor or hospital or after having just a check-up. Studies have shown that the placebo effect is real. Scientific studies indicate that the effect is caused by change of brain chemistry such as increased endorphins or adrenalin, which cause short term change in symptoms, though underlying disease persists. In addition, trust or distrust of the caregiver, past good or bad experience with a medicine can give rise to this response. John Haygarth was the first to study this effect and convincingly proved the existence of this effect in 1799.

While placebos can be a powerful tool for mitigation of pain or skin rashes etc, placebos cannot cure diseases that have pathological basis. Tumors would not shrink, heart diseases would not go away, diabetes would not be cured or a serious wound would not heal by taking placebos. But diseases like Parkinsons, depression, panic disorder and chest pain due to ischemic hearts (blockages) may be symptom free for as long as 6 months. Rheumatoid Arthritis obtains a high duration of benefit of up to two years, for patients who are susceptible to placebo effects. Since the placebo effect can be significant, in scientific confirmation of remedies, a remedy has to be shown to give beneficial result well beyond placebo effects, which may be, in some cases, as much as 40%.

***

Just as it is difficult to give a complete introduction to chemistry, physics, biology or mathematics, it is not practical to go beyond the above short introductory descriptions and cover more ground. However, since the connections between science, society and religion are of interest in this book, two topics, the disease of leprosy and mental diseases that illustrate these connections and dichotomies, are described below.

## LEPROSY

Leprosy is an emotional and scary topic for most people even today. It is intertwined with social stigma, religious faith related compassion and healing, and also with sin. The initial symptoms of leprosy are whitish lesions on the skin and there is less sensitivity and sensation at these locations. The lesions are accompanied by muscle weakness and numbness in the limbs. These symptoms persist and secondary infections, due to lack of care, develop at finger and toe tips and face. The infections in the presence of leprosy lead to nodules on the face, hands and legs, gradual crookedness and withering away of finger and toe tips and disfigurement of the face. The total numbness allows unfelt injuries which worsen into infections. This physical disfigurement and eventual stubbiness of fingers and toes with associated disabilities are what invoke so much fear as well as discomfort in someone who comes across an infected person. There was an unfounded fear too that it is highly contagious. This fear continues even when one is assured that the patient is not contagious. It is also easy to see that the crippling and disfiguring disease would be considered a curse or the wages of sin, because the disease is not terminal and can be construed as suffering and punishment. Much of this foolishness still continues in many countries like India. Armauer Hansen, the discoverer of leprosy bacillus stated "*There is hardly anything on earth, or between it and heaven, which has not been regarded as the cause of leprosy; and this is but natural, since the less one knows, the more actively does his imagination work*" (G. A. Hansen and C. Looft (1895) quoted in the article (Pandya 1998)). A very good review of the evolution of the disease and a good resource available in the web is an article by S. Dogra. (Dogra 2013).

Although it is not clear if early civilizations were referring to leprosy specifically or to all serious skin diseases, Leprosy or a similar disease was known in Babylonian and Egyptian civilizations and records of the disease are available in 3$^{rd}$ century B.C. in China. Every culture around the world has seen leprosy as a special disease requiring isolation and separate leper colonies or leprosaria were created during the middle ages in Europe, India, Japan and other countries. Some of them exist even today. Since the disease was considered to be highly contagious, some leper colonies were in isolated mountains and some even had their own currency, because people would not touch the money touched by a leper. Leper colonies broke up families and caused considerable suffering to the already suffering and stand as an example of the injustice of mass quarantines. At the peak of the disease, 20-25% of the population was afflicted. (Greene 1985). In 1225, there were close to 20,000 leprosaria in Europe. In Japan, there was continued harassment of lepers over the centuries and prefectures felt they needed to be purified and be rid of lepers, even though doctors had established treatment centers as early as in 1875. But after 1955, an enlightened policy has been pursued in Japan.

In India, where there continue to be as many as 130,000 cases every year (almost half of the world's leprosy cases), there is continued banishment of leprosy patients from villages and the government has done very little to provide proper solutions to the leprosy problem. In early days, Hindu laws prohibited contact with leprosy patients and marriage into the family of a patient was punishable by the King. When India was an English colony, the English feared the "Imperial danger" from leprosy and a commission was dispatched by the government which found that India had 5.3 cases per thousand. The commission deemed it a low rate, knowing that leprosy is not very contagious and advised the Government to disregard it. But, under pressure from the people in both countries, the British Government enacted a Leprosy Law in 1898, where the well to do could go to care on their own cognition while the poor and the indigent were taken to colonies (sanatoria run by charities and government grants). The Law was repealed only in 1983, when a multidrug treatment was well established. In 2005, India celebrated elimination of leprosy according to the definition of less than 1 case per 10000. But the stigma remains and till today beggars, crippled by leprosy, can be seen on some streets and temple sites. Jacob and Franco-Peredes state (Jacob 2008),

> "The history of leprosy in India offers insights into one of the world's most misunderstood diseases. Furthermore, leprosy control and elimination in India still faces many challenges. Although many of the theoretical and practical approaches of the past have been discarded, their careful examination provides insights for the future. Sustaining the gains made so far and further reducing the disease burden in India require an innovative, holistic approach that includes ongoing education, efforts to identify interventions to dispel stigma, and the inclusion of nonallopathic practitioners in disease control programs."

It is well known that both the Old Testament and Torah had strict laws on leprosy patients. The book of Leviticus, the 3rd book of the Old Testament, describes the duties of the Levite priests and includes instructions on diagnosing a leprosy patient, dealing with a patient and how to declare a patient free of disease. These show both a good knowledge of the disease and its transmission and a clear stigma around this disease. Leviticus 13 (King James Version) states-

> And the LORD spoke to Moses and Aaron, saying: "When a man has on the skin of his body a swelling, a scab, or a bright spot, and it becomes on the skin of his body like a leprous sore, then he shall be brought to Aaron the priest or to one of his sons the priests. The priest shall examine the sore on the skin of the body; and if the hair on the sore has turned white, and the sore appears to be deeper than the skin of his body, it is a leprous sore. Then the priest shall examine him, and pronounce him unclean….

The passage continues to state that such a person should be isolated and studied for seven days and if the priest sees that the patient has scabs that remain, he would not be isolated and declared unclean. It continues,

> "Now the leper on whom the sore is, his clothes shall be torn and his head bare; and he shall cover his mustache, and cry, 'Unclean! Unclean!' He shall be unclean. All the days he has the sore he shall be unclean. He is unclean, and he shall dwell alone; his dwelling shall be outside the camp."

The covering of the moustache is stated as covering of the upper lip in other versions. Both indicate that there may have been empirical knowledge that the disease could be spread by nasal discharges (droplets). But, there was an implicit understanding that this Zara'ath is a punishment for sins and is curable only through prayers and God's mercy. Jews were asked to reject physicians in favor of divine intervention, making an example of King Asa, who died of a foot disease but sought only physician's care. Christian societies too viewed leprosy in terms of the Biblical statements and the treatment was conflicted. On one hand, they were declared to be Nazarites (a type of warriors with special vows of isolation). But their actual treatment was that of treating them as dead. The 13$^{th}$ century Mass of Separations *states "I forbid you to enter the church or monastery, fair, mill, market-place, or company of persons... ever to leave your house without your leper's costume... to wash your hands or anything about you in the stream or fountain. I forbid you to enter a tavern...I forbid you, if you go on the road and you meet some person who speaks to you, to fail to put yourself downwind before you answer...I forbid you to go into a narrow lane so that if you should meet anyone he might catch the affliction from you...I forbid you ever to touch children or give them anything. I forbid you to eat or drink from any dishes but your own. I forbid you to eat or drink in company, unless with lepers."*(Brody 1974) (Moore 1990). The diseased were not permitted to touch children and were required to wear a bell warning of their approach. The special houses, they were housed in, were called Lazarettos after Lazarus, the patron saint of lepers. As a compassionate dispensation, Leprosy was a purgatory experience on earth itself so that the rewards of heaven were quicker to the sufferers.

On the counter side too, as a result of the Christian teaching that God loved the lepers, the caring for lepers at the risk of contracting the disease was and is a holy cause, following the example of Jesus, who " *...put forth his hand, and touched him, saying, I will; be thou clean. And immediately his leprosy was cleansed."*- Matthew 8:3. The spirit of charity and service would shine particularly in helping those who are shunned by common folks. The work of Father Damien in the Molokai settlements in Hawaii and Mother Theresa in Calcutta are legendary. It was the untiring work of Christian missionaries and priests that made life bearable for those afflicted. Leprosaria were also shelters for the poor and indigent syphilitics in some countries of Europe. By 1350, leprosy started declining and was nearly gone by 1500s. The cause of the sudden decline continues to be a matter of scientific study. One theory is that it was displaced by the TB bacillus. Ecole Polytechnique Fédérale de Lausanne in Switzerland believes that the intense death rate provided a natural selection of those with immunity and the whole population has become immune to the bacterium. With this decline in Europe, Christian missionaries continued their work in other countries such as in India and Hawaii.

In Muslim countries too, the diseases was dreaded. Al-Bukhari's Muslim Hadith (volume 1, 2.443) documented Prophet Mohammed's apparent dread of leprosy in his statement: "*Escape from the leprous the way you escape from a lion*". (Dogra 2013). In the Koran (*Quran*) Hadith, however, there are conflicting statements about the contagiousness of the disease (*'baras'*). In one version The Prophet is said to have asked an afflicted to stay away from him and yet receive his pledge of allegiance and in another, The Prophet is said to have partaken meals with the lepers and had them dip their hands in his dish. However, the medieval Islamic scholars, such as Ali Ibn Rabban al-Tabari, knew about the transmissibility of the disease. Avicenna advanced an additional theory that the disease was inheritable (Isaacs 1991). By and large, Islamic countries also isolated leprosy sufferers. 'Umayyad Caliph al-Walid ibn 'Abd al-Malik ibn Marwan, Caliph of Syria, isolated lepers but was compassionate and provided for their comfortable living. Muslim Spain created a separate colony. The autonomy and the care given to them despite the serious fear of contagion resulted in half measures that did not strip the sufferers of their human rights. There is some evidence that lepers of other faiths were treated even more liberally by the Muslim authorities.

## Leprosy Types

After the decline of the disease in Europe in the 16th century, there was an endemic wave of the disease in Norway in the 1800s. This led to the first scientific study of the disease by Daniel Danielssen and Carl Boeck in 1847 and a detailed report showing the distinct characteristics of leprosy, distinguishing it from other lesions such psoriasis, scurvy and syphilis. These scientists believed that the disease was hereditary. In 1873, Danielssen's son-in-law, Gerhard Armeur Hansen, identified the bacteria responsible for the disease when examining a nasal biopsy, visible as rod shaped under a microscope. He named it Mycobacterium Leprae. (Since this could be stained only by the special Ziel- Neelson method, this was identified as a Mycobacterium). He found that this bacterium bore considerable similarities with the tuberculosis bacillus (TB), but had a variety of shapes unlike the TB and formed intracellular connections. There are also differences in which dyes could stain each. With the decline of the disease in Europe and no urgency in finding vaccines etc, there was no further progress for a long time, beyond this discovery on the now termed "Hansen's Disease". Management of the disease was still based on the European model of isolation till death. A year after the discovery by Hansen, Henry Vandyke Carter of Bombay Medical Service, India, and the famous illustrator of anatomy in Gray's Anatomy, wrote an illustrated book on the disease showing the two types of the disease- *Elephantiasis Tuberculata* and *Elephantiasis Anaisthetos.* Carter is also the first physician to diagnose the details of the disease by dissections at the Jamshedji Jeejeebhai Hospital in Bombay and describe it as a disease of the peripheral nerves. The progress in leprosy treatment was slow for a number of decades. After several conferences, an international system of classification of strains of leprosy and a lepromin skin test for leprosy were adopted in 1953. This is similar to the TB test in which deactivated leprosy bacteria are injected under the skin of the forearm and an indication of significant swelling shows the presence of antibody response to a dormant or active disease in the individual. Figure 3.11 shows the prevalence of the disease in 2003, around the world.

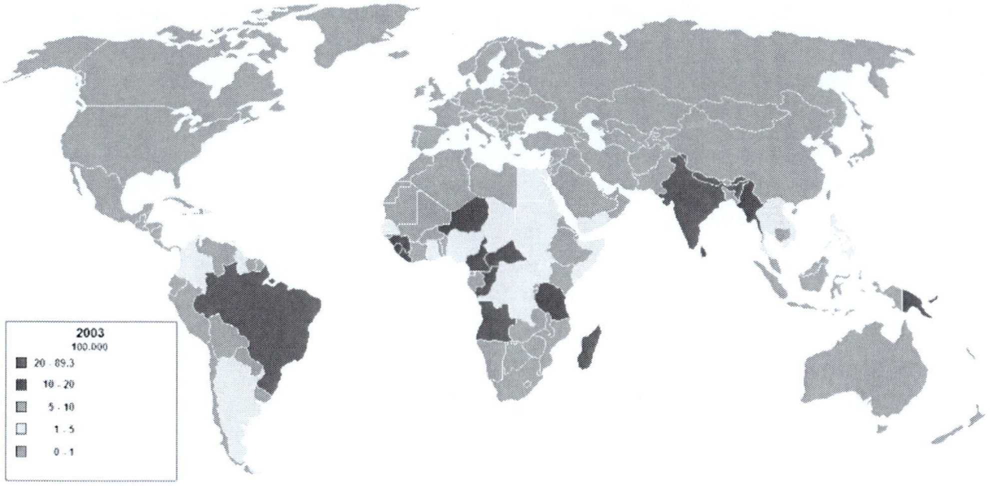


Figure 3.11. Distribution of leprosy around the world in year 2003. Source: Wikipedia.

There are two forms of the bacteria, one Mycobacterium leprae and Mycobacterium lepramatosis. The second one which was recently identified in 2008, is not common but causes diffuse lesions. The M. leprae itself may be tuberculoid, paucibacillary (PB), which cause few large white skin patches and loss of sensation due to damaged peripheral nerves. These are rod shaped bacilli growing at the nerve tissues and give a positive result for the lepromin test, which is indicative of the well developed immune system of the patient. The M. leprae can be lepromatous multibacillary (MB), which gives several smaller symmetric skin lesions, nodules and thickened skin. The nasal mucosa is infected and the lepromin test gives a negative result. The site of infection shows microphages and multiple shapes of bacillus at the nerve tissues. There are borderline cases involving a combination of symptoms and these may later become tuberculoid-like or lepromatous-like.

## The Leprosy Mycobacteria and Mechanism of Infection

The M. leprae has lost many of its gene functions, with 27% of genome being inactive pseudogenes, while these same genes are active in M.tuberculosis and therefore both appear to have evolved from the same source, with M.leprae in reductive evolution not requiring use of these genes. 23% of the genes are genes that do not code for any proteins. Because of this loss of functional genes, M. leprae cannot be cultured in a laboratory and needs a host. This is one reason why the discoveries relating to this bacterium are slow. The bacteria absorb nutrients very slowly due to its thick waxy wall made of lipids. The loss of the gene functions had also reduced its absorption of oxygen (this is an aerobic bacterium) and it cannot transport electrolytes on its own. As a result, the bacterium's doubling time is 14 days! Interestingly, the only other animal that can be infected with it is the armadillo, which has a low body temperature, since M.leprae requires less than 33 deg C to thrive and is eliminated at above 37 deg C. This explains why the cooler tips of the nose and fingers tend to be hosts for the bacteria. The detailed pathways for the functioning of the bacteria have been found. Carbon cycles are also similar to the TB. But the pathways for energy production through oxidation of acetates, derived from carbohydrates, fats etc. (Kreb cycle) are closed off and the organism can only produce energy from glucose. The organism therefore needs a host to provide the glucose. The organism cannot assimilate vitamins and this too is one reason why the bacteria grow slowly.

The M. leprae uniquely infects and survives in the Schwann cells of peripheral nerves. The Schwann cells perform the same function that some brain cells do, namely create the support chemical myelin for the neurons. The Schwann cells also have other functions such as, creating a matrix to support neurons and arranging nutrients for it etc. (See figure 3.12).

Essentially, the M.leprae bacterium strips the cells of the myelin sheath that insulates the nerves, prevents dissipation of the nerve signals to the immediate vicinity and thereby helps better conduction of the signals. (The insulation is a wrapping of membrane around the nerve cells for about 100 layers and impregnation of the layer with myelin. This is amazingly similar to the modern electrical insulators made of fiberglass reinforced resins). The actual mechanism of penetration of the bacterium into the host cell is not yet understood. However, a very recent study has found that the M.leprae performs a feat that no other bacterium does: It converts the Schwann cells into the so called "neural crest" stem cells, which are the cells that are found in embryos and are used to make the physical body of the infant. (Masaki

2013). The infected stem cells then migrate and make different cells such as muscle cells. (Author's note -This is perhaps only true for the type II response, since in the type I, systemic disease is not seen- see below).

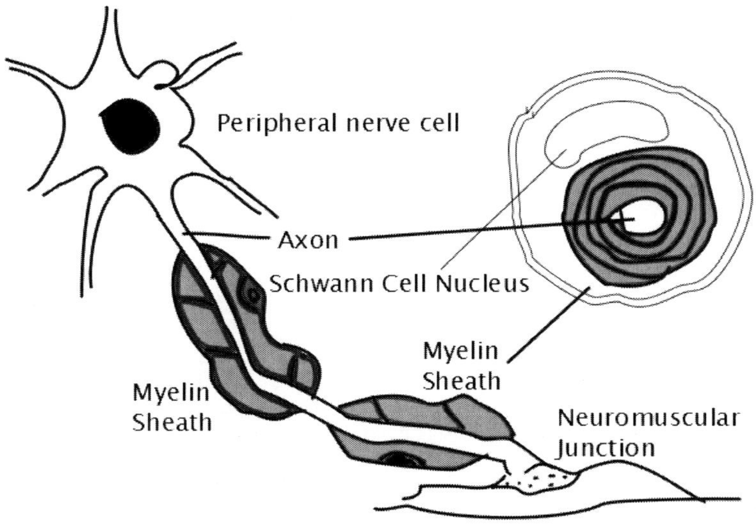


Figure 3.12. Schwann Cells and Myelin sheath.

The other mechanism of the bacteria is the incorporation into the macrophages that rush to the infected site to seclude and destroy the disease organism which might have been rendered inert by the action of the antibodies. The macrophages arrive in circular structures of granolumas and the M.leprae bacteria hide in these granolumas thus evading further detection.

Much of the damage to the body occurs because of the immune response of the host cell. The immune response itself is quite varied. Type I immune response is called a "reversal reaction" in which the immune system suddenly changes to a different status and mounts a strong defense that leads to a severe tissue damage. This happens over two to three days and is therefore called a Delayed Type Hypersensitivity. In these cases, extensive inflammations and skin lesions that eventually cause scales and nerve damage are seen. Further diseases such as peripheral lymphoma (a type of cancer of the blood) may occur. The type II response, known as erythema nodosum leprosum- ENL (somewhat specific to this disease), is an immune complex disorder and results in continued and widespread symptoms of inflamed nodules, both surface and deep. This creates cells that are quite diseased and may be necrotic. The Type II response results in systemic illness of fever, weakness, joint tenderness and swellings etc and most of the patients show a chronic disease.

## Treatment for Leprosy

In the early Indian medical text of 600 B.C., Sushruta Samhita, the *Chaulmoogra* oil (oil from the nut of the *Hydnocarpus wightiana* or the *Hydnocarpus anthelmintica* tree) is suggested as the treatment for leprosy. This has been used with limited success in India and

China for a long time. The therapeutic seeds, sold in bazaars of Calcutta, India, were found by Sir David Prain in 1901 and in 1904 and Fredic Power isolated the chulmoogric and other acids that seemed to confer the cure. Opinions on its efficacy were divided among physicians. The British Government conducted a therapeutic trial in 1915 and ordered its use in hospitals in India. Despite its limited success, this was the only treatment available till 1941, when Guy Henry Faget at the U.S. Public Service Health Hospital, in Carville, Louisiana discovered the drug Promin, which has been hailed as the “miracle of Carville”, in line with the Christian concept of miraculous healing of lepers by saints. While the chemical itself had been synthesized by others, the discovery of Promin as a cure for leprosy was made in the atmosphere of the excitement of curative powers of sulfanamides and sulfones. Faget obtained the drug from Parke-Davis. Faget quickly started drug trials with six patients and despite its toxicity, the curative powers of the drug was established. Today, Dapsone, a drug similar to Promin, is the drug of choice because of its lower toxicity.

However, there has been increasing resistance by the organism to Dapsone and therefore, the World Health Organization has instituted a multi-drug therapy which includes antibiotics rifampicin and clofazimine. The actual treatment and dosage vary depending on whether these are for paucibacillary (such as tuberculoid) or multibacillaory (negative skin test) disease. The treatment is to be 6-12 months for the paucibacillary (no clofazimine is needed) and 2 years for the multibacillary type. Monthly supervised doses as well as daily self administered doses are suggested. These drugs work along the lines of other sulfa drugs and antibiotics.

Today, while there is a general decline all over the world in leprosy cases, the efficacy of the drugs is coming into question and the increase of antibiotic resistance (up to 0.65% in PB and up to7% in MB) is a matter of deep concern. A sociological approach through accredited social health activists (ASHA) is being recommended to ensure adherence to drug regimens and patient follow ups. New drugs, currently being discovered, may also be needed.

Despite the fact that leprosy is a slow developing disease requiring special conditions, is not a very contagious and virulent disease and does not cause prompt deaths, M. leprae is an old organism which multiplies slowly but surely and appears to be in tune with special aspects of human physiology and evolution. The scientific understanding of its emergence, its transmission, its infection mechanism, the drugs to counter and patient care are phenomenal areas challenging the society and medical science is well poised to address these on a continuing basis. From the days when the disease was considered a curse to the modern day understanding that it is just an infection caused by an organism and the drastic reduction of the incidences of the disease are a testament to the organized methodology of science and the continual knowledge building where society itself is enriched. Leprosy also provides examples of the inhumanity, fear and superstition that a lack of scientific knowledge can create in an otherwise compassionate society.

## Mental Diseases

In sections of traditional societies as in India, China and many other places, mental illnesses were and are even now, attributed to spirits, sometimes evil, possessing the mind of the person. This was also the case in Europe, Americas and other countries during the previous centuries. The interpretation of whether it is evil or a goddess depends upon the behavior of the patient and the advice of the priest. In India, such possessed individuals predict, interpret and judge events. If the spirit that possesses them is deemed evil, the priest would throw ashes, water and other things and beat the individual with sprigs of the *neem* (Margosa) tree, while using threatening chants, to drive the demons away. This exorcism is very similar to the ones practiced in Christian societies and has thrilled many a filmgoer. Ancient Greece held such notions of possession, both by good and bad mystical things. However, by the time of Hippocrates, whose idea that imbalance of humors caused disease, was widely accepted, mental illness was included in such causation. The rational thoughts behind this placed mental disorder in the same category of illness as physical disorders and treatments were at par with physical disorders, except for any controls that might have been necessary for safety. Epilepsy was due to congestion of phlegm in the brain and depression was due to the accumulation of black bile with the term melancholic describing such a state. Women had the additional category of a wandering womb which is the source of the word 'hysteria'. An example of the "sympathetic" method of humoral therapy was the prescription of balancing rest and exercise, abstinence from sex and alcohol etc. for the mentally ill.

In 600 B.C., the Indian text *Charaka Samhita* also describes mental disorders as *Doshas* (faults) in the balance of the three fluids *(kapa, pitta, vayu)*. As is common in India where multiple contradictory concepts can coexist, these ideas did not displace ideas of possession. In China, the emphasis was on relationship between body and mind, following Chinese traditional philosophy, and the treatment was holistic. Yet, in all these traditions, it was believed that mental illnesses involved metaphysical, emotional and moral disturbances, such as disrespect to elders and gods, not performing one's duty, shock due to being overjoyed or extreme sadness etc. Occasionally, the effect of a serious injury to the head or imprudent sexual activities, were recognized to be the cause. In Rome, before Galen, there were multiple approaches, some humane, using diet, exercise and rest, following the Alexandrian physician Asclepiades. Asclepiades classified mental illnesses into hallucinations and delusions, differentiated between acute and chronic disorders and suggested that emotional disturbances rather than supernatural causes were behind the diseases. Aulus Cornelius Celsus (see above) proposed, in the first century C.E., that mental illness was mental imaginings run out of hand. He was the opposite of Asclepiades and suggested more typical method of restraints or even frightening the patients with deprivation and darkness, implying that the patient needed to be distracted and shocked away from his imaginings (Hinshaw 2006). (The latter idea would resurface in the form of physical treatments in the $19^{th}$ and $20^{th}$ centuries). In later centuries, with Galen repopularizing the Hippocratic tradition of humoral therapy, symptomatic treatment of sadness, and manic excitement were each attributed to excess or depletion of one or more of the humors. According to this advice, bloodletting was common. Since mental illnesses are confusingillnesses, with unclear and sometimes nebulous symptoms, the therapy combined the humoral theory with possession by spirits, as in India, but with a greater appeal to philosophy. An analogous situation continues even today in many traditional societies, where Western medicine has not taken hold, both because of distrust and for reasons of cost and availability.

Overall, in the early history of medicine, there was no coherence in theory and practice. Intolerance of the mentally ill with punishments for misbehaviors and humane treatment suggested by various prescriptions, existed side by side. This is, in no way, different from what we observe in many places even today. The 'softer' science of mental illness proves to be harder than other sciences.

There was a bright spot in the ninth and tenth centuries when Muslim societies had so called *mauristans*, institutions for caring for the mentally ill, in large cities like Damascus and Baghdad. The insane were deemed to be inspired by the grace of God and therefore were held in esteem and cared for in a relatively high level of comfort (Duffin 2000). The science of *Nefs* guided Islamic philosophy on psyche, mind and related issues. Muhammad ibn Zakariya al-Razi, and Avicenna provided descriptions and treatments for mental illnesses, which, like Asclepiades' treatments, included music, rest and exercise and occupational therapies. During the middle ages, with the emphasis on divine explanations, ideas of sin and mercy of God, the natural explanation of mental disease retreated in the Western world. Although Galen's theories still dominated the treatments for physical ailments, it is easy to understand that mental illness would come under the purview of religion, since it is closely related to behavior, thoughts and speech which are affected in the mentally ill. It is also conceivable that those religiously charged atmosphere in those periods would create a greater fear of possession by the devil. Monks and priests became caretakers and healers of the mentally ill. The term lunatic was coined connecting episodes of lunacy with phases of the moon. The mentally ill were thought to have had a weak mind thus allowing possession by Satan and his minions. The treatment was obviously punitive with exorcism rituals and isolation in remote houses. This gave birth to the concept of institutionalizing the individual, now in colloquial parlance all over the world, such as that a person deserves to be in Kilpauk (Chennai, India), Bellevue (New York), or 999, Queens St. (Toronto). This institutionalization freed and continues to free families and society from the obligation of caring for the difficult, unruly, depressing individuals, whose mere presence could ruin a happy household and send people scattering on the streets. The holding of violent individuals in chains was common and justified in many cases, but it is easy to see that such treatments would be misused to make patients obedient. The exceptions to these institutionalizations are when the mentally ill were given refuge in a monastery or if they were rich or distinguished enough that they could be cared for privately and kept away from public knowledge.

The 15$^{th}$ century saw mental illnesses in even darker terms. The illnesses were now known as Devil-sickness and there was a violent fear against these "creatures". In 1486, the German Dominican friar Heinrich Kramer wrote *Malleus Maleficarum*, commonly known as the Hammer of the Witches, which reaffirmed that witchcraft did exist, particularly among women, and that it should be dealt with severely. The treatise became very popular and influenced people into thinking that witches cast spells on innocent victims and turned them mad. The presumed witches were themselves evidence for this. The brutal acts in trying to extract confession caused delusions in the accused and proved the accuser's case. (Stephen Hinshaw, see above).

In the 16$^{th}$ century, Johannes Weyer, a Dutch physician to the Duke of William of Rich, wrote one of the early treatises refuting the practice of persecution of the so called witches and holding generally a skeptical view of the supernatural cause of mental disease, even though he never denied the existence of the Devil or its power to cause suffering. In fact, Weyer was an inspiration to many writers, occultists and demonologists who believed that magic could fight the Devil and his demons. Weyer, while using principles used by the persecutors themselves, was the first person to criticize the *Malleus Maleficarum* and coined the term "mentally ill". (Clarke 1997). Weyer did not accept the guilt of the witches, because they themselves were not willful like heretics and lacked the will, rational thinking and malice to commit evil acts. He is now considered the father of legal application of psychopathology, by introducing insanity defense and producing various evidence, including framing of individuals by mobs. Following him, the celebrated and sainted Vincent de Paul opposed the branding of the mentally ill as witches or victims of witchcraft. The words of a holy man had much influence.

The gradual deliverance of Europe, England and the Americas from the clutches of witch hunts, after the horrid past with events like the Salem witch trials in 1692, occurred during the 18$^{th}$ century. But the progress was very slow. Patients continued to be treated cruelly, imprisoned and isolated in squalid conditions. Diluted ideas of Demonic possession still persisted. This is not surprising, given the condition of the poor, particularly the urban poor, in countries like England and France. But the public regarded the mentally ill as less than human, because of their diminished reason. Even King George III, who ruled in the last decades of 1700 and early 1800, was chained, starved and held inert in mechanical devices, making him go through crises that would shock him out of his madness. The various descriptions of the treatment of mentally ill, though differing from the witch hunts, were equally unwholesome and let the mentally ill to languish in the so called mental asylums. Towards the early 1800, French revolutionary concepts demanded human liberation and naturalistic thinking which included the humane treatment of the mentally ill. During this period, despite the negative attitude of the public, there was progress in discriminating illnesses. The Scottish Doctor William Cullen coined the term 'neurosis' for the impairment of sense and motion and therefore identified these diseases as the disease of the nervous system. Phillippe Pinel, the physician who identified *dementia precox*, now known as Schizophrenia, was at the forefront of people who literally unchained residents of mental asylums and this took on a symbolic meaning as well. These incidents echoed in England and the United States. Pinel instituted a new paradigm for treatment based on his theory that while mental disorders damaged the moral sense, the ill retained a semblance of reason and therefore had to be treated humanely. (Pinel and his colleagues developed symptomatic discrimination between mental retardation, senility, melancholia and what would be neuroses of today-monomania- pathological preoccupation with one thing). While, the idea caught on, the implementation was far from universal. In 1840s, Dorothea Dix, who went on to become the Superintendent of Nurses in the U.S. Army, started a crusade to bring legislation for the humane treatment of mentally ill and was instrumental in the construction of the first proper mental asylums in the U.S.

## Psychiatric Treatments

In the 19$^{th}$ century, anatomical explanations for diseases were popular and mental diseases were no exceptions. Epilepsy, vasculitis, induced mental illness (tertiary syphilis), stroke and allergies were found to have organic origin and were no longer diseases of the mind. Hope grew that other mental ailments would also be found to have organic roots. But no true physical findings were made beyond what Cullen had found. A new term 'psychosis' was coined by Karl Freidrich Constatt to signify psychic neurosis to describe all mental illnesses that could not be attributed to neurosis. In 1802, Johann Christian Reil, a German physician of high repute coined the term psychiatry which covered psychosis. In 1803, he published a highly influential article *"Rhapsodies on the Application of Psychological Methods of Cure to the Mentally Disturbed"*. Reil came to regard insanity as due to the fragmentation of the self, an improper functioning of the consciousness and a poorly constructed personality that is unable to make sense of the world. Reil ascribed mental powers to that arising out of the brain, which is composed of many "sounding bodies" having a tight and mutual relationship to give meaning. (This relationship would be the nerve connections in the brain). In this way, he conceived of the brain as the seat of the soul. The dynamics of this relationship determined the "orchestration" of the whole brain. The diseases, according to him, then arose because of the disturbance in these relationships and the soul then would be unable to express its will. He advocated psychological treatments to realign and readjust these nervous connections in order to restore the mind. One can see that Reil's description of mind is quite logical and it is clear that Reil was a forerunner to modern psychiatry.

Main stream German psychiatry was established by Wilhelm Griesinger in middle 1800s. Griesinger was an adherent of the materialistic interpretation of physiology and diseases and firmly stated that mental illnesses are diseases of the brain. He cited cases of brain inflammation, heredity and head injuries which cause mental illnesses as proof of the physiological nature of these illnesses. Pupils of Griesinger continued the work of neurology and neuropathology and their relationship to mental diseases. As stated before, such a materialist view did not bring cures. A bad interpretation was that if madness and mental weaknesses were inherited diseases, then these would get worse with each generation (degeneration) and these triggered fears in the upper echelons of the public, of proletarian unrest, increased sexual perversions and criminal tendencies among the lower masses. Reversing this logic, criminals and the insane were considered to be degenerates, typified by facial features which did not correspond to the standards of a handsome European face.

## Freud and Jung and Birth of Modern Psychotherapy

Unless we ourselves have been through the experience, we often wonder what really happens when we witness someone being hypnotized on stage for a show or in a movie. However, one does really know that the hypnotist can indeed help alter the mental state of the patient. The layers of these states were discovered just when physicians were becoming frustrated with their own theories and asylum treatments, which were yielding positive outcomes only sporadically. One of the first, though falsely interpreted method of treatment was developed by Franz Mesmer. Mesmer believed that people possessed "animal magnetism" to various degrees and that they could transfer and interact with animate and inanimate objects through this. Therefore, people with a high degree of animal magnetism could heal mental illnesses in others. Though his ideas were very controversial, he did have pupils and supporters. In his treatment, he would sit in front of the patient with his/her knees touching his, holding hands, sometimes for hours and the patient would relate a flow of sensations through the body, which Mesmer interpreted as crises and resolution of crises. He introduced a group therapy version of this, where he used objects of iron of various shapes and a hanging rope for communicating. Some patients felt better after these sessions. When the Royal Academy, at the request of Mesmer, looked into animal magnetism, they found no evidence, but the arguments that surrounded it gave good publicity to "mesmerism". In 1841, a successful Scottish surgeon James Braid witnessed a demonstration by Charles Lafontaine, a show magnetism demonstrator and was convinced of the effects of the transformation of going into and coming out of states. He rejected the magnetism explanations and decided to investigate it as an operator assisted procedure. By using the dynamics of gaze (looking at the brow or tip of one's nose etc.), which has been practiced in Indian traditional exercise of yoga, he was able to "auto hypnotize" and therefore demonstrated that there was no mechanism of operator magnetism. James Braid practiced, what are well known methods in yogic transcendental meditation, concentration on an object and breathing in a controlled way. With these experiments, he determined the psycho-physiological effects and the feedback mechanisms involved. He first demonstrated hypnotism on a subject at his own home, then delivered a public lecture and demonstrated his discovery. Though the word hypnosis was introduced later, Braid used the word hypnotism to imply that it had nothing to with the operator exerting his will to control the subject. In 1843, he published his ideas in a book, *Neurypnology; or the Rationale of Nervous Sleep*, which became hugely popular. Soon, physicians such as Jean-Martin Charcot started making hypnotism their central method, noting its efficacy in diagnosis of ailments like hysteria and other neuroses. Charcot's experiments for clinical determination of a variety of illnesses, including headaches, strange visions and other impairments of sensory and motor functions, yielded the so called "hysterogenic points" or zones of hypersensitivity, which when touched could provoke a strong response. (One might speculate that this might be the basis for the Chinese treatments of acupressure and acupuncture). Though Braid was well intentioned and theorized that the patient response was all her own, he neglected to take account of his own influence and suggestion on the patient and the charged atmosphere of a theatrical demonstration. The fact that the truth lay between the concepts of Mesmer and Braid, was revealed by the work of Sigmund Freud, a Viennese doctor and physiologist.

### *Sigmund Freud*

Freud's education was his realization that his own early ideas were false. Freud's work, with Charcot and another physician Josef Breuer, on patients under hypnosis, had convinced him that neuroses stemmed from early sexual traumas. In particular that hysteria was due to the repressed memories of prepubescent sexual seduction by the father. He went public with this theory in 1896 and this still reverberates today. However, in 1897, Freud realized that the patient's narrations were fantasies and rejected his seduction theory. However, reconsideration of the patient narratives convinced Freud of childhood sexuality and Oedipus complexes as the underlying causes. (Some psychiatrists now believe that Freud got it right the first time). He placed his discovery of repressed Oedipus complex among the most important in psychoanalytic theory. The psychoanalytical theory of Freud then is centered on repressed childhood sexuality. Revealing of the associated repressed memories and feelings from childhood then formed the psychotherapist's methodology. Freud abandoned hypnosis as a method and took to developing psychoanalysis, the completion of which is what we see today. His theories include, in addition to sexuality, unconscious mental states, interpretation of dreams, symptoms in neuroses, etc. Free association, in which the patients could speak freely of their own accord and interpretation of dreams, became very important tools of psychoanalysis. Freud came to believe in the concept of free association when a patient, "Miss Elisabeth" chided him for obstructing her flow of thought during a session. Free association continues to be a central pillar of scientific psychoanalysis today. Following

Freud, psychoanalysis is often based on resolving unconscious conflicts in the mind and the subconscious. Though Freud believed in biological origins, his theories only addressed the dynamics of the unconscious mind. Freud's theories had three parts (1) Theory of the unconscious mind (2) Infantile sexuality (3) Structure of mind and psychoanalysis.

The theory of unconscious mind is an expression of determinism, in the sense that all behaviors and actions are meant to be. For example, a slip of the tongue is a Freudian slip in that it is an expression of an unexpressed and unconscious mind state. Freud believed that such a state could be brought out in the open in an analysis session. The theory also implies that the conscious mind is only the visible part and there is a much larger unconscious (often termed as subconscious now) under it. The unconscious mind possesses and drives actions through instincts of different kinds. But the primary ones are the *Eros* (life preserving and sexual instincts) and *Thanatos* (death, aggression, self-destruction instincts, so named by one of his students). While sexual instincts are well known to be Freudian, the non-sexual death instinct is also a crucial aspect of Freudian theory. Infantile sexuality drives all bodily pleasures from birth and many actions are tuned to express this instinct, starting from suckling- the oral stage, defecation- the anal stage and interest in the sexual organ- phallic stage. The last stage leads to the Oedipus complex, attraction to the parent of the opposite sex and fear of the parent of the same sex. It leads to anxiety of castration by the father and penis envy in male infants. He theorized that the child resolves these dilemmas by the age of five and reconciles with the father. Then the child enters a latency period until puberty. Freud believed that adult mental ailments result because the normal progress of the mental state of the child is arrested by parental and social controls. He opined that obsessive compulsive disorder of cleanliness and germophobia are caused by conflicts in the anal stage and that the lack of resolution of Oedipus complex leads to homosexuality. Freud posited three substructures of the mind- id, ego and super-ego. The sexual drives and pleasure centers are situated in the id at one end and at the other end is the super-ego, in which the conscience, altruism and social instincts of self control are situated and fostered by parents. The ego is then the one which is caught in between the id and super-ego tensions and also has to deal with external realities. So, the ego is the conscious mind making decisions, the id and super-ego are hidden. Id is permanently in the subconscious mind and imparts to the ego the pleasure drive while the super-ego applies caution and restriction. While these are theoretical, Freud seems to have applied them in the literal sense. He stated that the ego's lack of resolution of the push-pull of the two extremes, leads to problems in which defense mechanisms of repression, sublimation (channeling the sexual energies into ambitious accomplishments), fixation (arrested development at one stage) or regressing into a development stage, occur. Repression, then, is a central component of the defense and leads to the super-ego relaxing its vigilance, which then leads to behavioral flaws. The Freudian death drive is in opposition to the *Eros* drive and replays trauma, relives distressing situation and overcomes the pleasure principle. He believed that war trauma (and similar disorders, now termed PTSD- post traumatic stress disorder) is caused by conflicts due to this part of the drive. One aspect of Freud's theory which is uniformly criticized is how sexist and male-centric it is.

Taking the cue from his mentor and coworker Josef Breuer's idea of free association, Freud's method involved the patient talking freely in the absence of other distracting sensory inputs and an unobtrusive analyst's assistance. The therapist enables disarming the super-ego and preventing the inhibition of the ego. This talking, over many sessions, would bridge the three elements and restore the harmony. (This is reminiscent of the recurring theme of Greek humoral theories and eastern mystic theories on harmony between elements. It appears that the push pull of differing components in life and in science is the essential part of order and disorder and chaotic phenomena, be it our minds or the Universe. One philosophy that emphasizes this is the Daoist yin and yang theory, though invoked in a positive context of balance). The dreams and unconscious gestures and slips are also symptoms. Freud presumed that the super-ego does not impose control during sleep and the latent meanings behind manifest dreams are symptomatic of any conflict. The interpretation of these dreams would let the patient realize the conflicts behind them.

While Freud's genius in arriving at his theory is not in doubt and each component of his theory is appealing to the average observer, the scientific accuracy and applicability of these theories is controversial. This is important because Freud and his followers like Carl Jung believe that they are foremost, scientists. The success and the failure of Freud's theories lie in the fact that they are compatible with all possible observations and therefore cannot be proven to be false. The theories explain all cases of people from those who are normal to the extremely distressed. This renders the theories to be non-scientific. On top of it, the cause is revealed by the symptoms and behavior, which in turn are caused. This is a circular argument, giving the impression of a religious idea. This criticism becomes particularly relevant when we consider that Freud believed religion was an illusion, created to defend oneself against the superior crushing power of nature. While he felt that the educated and the intelligent did not need religion to be good, the great mass of the uneducated and oppressed class needed a God to forbid criminal acts. Whatever the scientific value of Freudian theories, his methods have found favor both with patients and practitioners.

### *Carl Jung*

Carl Gustav Jung, the famous Swiss psychiatrist, was the son of a Swiss pastor of the reformed church. His religious upbringing came to naught when he realized that he felt nothing at the time of his Christian initiation ceremony. It is said that much of his work was related to this and his attempt to find an alternative to this faith. After he took up his profession in 1900, he published the book *"Studies in word association"* in 1906. He became well known for this field, where a person's mental complexes would be revealed by word association using a so called psycho-galvanometer, which measured the skin's electrical resistance. The effect is known to be due to the "cortical arousal" that reduces skin resistance. Jung had known about Freud's ideas on unconscious action and slips and relating this to the topic of word associations, he sent a copy of his publication to Freud. This resulted in a 13 hour conversation at their first meeting and a six year long friendship and collaboration between them. Freud even called Jung, his eldest adopted son. But this ended in 1913, when Jung resigned from International Psychoanalytical Association, where he had been elected with Freud's support. A year earlier he had published the "Psychology of the Unconscious" that departed from Freud's theories. Jung felt that Freudian theory of the unconscious was negative and theorized that the unconscious was a reservoir of repressed memories, with both negative and positive values such as creativity. He also thought that Freudian theory focused too much on sexuality.

Jung's contribution to psychiatry, in addition to providing the technique of word association, was (1) Ideas of individuation, complexes and persona (2) Concept of Archetypes (3) Concept of personality types. All these ideas are in current use and the popularity of these ideas is proven in colloquial descriptions. Jung was persuaded by Freudian theory to believe in the existence of the unconscious mind and its expression through the ego. His word association test revealed complexes, such as inferiority or superiority complex, a guilt complex etc. The complexes were nodes in the unconscious mind, a knot of unresolved issues. Jung's concepts on inferiority complex and guilt are now well recognized characteristics in psychiatry.

The psycho galvanometer was clearly too simple a tool for detailed diagnostics and Jung discontinued its use. He launched his idea of personal advancement and growth through individuation, the integration of the different parts of the psyche. This description of parts differed from Freud in significant ways. He identified the parts as the persona, the ego, the shadow, the anima and the self. He identified an archetype or a symbol for each of these parts. The Persona is what we present ourselves to the external world, a mask so to say, such as employee, manager, father, mother, son, daughter, law abiding etc. Shadow is part of the hidden personality. At the end of this persona is Ego, conscious understanding of the self, that lets us feel different or distinct from others. The inner layer is the Shadow, which is, in essence, Freud's Id and contains the animal instincts and drives. Further inside is the Anima (male) and Animus (female) that is the node of feelings with respect to the opposite gender. The complexes associated with this correspond to Freudian concepts like Oedipus complex. The Self then is the whole psyche. It is both within and the whole. Freudian Ego would correspond to parts of persona as well. Jung states that the Ego perceives reality in four ways – Thinking, Feeling, Sensing and Intuition. (In modern personality test, the two more ways, Perceiving and Judging, are added). Jung theorized that while conscious, we use our strong function and in unconscious mode, the weakest function is expressed. While not underemphasizing the importance of sexuality in development, Jung believed that the unconscious deals with other things as well and expressed these in dreams and involuntary expressions. Jung believed that while the first half of life was development of the mind itself, its integration – the individuation occurred in the second half of life. In his theory, the Shadow is what the Persona is not and the Ego refuses to acknowledge it. The individuation process then brings reconciliation between the two, realizing the self. This reconciliation and integration into the self is the major goal of the Jungian therapy. Jung also introduced the concept of personality types, the introverted and extroverted. The former is wholly interested in one's own mental state and therefore is deliberate, reclusive and low key. The extrovert is more interested in gratification obtained from outside. These terms are now used to express shy or bold individuals, whereas Jung deemed them in terms of mental makeup.

Another of Jung's significant ideas is that of the collective unconscious, which is derived from memories of ancestors. This was a bold idea, not based on any of his predecessor's ideas. He postulated that an infant is not a blank slate but has the collective unconscious of people from all times. This is a very valid concept which is borne by the fact children are born with knowledge of the structure of the mother language and the inheritance of genes also corresponds to some aspects of the brain. Jung stated *that "In addition to our immediate consciousness, which is of a thoroughly personal nature and which belief to be the only empirical psyche,...there exists a second psychic system of a collective, universal and impersonal nature, which is identical in all individuals. This collective consciousness is not developed individually, but is inherited. It consists of preexistent forms, the archetypes, which can only become conscious secondarily and which give definite forms to psychic contents."* (Jung 1981).

Jung's archetypes idea is a difficult but interesting one. He derived these archetypes after listening to the imagery of his patients. The mother complex is symbolized by the mother archetype. The complexes expressed by the patient are the expression of an archetype that dwells in the psyche. Jung cites the heroes, villains and saviors of mythology as archetypal representations and he theorized that these archetypes are derived from the collective consciousness and therefore borne with the individual. Jung states in his book *The Structure and Dynamics of the Psyche* (Jung 1960), that all ideas go back to archetypes, particularly in religion, but also in science, philosophy and ethics. A true understanding of the Jungian concept of archetypes borders on philosophy. The overall theories of Jung show the strong influence of Christianity, Hinduism and Taoism. He saw eastern ideas of harmony represented in the unitary nature of the collective unconscious. Though separated from Christianity, Jung felt that spirituality was a necessary goal and life's purpose was ultimately spiritual. He felt that therapies such as for individuation were at once, a journey towards the self knowledge as well as the spiritual. The yearning of the mind, to integrate the conscious and the unconscious, was itself a spiritual quest. While many of Jung's ideas are very much in use today, his tools on personality assessment are very popular, such as in the Myers-Briggs test. The Jungian Type Index also guides the psychoanalyst towards the subjects' preferred mode of examination.

While advances have occurred in the details of psychiatric practice today, the influence of Freud and Jung, on the practice of psychiatry and psychological counseling, is very significant and the popular expressions in ordinary conversation reflect the deep ingraining of the ideas of these two giants.

## Modern Medicine and Mental Illness

The experiences during the war and post war economies demonstrated the usefulness of psychotherapy for neuroses for all people and its use, sometimes unnecessary, became prevalent particularly in the United States. The treatment of the truly ill, insane, schizophrenic and patients with dementia was still in flux. The non responsiveness of the diseases led to extreme therapies. In 1880s, it was accidentally found by Julius Wagner-Jauregg that a high fever due to an infection cured a certain mentally ill patient and in 1917, he used malaria and TB to cause fever and reverse the mental illness with some success. The shock treatment with camphor as a convulsive agent was also tried in some places. An even more risky treatment was the injection of insulin in 1943 to cause coma in a patient for 20 minutes and then revive the patient with the expectation that his mind would be reset. These treatments were said to be very successful only to be later assessed as not worth the risk. In 1938, the notorious but sometimes effective method of electroconvulsive therapy ECT was developed in Italy by the psychiatrist Ugo Cerletti and his assistant Lucio Bin. Though it caused some memory loss, was fearsome to the public and had notorious cases of misuse, ECT proved to be useful for treating schizophrenics and cases of serious depression. Today ECT is still in use for some psychotic disorders. Brain surgery, in particular, lobotomy, too is an extreme measure that can calm cases of strong convulsions and violent disorders. These extreme and invasive measures indicate the desperation of caring psychiatrists who want to bring normalcy to asylum bound patients. But these methods could also amount to torture in the hands of a callous doctor. These days, Psychosurgery targeting the limbic system is practiced in very limited ways to cure very specific and well justified cases of mental illness.

In modern medicine, psychiatry has been separated from psychological counseling, the latter requiring a Ph.D. degree but not a doctor of medicine degree. Serious mental illnesses require the intervention of medically trained psychiatrists, who combine free association therapy with medications. The delicate nature of the patient's illness requires highly ethical practice and the scrupulous observance of the Hippocratic Oath of not doing any harm is especially applicable to psychiatry. Psychiatry has numerous fine divisions ranging from pediatric psychiatry to addiction to biological psychiatry to forensic psychiatry and so on. (The last, popular in TV series, uses the psychological profiling to determine criminal behaviors). A psychiatric diagnosis takes place by taking into account anatomical anomalies (brain scans, injuries etc.), psychopathological history (symptoms and syndromes recognized by medical field) and neuro-physiological manifestations (ticks, tremors, convulsions etc.) There are 4Ds that are recognized in psychopathology, Deviance- socially unacceptable behavior, Dysfunction- inability of the patient to carry out normal activities, Distress- patient is deeply troubled and Danger –violent or suicidal. The previously defined and further refined personality tests, word associations test, cognitive tests to assess the level of reasoning, dissociation etc. are given. A mental status exam (MSE) includes the above tests, physical examination of appearance, study of attitude, mood and behavior, speech, thought process and thoughts themselves and the assessment of personality in terms of the four Jungian categories of cognition. Inpatient treatments are uncommon and are required only in case of danger or for those requiring constant observation. Involuntary commitment is highly regulated by laws in most countries, since it violates the civil rights of the individual.

The major therapy that has evolved, since the days of confusion, is the development of drugs for the treatment of mental illness. It started when Austrian pharmacologist Otto Loewi discovered that a chemical called acetylcholine is involved in the signaling between nerve cells. These neurotransmitters became the clue to further developments in "psychotropic" drugs. John Cade, an Australian psychiatrist working at the Bundoora Repatriation Mental Hospital in Melbourne during the late 1940s, was experimenting in the crude kitchen, by injecting the urine of mentally ill patients into guinea pigs and testing if an excess of uric acid content in mental patients caused illness in the guinea pigs. There was a high mortality rate in the guinea pigs when they received this injection. In order to increase the concentration of the uric acid and confirm that it was the uric acid that caused the pigs to die, he used additional uric acid in the form of lithium urate in his injection. The pigs actually got better. His realization, that lithium was the beneficial content, led to treatment of bi polar disorders and manias with lithium, which is still in use. Spurred on by this discovery, others got into the field of psychotropic drugs. Chlorpromazine, previously known as 4560RP, was discovered by Henri Leborit, a French naval surgeon, as a drug to calm even the most violent patient prior to surgery and this became the favored treatment for violent patients. As research into these drugs revealed considerable promise, private pharmaceutical firms got involved in the research and development and by 1950s, these drugs started becoming available for not only seriously ill patients, but also for milder problems of anxiety, anger, depression etc. Tranquilizers like meprobromate and benzodiazepine relieved anxiety and stress. Valium (diazepam) was discovered by the Croatian-Polish-American Leo Sternbach in 1963, became hugely popular and was even abused by many people. A new culture of drug taking for mental discomforts, like aspirin for headache, started. Such medicines relieved symptoms quickly in ill patients while psychotherapy would take years. This has made the chemical therapy very popular.

In the 1960s, three important brain chemicals- serotonin, dopamine and norepinephrine were discovered. Serotonin is a chemical, most of which resides in the gut, helping in intestinal movements. Some of it is held in platelets and aids in clotting. However 5% of the chemical resides in the central nervous system and regulates mood, appetite and sleep. In the brain, it is a neurotransmitter that carries signals between brain cells. When a signal is received, the recipient cell passes it on to the next cell in the communication chain. About 10% of the chemical is lost in this process and 90% of it released by the recipient cell and taken up by the serotonin transporters. Serotonin is then implicated in depression when it is deficient in the receptor cells of the brain. In the 1970s, serotonin selective reuptake inhibitors (SSRI) like Prozac were discovered to slow down the process of the release of the serotonin by the receptor cell during synaptic activity. These antidepressants have proved their efficacy for serious cases of depression. But for milder cases, the jury is still out. Paradoxically, for some patients, a reuptake enhancer helps depression in patients who are resistant to SSRI therapy. This dichotomy points to the fact that depression is not a well defined illness. The poorly understood antidepressants have also found other applications such as in pain control and in reducing the severity of premature ejaculations in men.

The 20$^{th}$ and 21$^{st}$ centuries have continued to see development of different SSRIs and other medications. The drug discovery has been spurred both by urgent medical needs and also because of demand by people seeking relief from the stresses of day to day life. The categories of psychiatric medications are as follows:

- Antidepressants treat clinical and neurotic depression, eating disorders, chronic pain, premature ejaculation. Also (borderline personality disorder where a person has considerable distress due to emotions, with marginal efficacy)
- Antipsychotics overcome psychosis in schizophrenia and mood disorders.
- Stimulants treat attention deficit disorders, sleep attacks (narcolepsy) and are also used for dieting.
- Mood stabilizers treat bipolar disorders and the so called schizoaffective disorder which is a combination of psychosis and mood disorder.
- Anxiolytics relieve anxiety disorders.

Generally, the term mental or psychiatric disorder is applied in a social context, where a subject displays behavior, feels emotions and has perceptions that are well outside the social norms. The classification of disorders follows an international system of diseases, which includes standards for diagnosis. Some examples of disorders are anxiety disorder, phobias, bipolar disorders with heightened happy and depressed moods, post traumatic stress disorder, panic disorder and psychotic disorders like schizophrenia, delusional disorders and schizoaffective disorder. Personality disorders are categorized under conditions, where the subject is rigid, antisocial, paranoid, unable to adapt to normal conditions, overly dependent etc. and symptoms exist from childhood. There are also sexual and identity disorders, including the famous multiple personality disorder. Additional categories are syndromes where symptoms are clear, but underlying pathology is unknown. The abilities and disabilities associated with any disorder, vary over the complete spectrum. Of these, schizophrenia, depression, autism, bipolar disorder, obsessive compulsive disorder and attention deficit disorder receive much discussion and attention. A brief description of two of these, are given below.

## Schizophrenia

Schizophrenia is a serious disease affecting 1% of the people across all classes of people but favoring males. It impairs thinking and emotional processes, creating delusions, paranoid beliefs and fears, hallucinations, significant lethargy and lack of motivation. The disease results in dysfunctions and inabilities such as in speech, socialization, hygiene and experiencing pleasure. It is usually diagnosed during late teen or early adult years, though most families would testify that there were earlier signs. No single organic cause has been isolated, but the focus on neurobiology has increased since the discovery of reduced brain volumes in the frontal and temporal areas. It is not clear if this is a cause or an effect. PET scans reveal differences in brain activity in the frontal cortex, hippocampus and temporal lobes. However, many factors, environmental, genetic, psychological, and developmental, appear to be involved in the appearance of the disease and possible involvement of drug use has been found to be important. Recently the disease has been linked to a retrovirus incorporated in human genes, which separates from the DNA to reproduce. There are many versions of schizophrenia and the disease is categorized under psychosis (approximately, loss of touch with reality). The disease is managed with antipsychotic drugs chosen on the basis of the differentiation, with family and social support. While there are a few cases, Schizophrenia patients are rarely violent or a danger to themselves. Though occasional institutionalization may be necessary, most treatments occur in the home environment. Most patients are able to have normal or near-normal lives, but may require treatment for a long time.

## Depression

This is the most common mental illness that might afflict a person at all stages of life. Some last for a short time and others are hard to cure. According to the National Institute of Mental Health, symptoms of depression are following:

- Poor memory, difficulty in concentrating
- Fatigue and low energy
- Feelings of guilt
- Feelings of worthlessness, sadness
- Pessimism
- Sleep disorders, lack of sleep or excessive sleep
- Irritability
- Restlessness
- Loss of interest in activities and sex
- Eating disorders

- Headache, cramps, indigestion – unresponsive to treatment.
- Thoughts of suicide and suicide attempts

One can see that these symptoms (except the last one) are commonly experienced by most people at one time or another. What distinguishes depression is the duration and intensity of these symptoms that do not respond to normal stimulus. The major category of Clinical Depression, which affects 5-10% of the populations and women disproportionately, has a constant and very strong symptom of despair and worthlessness. Depression is triggered by a personal loss, social isolation, major life change, emotional or sexual abuse etc. As stated above, in one hypothesis, the imbalance between the neurotransmitters serotonin, norepinephrine and dopamine, is the source of depression and SSRIs or a similar antidepressant to boost serotonin or norepinephrine are used to treat the disorder. The monoamine hypothesis of the neurotransmitter is clearly not fully explanatory because of other contradicting characteristics of the disease. For example, the symptoms decrease within hours whereas changes in serotonin take a long time. Also, moods of healthy people do not seem to be sensitive to changes in serotonin levels. Studies are in progress.

Although mental illnesses are as old as physical diseases, the factors involved and the spectrum of the diseases with similar symptoms is so continuous and the diagnosis of the diseases and determination of underlying causes are so difficult that treatments for these diseases are progressing only slowly. While serious mental illnesses of psychosis and schizophrenia are now managed and even cured, milder forms of the diseases like depression and behavioral disorders are still difficult to treat. If the sudden spurt of progress in the recent decades and developing understanding of psychology and neurology are any indication, one can be quite optimistic about the medical treatment for mental illnesses. But considering that social ills, inequities, familial dysfunctions, decline in support from extended families are factors and that care of mentally ill persons is a social responsibility, an enlightened public continues to be the best bet for the safety and welfare of the mentally ill.

## MEDICINE, SOCIETY AND RELIGION

Unlike other sciences, medicine is a scientific field that is easily understood in terms of its purpose. In that sense, it is supported wholeheartedly. But, at present, the expectations heaped on the field of medical research and science, exceed the status of the field. Medicine has been a victim of its own success. It is easy to forget that the present state of medical science and practice is less than a century old, after a history of millennia of superstitions, poorly understood palliatives, ad hoc theories of diseases and cures and patient care that were both undeveloped and discriminating against the poor. It is also not well acknowledged by the society that modern medical science has developed at a relatively faster pace in terms of its applications, than other sciences and has saved and bettered lives of more people than have lived in the past millennia. In fact, more lives could have been saved by the society and a better quality of life could have been accorded to the world's population, if medical resources had been shared more equitably and if the Governments had taken public health as one of the priorities in the past, as they do now. Also, society does not demand better of the Government, in supporting higher levels of medical research and according it even a higher priority than given now. The public and Governments must realize that society can advance only if the physical and mental health of its people does. In other words, health and medical facilities are not a measure of the quality of life, but are prerequisites to a good quality of life.

An important aspect of research, not concretely understood by the public, is that medical research should not be wholly tied to immediate applications, such as curing of diseases, transplants, increasing life expectancy or addressing ageing. For too long, medical research has been tied to the public clamor for cures and higher quality of life. Fundamental research, which is not tied to needed applications and which has been the foundation of all other sciences, should be a foundational basis of medicine. Some part of medical research should be free from pressures of applications and should be pursued solely for understanding nature. While one may think this work is carried out in biology, biochemistry or biophysics, in reality, the actual knowledge and training of medical researchers provides better context for broadening this research. An example is the development of imaging by MRI and X-ray machines. The advances in these modalities are instigated and supported by practicing physicians. Public should give a carte-blanche support for this type of research, where the areas of work are motivated solely by curiosity and the need to understand the functioning and malfunctioning of the animal bodies. Governments would do well in participating in the identification of fundamental questions in medicine and fund these. While these questions (examples are DNA damage, evolution of humans through viral DNA recombination, stem cell research, immunity) would have definite applications, but those should not be the primary motivation.

Education is the most important tool the public can employ in managing and advancing healthcare. It is insufficient to look up the disease and a set of cures on the internet and form opinions. While helpful, such attempts cannot truly reveal what lay behind these discoveries or how a certain procedure or drug is applicable in a given case. A public, armed with the knowledge of medical methods, is the one that can ask the important questions of the doctor and obtain satisfying answers. The public and private enterprise, particularly in developing countries, are on a different track than science as far as medicine is concerned. The advancement requires both trust in science, one of the key themes of this book, and concern for the broad society's health. While medical researchers and practitioners work out of concern for the sick, the public needs to respond in kind by playing a supportive role. Often, the public takes the easy path of blaming the medical institutions, their failures, and the unavoidable fraction of callous and greedy practitioners, when something goes wrong. In addition, there is a tendency to believe in unsubstantiated rumors, such as believing that the public health measure of fluoridation of water is extremely harmful. The public must know that the broad cadre of medical researchers and practitioners are in it for the love of the field and for making a difference to people's lives. Doctors have a grueling training, far longer than most professionals do, working for 90 demanding hours a week for 8 to 12 years beyond their college years and one may excuse the occasional arrogance or impatience some doctors show. The nurses and other support staff deserve equal support, because they are always there dealing with day to day patient issues. Patients, working as hard as their doctors and nurses in cooperation and trust, can improve healthcare.

General surveys indicate that in Western countries, the public shows a little less ambivalence towards medical discoveries than in other scientific fields, mainly because of the direct impact they can feel. This greater interest has, however, not translated into a sustained support for the research nor the practice. The poor understanding of the nature of medicine and its methods, have resulted in the above noted expectations. A failure of the profession in meeting these expectations is taken personally by the individuals as lack of knowledge, incompetence, callousness and working out of greed. An example is the lack of clear understanding of how medicines are discovered and approved for use, their side effects and potential interactions. The personal concern of the patient and expectation from the system is not matched by the patient's initiatives to learn about the system. Idle talk, rumors and media versions of medical developments seem to be more prevalent than hard facts. A drug's long term effects such as potential for cancer are likely to be exaggerated in people's mind, compared to the harmful effects of their daily habits of food, water and lack of exercise. A good example is the lack of attention to the fact that sugar is harmful in the immediate while the long term harmful effects of sugar substitutes are likely to be better known.

An aspect that is quite troubling to the medical scientists and practitioners is the lack of balance in patients' expectations. The physician, like any professional dealing with clients, deals with a patient at two levels. At a personal level he/she acts with a healing touch. (For example, when a doctor asks a patient, "How are you doing?", he or she is not asking a question of diagnosis, but a simple social question and therefore, this is not a reflection on the knowledge of the doctor). The expectation that physicians should be well honed in their practice and should be knowledgeable is also correct. But, when the patients find out that medical science and therefore the physician does not have an answer to a particular question, they are likely to believe that all of medical science is flawed and undeveloped. Some might complain that "medical science cannot even figure this out", without knowing how complicated and difficult the issue is. (In some cases, this may be because they are used to getting ready but unproven answers from traditional systems of medicine). Modern medicine is a relatively young field and it is growing very fast under the pressure of the public's urgency in healthcare and this leads to gaps in knowledge. In this empirical field, some side effects, interactions and long term efficacy will only become known after some time has elapsed, and if the results show harmful effects science does reverse itself.

The system of medical insurance, with non medical professionals deciding on allowing or disallowing treatments, has disrupted patient-doctor relationship and trust. The public gets medical and healthcare news from the media, which may have its own spin or incorrect representation, instead of getting it directly from the doctors. A system that minimizes barriers in the scientific and humane application of healthcare and improves patient access to education, information, doctors and treatments, is essential. Medical practitioners, on their part, have been lethargic or even apathetic about the equitable use of healthcare resources. Many doctors like Atul Gawande in the U.S. have done well in pointing out the shortcomings in the U.S. healthcare, but doctors must devote some time to educate the public at large about the methodology of medicine too. In the long run, this will help the profession.

Drug companies that have been instrumental in many of the discoveries, are increasingly becoming corporate minded in a field where public welfare should reign supreme. The societies harboring such greedy ventures, whatever their attitude towards capitalism is, should not link profits and health systems. The bottom line is that crucial curative drugs, whatever the cost of development, should be borne by the society with corporations making no profit. The choice of availability should not be made by profit conscious corporations alone. If necessary, Governments may contract the production of drugs that the companies deem unprofitable. Other drugs should be provided at reasonable cost.

A somewhat worse situation exists in the Eastern countries, where modern advancements have rushed in at great speed in a matter of decades without the background of historical events marking modern scientific discoveries. While modern medicine was forced in for good or bad (mostly good), by colonization in most countries, the assimilation of these practices has not found equilibrium. One reason is that there is a direct distrust of the system- it is foreign and originates in the exploitative and colonialist countries, whereas the local system is indigenous, traditional and arrived at with centuries of wisdom and therefore "Good". The other reason is the same as in the West, a poor understanding of the methods of science, in particular, how drugs are approved. In Eastern countries, where home remedies of herbal medicines, oil massages, skin applications and hot and warm compresses are common, the taking in of "chemicals" is an anathema. This is despite the fact that inclusion of gold, copper and silver in food, water and medicine is considered to be good, even though these too are chemicals. The fact that traditional use is a long form of drug trial, and modern medicine is based on the understanding that statistical and epidemiological methods can produce quicker and more conclusive answers, is poorly understood. In countries where there is a strong and evolved tradition of indigenous medicine such as India, China and Middle East, it is very frequent to hear the argument that Western approved drugs have been proved to be very harmful (often omitted phrase- over long periods of use). Apart from the fact that, the number of wildly successful medicines far exceeds those that caused harm, what they fail to realize is that it is the same Western scientific process that actually discovers the harm, admits the facts and advises against the continued use. They also fail to note that there is no equivalent system that would bring out the harmful effect of the versions of traditional medicines.

The increasing problem is that traditional medicine and the so called nutritional supplements are carelessly prescribed and suggested by untrained individuals, who themselves may have learned about these in a poorly scripted and unaccountable television programs or from anecdotal cases. Relatives and friends often suggest these remedies to patients, who comply out of trust or because they feel they must obey. What is not understood here is that in such narratives, only success stories, which may or may not be true or which may have been due to the well demonstrated placebo effect, are told while many harmful results or failures would not be. In some of these cases, the benefit might have been symptomatic due to placebo effect. In this sense, traditional medicine is like religion, not testable and faith based, the harm or good done by it is poorly and subjectively assessed and anecdotal evidence is enough to support a thesis. The unregulated use of traditional medicine (both pharmacological processing and inappropriate use by poorly trained practitioners) can cause harm. While good traditional medicines were highly prescriptive in drugs and habits, the modern user may mix medication, causing undiscovered drug interactions. The pity is that one would never know that such harm had been caused by a traditional medicine, because there is poor accountability nor is there a follow up. The value of modern drug trials in this regard is also not appreciated by the public in these countries and a regulation to subject traditional medicines to such trials will have a salutary effect.

***

Healing is a subject dear to religion, particularly Christianity. In all religions, there is a god, goddess, angel or saint of health and praying to such powers often provided healing. On the contrary side, most cultures have the concept of the 'Evil eye' as being the reason for illnesses and deaths. *Voodoo*, spells and punishments for sins are still deemed causative in some cultures, sects and classes of societies and superstitious counter measures are instituted. A large number of these, both cures and diseases, are founded in the suggestive power of one's own mind, particularly for psychosomatic illnesses.

In the Bible, Matthew 8:1-4 states, *When Jesus had come down from the mountain, great crowds followed him; and there was a leper who came to him and knelt before him, saying, "Lord, if you choose, you can make me clean." He stretched out his hand and touched him, saying, "I do choose. Be made clean!" Immediately his leprosy was cleansed.* (See figure 3.13). Even healing from a distance was described in the Bible. Matthew 8.5.13 states, *"When Jesus had entered Capernaum, a centurion came to him, asking for help. "Lord," he said, "my servant lies at home paralyzed, suffering terribly." Jesus said to him, "Shall I come and heal him?" The centurion replied, "Lord, I do not deserve to have you come under my roof. But just say the word, and my servant will be healed.. ".. Then Jesus said to the centurion, "Go! Let it be done just as you believed it would." And his servant was healed at that very hour."*

Figure 3.13. Jesus Christ Cleansing a leper, by Jean-Marie Melchior Doze, 1864 (Source: Wikimedia Commons).

Hindu scripture *Bhagavatam* narrates that a princess Uttara delivered a stillborn son of Abhimanyu. Krishna's relatives and friends begged him to revive the child. Krishna, invoking his *dharmic* (righteous) character sprinkled drops of water on the child and touched him on his chest, reviving the child (See figure 3.14). There is also a tradition of faith healing in Islam, although not as strong as Christianity and Hinduism, particularly among the Sufi Muslims. In religion, saints are also said to have healing powers and there are many mythologies about this. In India, faith healer Swamis are many and some are very famous, although there are no official records of their healing. Most of these stories survive by hearsay. The evolutionary process of messages ensures that the ones that are exaggerated survive and spread.

Figure 3.14. Lord Krishna saves Uttara's son (Source: Wikimedia, Author: Ramanarayanadatta Sastri, Volume: 6, Gorakhpur Geeta Press, Gorakhpur, India).

Healing is one of the motivations for religious belief. In each culture, different religious observances mark a prayer for a loved one or oneself to get better. In India, a woman would tie a coin in her *saree* front and promise to deliver it to a certain temple after the person recovers and a Catholic makes an offering of a charm of a metal eye (called exvotos in Europe) to the altar of Mary, for her relative or friend to recover his eyesight. In the Shinto religion, a prayer for the sick to be healed is written on paper and tied to a tree at the temple site. There is no doubt that such offerings and prayers provide hope and courage to the people. It should therefore not be looked down upon, so long as they are also doing everything within their means to provide the sick with medical care. There is nothing more disheartening or frightening than to see one's child or spouse or other dear ones being seriously ill and any means that can console and give encouragement and strength should be used. But, where good treatments are available, prayers should not replace medical treatment. Some religions like Scientology harm their followers by insisting that they should not use medications and forbidding modern treatments.

Healing by touch is a common anticipation among the religious. The previously quoted case of Mesmer's animal magnetism theory is a case in point for touch healing. It is evident that the psychotherapy of a patient, touching the practitioner knee to knee and finger to finger and listening to the "mesmerizing" sounds of the practitioners had, in many cases, cured patients of their illnesses. Even Mesmer attributed this to a scientific phenomenon and now we know that this healing is a form of therapy. Faith healing is, by and large, of this type. First, at a logical level, it should be noted that faith healing has not been particularly effective for millennia in that era when there was much faith in faith healing and it was one of the major methods. If they had had significant effect, scrupulous records kept by ancient and more recent civilizations would have recorded the great successes and would not have advanced theories of natural causes and treatments based on scientific methods. These civilizations would not have gone away from such systems. Again logically speaking, even if faith healing was only a partial success working for the most deserving, it is not a reliable method based on clear prescriptions for successful treatment.

The American Cancer Society website states, *"Although it is known that a small percentage of people with cancer experience remissions of their disease that cannot be explained, available scientific evidence does not support claims that faith healing can actually cure physical ailments. When a person believes strongly that a healer can create a cure, a "placebo effect" can occur. The placebo effect can make the person feel better, but it has not been found to induce remission or improve chance of survival from cancer. The patient usually credits the improvement in how he or she feels to the healer, even though the perceived improvement occurs because of the patient's belief in the treatment. Taking part in faith healing can evoke the power of suggestion and affirm one's faith in a higher power, which may help promote peace of mind. This may help some people cope more effectively with their illness."* (American Cancer Society 2013)

Systematic surveys and tests on prayers have yielded no positive correlation. The largest study on the role of prayer in healing was conducted in 2006 by Herbert Benson, a professor at Harvard medical School and Director-Emeritus at Massachusetts General Hospital. The study (Benson 2006) used double-blind protocols, in which patients were distributed into three groups. The experimental Group 1 and control Group 2 were informed that they may or may not receive prayers, while actually only Group 1 received them. Psychosomatic effects were tested for in Group 3, which was told that they would receive prayers and did receive such prayers. Three congregations of three churches prayed for a successful surgery and a healthy recovery of patients. On assessing the results, it was found that 51% of the patients in Group 1, 52% in Group 2 and 59% in Group 3, died of major complications within 30 days. (Essentially there was no significant differences in survival rate between the three groups). Some of the patients for whom prayers were offered did worse than those who did not receive the prayers. While this is not necessarily the last word on the efficacy of prayers on healing, it does show that prayers are not the ready answer, where science is not faring well. 50% survival rate is considered poor in medicine. But if religious persons observed a 50% survival with prayer and no control group, a 50% survival rate would be considered miraculous, since other 50% died and would be attributed to prayer.

In poor countries and even in advanced countries where the poor have no access to medical care or are not aware of its availability, prayer and surrender to the Divine is the last resort. Often, hardships such as these become the reasons for religious belief. This is a heartbreaking situation that the inequities in the world create. Proper availability of healthcare with reliable recoveries would give hope and happiness to such people and religion would then be reserved, if needed, for personal spiritual upliftment and not for appealing to a God in conditions of misery.

Miraculous or not, scientific tests are equally applicable to all phenomena. If a cure is found or happened for a sick person under a scientific study and no scientific cause can be found, science will admit that it does not have an explanation and will go on and work further to discover scientific explanations. In some cases, where there is no predominance of such events, science might even ignore those and move on, because the relevance would be deemed to be poor. But if there are important scientific principles in these cures, these would eventually be found out, as seen in accidental discoveries in etiology and cures for diseases.

A final word on the topic of healing in this chapter: The history of medicine shows that recent advances in other sciences are accompanied by even faster advances in medicine. It is correct to be optimistic about a future where we will have greater control over diseases, ageing and our overall health. The future advances in the fields of biology, chemistry, physics and computer science and increasing understanding of statistical principles in assessing diseases and treatments will revolutionize medicine, and in future decades, medicine might be extraordinarily different. Medical science has advanced enormously and yet has barely started. In the meantime, it should be understood that healthcare is also a social justice issue. It is most important for individuals and societies to be as deeply involved in advancing healthcare and to support necessary research without bringing in extraneous and contentious issues.

## REFERENCES

American Cancer Society; *Faith Healing*. EText: http://www.cancer.org/treatment/treatmentsandsideeffects/complementaryandalternativemedicine/mindbodyandspirit/faith-healing (2013).

Benson, H.; *"Study of the Therapeutic Effects of Intercessory Prayer (STEP)", American Heart Journal* Vol. 151, No,4, (2006): pp. 934–42.

Brody, S.N., *The Disease of the Soul, Leprosy in Medieval Literature*, Cornell University Press, NY (1974) pp.132-146.

Cattermole, G.N.*; "Michael Servetus, Physician, Socinian and Victim", J. Royal Soc. Medicine*, Vol. 90, No. 11(1997)p.642.

Clarke, S.; *Thinking with Demons: The Idea of Witchcraft in Early Modern Europe*, Oxford University Press, Oxford, U.K. (1997) p.199-203.

Dogra, S., et.al., *Indian Journal of Medical Research*, Vol. 137, No. 1 (2013)pp.15-35. EText: http://www.ncbi.nlm.nih.gov/pmc/articles/PMC3657879/.

Duffin, J.; *The History of Medicine*, University of Toronto Press, Toronto, Canada (2000).

Emsley, J.; *Oxygen, Nature's Building Blocks: An A-Z Guide to the Elements*. Oxford University Press, Oxford, UK: (2001) pp. 297–304.

Greene, L.W.; *Kalaupapa-An Exile in Paradise*,, U.S. National Park Service, Molokai, Hawaii, Denver Publishing Center, Denver, CO (1985), EBook: Hathitrust, http://babel.hathitrust.org/cgi/pt?id=pur1.32754073195046;view=1up;seq=3.

Hinshaw, S.P.; *The Mark of Shame of Mental Illness*, Oxford University Press, Oxford, U.K. (2006)).

Hippocrates, *Aphorisms*, Kessinger Publishing Co., Whitefish, MT (2004) p.1. Etext: Massachusetts Institute of Technology, http://classics.mit.edu/Hippocrates/aphorisms.1.i.html.

Isaacs, H.D.; "*A Medieval Arab Medical Certificate", Medical History*, Vol. 35, Issue 4 (1991) p. 250-257.

Jacob, J.T. and and Franco-Paredes, C.; *The Stigmatization of Leprosy in India and Its Impact on Future Approaches to Elimination and Control,* Neglected Tropical Diseases, PLOS, Vol 2, No.1, (2008)e113.

Jung C.G.; *The Archetypes and the Collective Unconscious,* Volume 9, Issue 20 of Bollingen series, Editor: R.F.C. Hull, Princeton, University Press, Princeton, NJ (1981)p.43.

Jung, C.G.; *Structure and Dynamics of the Psyche*, Collected Works, Vol. 8, Pantheon Books, New York, NY (1960).

Masaki, T., et. al.; *Cell,* Vol 52, No. 1 (2013), p.57.

McDonald, K.; *Biologists Uncover Mechanisms for Cholera Toxin's Deadly Effects*, News Center, University of California San Diego, (Sept. 11, 2013). EText: http://ucsdnews.ucsd.edu/pressrelease/biologists_uncover_mechanisms_for_cholera_toxins_deadly_effects.

Moore, R.I.*;"The Formation of a Persecuting Society: Power and Deviance in Western Europe",* Blackwell Publishers, Oxford, U.K. (1990).

Nester, E. N., et.al.*; Microbiology: A human perspective*, (5th edition) WCB McGraw-Hill, Boston, MA (2007).

O'Malley, C.D.; *Christianismi Restituto and other writings by Michael Servetus-1553*, translation, Classics of Medical Library, Birmingham, AL (1989) pp.207-209. EText: *Christianismi Restituto -1553,* Servetus International Society, http://www. servetus.org/en/miguel-servet/writings/63-christianismi-restitutio-1553.html.

Pandya, S.S.;"*Anti-Contagionism in Leprosy, 1844-1897", Int. J. Leprosy and other Mycobacterial diseases*, Vol. 66, No.3 (1998) pp.374-384.

Pasteur, L.*;* address delivered by Louis Pasteur at the*"Sorbonne Scientific Soirée"* in 1864, Etext: University of South Florida, Tampa, FL. http://www.rc.usf.edu/~levineat/pasteur.pdf.

Pasteur, Mm et. al.; *The Germ Theory and Its Applications to Medicine and Surgery,* Comptes Rendus de l' Academie des Sciences, Vol. 86, Issue 5 (1878) pp. 1037-43.

Porter, R.; *Greatest Benefit to Mankind*, W.W. Norton and Co., New York, NY, London, U.K. (1997).

Tarlow, S.; *Ritual, Belief and the Dead in Early Modern Britain and Ireland,* Cambridge University Press, Cambridge, U.K.(2010)p.64.

Voltaire, F-M., A.; *Dictionnaire philosophique portatif ("A Philosophical Dictionary")* (1764), University of California, Berkeley, CA(2007). Etext: Wikiquote: *en.wikiquote.org/wiki/Medicine.*

Worboys, M.; *Notes Rec Royal Society of London,* Vol. 67, Issue 3 (2013), p. 199.

*Chapter 4*

# ORDER, DISORDER AND CHAOS: COMPLEX PHENOMENA AROUND US

"In all chaos there is a cosmos, in all disorder a secret order."- Carl Jung (Jung 1968).

When do we not deal with order, disorder and complex situations, phenomena or behavior? It is all around us; sometimes clear as a bell, sometimes beautifully complex, sometimes maddeningly chaotic and sometimes frustratingly complicated. Order is what religion is said to bring to our lives and order is what science is said to bring to our thinking and knowledge. In our society, order has a high value. "Law and Order" invest us with control over situations and behaviors so that the goals of society are met. The laws allow us to navigate around our worlds and we see order as the expression of this law. Religious commandments are said to be derived by "God's Laws" and order is established and maintained by these laws. In a society too, the legal Law is the theory of desirable behavior and relationships and Order is the enforcement and observance of these laws. In the case of religion or in a society, there are usually laws that need to be obeyed if judgment and punishment is to be avoided, but a choice remains. In science, however, laws are implicit and implicate in our lives and we cannot evade the natural laws, such as escape gravity on earth. The order that scientific laws entail too are evident in fundamental ways, as we see in Newton's or Einstein's laws of mechanics or Maxwell's laws of electromagnetism or laws of chemical interaction, which result in predictable relationships between properties of objects and outcomes in an event. In religion and society, disorder is an undesirable state to be avoided while disorder in science is a state to be noted, explained, understood and incorporated in the science of nature.

While the major branches of science dealt with an orderly universe, its various branches such as mechanics, nuclear physics, high energy physics, condensed matter physics, fluid physics, organic and inorganic chemistry, cell and molecular biology proposed and confirmed natural determinism. Most of what we generally seem to recognize as science, deals with predictable behavior of particles, atoms, objects, stars, cells, plants and animals. As an example in physics, one could write an equation for an object in a given situation and the object's trajectory (future) in the relevant phenomenon would be determined. This determinacy encouraged Laplace to make a statement that given a powerful calculator of the behavior of every atom in the Universe one could predict exactly what would happen in the future. But the modern day development of the field of Chaos has changed this thinking that the world, according to science, is deterministic and predictable. It must be noted that this is, in addition to the statistical nature of quantum physics which has the underlying “uncertainty principle”.

We see that our lives are orderly in many cases and seem to be unyieldingly disordered in other cases. In yet other cases, the complexity of it all hides the order. We negotiate a wage for some work, we do it and we get the money from the employer. This is the order we like. We buy a lottery ticket and the chance of getting even some money from the lottery is totally random. But when an unknown uncle leaves an inheritance to us, it seems to be a matter of luck, but it might have had a complex set of orderly events in them. Since the 1970s, it is becoming increasingly clear to scientists that, like in the above cases of life, the science of deterministic cases is an incomplete description of nature. The foundation for these ideas had been laid in earlier periods, but the importance of that had not been realized till recently. It is now fully known that even simple scientific descriptions for particles in orbits, planets in orbits, water dripping from a tap and water boiling in a pot, though may seem orderly in one sense, are incomplete and these systems have other hidden orders and may even go into disorder. Smoke from a small fire rises lazily and seems to have its own mind, even though it obeys laws of physics. We see this in our daily near-orderly lives. In a different domain, the way a leaf forms, evolution proceeds, an asteroid moves through the solar system, the weather is established in our region, and an embryo grows, have multi parameter dependencies resulting in complexity, but often have simple rules that govern the development of this complexity with profound results. In yet another regime, the way an ant or bee colony forms and maintains or how pedestrians walk along a crowded street in opposite directions, result from collectivization of seeming disorder into order. So science and life mirror each other. Given this, it would be hard to find a case of pure order or disorder in nature, though there are some cases, like the orderly binding of quarks into neutrons and disorderly behavior of hot gas molecules in a highly insulated container. But even more importantly, as we approach from orderly behavior to a combined regime of order and disorder, the order is hidden in some cases and the system is unpredictable enough to appear random. Such a state is the state of Chaos, a term used in literature implying confusion and lack of control, but in science, for a state of multiply unpredictable order.

The relationship between orderly states, chaotic states and complexity is intricate and is not amenable to solutions in physics, while mathematical descriptions do exist. Typically, one would have to use a long computation to illustrate the effects. This is a lot like the fact that one has to live through a life in order to understand its complexities. The scientific ideas on order, disorder and chaos, which result in complexity are lucidly laid out in two exceptional books, “The Quark and the Jaguar” by Murray Gell-Mann (Gell-Mann1994) and “Deep Simplicity”, by John Gribbin (Gribbin 2004). The book “Chaos- Making a New Science”, by James Gleick (Gleick 1987) is highly readable for its narratives on the people behind the discoveries in the field. These books and other descriptions of this field are meant to give an understanding of the development of these fields and their description in science.

Here we give examples from this field at an introductory level to once again demonstrate that science proceeds with caution and integrity. But this topic does something that other topics in this book do not do. The theory of chaos applies to economics and finance, social science, political science, populations, all fields of science and even the process of scientific discoveries. Every field of human endeavor and inquiry involves one or several aspects of the theory of Order, Chaos and Complexity. So, the topic can be broached starting from any branch of inquiry and we would find the phenomena occurring. So, as we present a specific aspect of the order, disorder, chaos and complexity, we will also draw examples from different walks of life and various phenomena in this world. This is not to say that we definitely know that the scientific phenomena are what govern our lives, but the parallels do make us wonder. Very importantly, this chapter shows clearly that science rules our lives, whether we recognize it or not. (Most of the time we only recognize the science behind objects but not events in our lives). By reading this chapter and the books mentioned above, the profoundness and trivialities of our lives would become clear and the readers can form a philosophical understanding of our lives.

It is normal for one to wake up sometime in the morning and gradually or hurriedly finish the morning routine, dash off to work or school or wherever one has to be do, finish what is planned for the day, evening and night and then go to bed. A 'normal' day is when all one intends for the day happens more or less in expected ways with insignificant differences. Also, such a day may not make a deep difference in one's life. But, let us take the example of a man leaving for work like many days in his work life. Let us say, he usually takes the 8:14 AM train and reaches his office building at 8:55 AM and then takes an elevator up to his floor and is usually at his desk at about 8:58 AM. On most days, within a certain margin of error, he is able to do that and nothing much different happens. On a certain day, he is a little behind in getting ready for work, dresses, but in his hurry he spills coffee on his shirt. He has to change his shirt which causes more delay. Because of the delay, he misses his 8:14 train and arrives at the building only at 9:10. A woman, visiting one of the offices, also arrives precisely the same time at the elevator. They exchange pleasantries and as they ride together on the elevator, they become acquainted. That acquaintance later turns into romance and he ends up marrying her and has children and a whole set of generations with its own set of remarkable events arise and change the world in its own ways, far into the future. All this is because, unlike on the usual days, this day the man spilled coffee on his shirt. The proverbial poem illustrates this for a different situation.

*For want of a nail the shoe was lost.*
*For want of a shoe the horse was lost.*
*For want of a horse the rider was lost.*
*For want of a rider the message was lost.*
*For want of a message the battle was lost.*
*For want of a battle the kingdom was lost.*
*And all for the want of a horseshoe nail.*

We wonder at these miraculously connected yet not so rare events. But, through an astounding array of discoveries in the field of mathematics, study of dynamical systems and simulations, we find that these types of events can be found in natural science and therefore we should not be surprised that these happen in our lives. It is tempting to ascribe these to miraculous arrangements specially designed for one's own life and assign it to a category of intervention by God, a superior being or Dame Luck etc. But the fact is that our own lives are a string of natural events, whatever additional qualities, such as emotional, religious or spiritual feelings that we may find in our lives. This means that our lives also proceed along the wonderful and often naturally unpredictable paths, precisely according to laws of nature in the domain of Chaos.

The above example in a man's life has the feature that a small change in his daily routine caused a major change in his life and other's lives. (We can speculate on what string of events brought his would be wife to the same elevator). The process involves many interactions (many parameters and objects) and some are the so called "non linear" interactions (the response to a small stimulus is very strong in result) and some situations are in a critical state (phase transition, like water about to boil). This story is characterized by several features – (1) High sensitivity to initial conditions, where a small change in initial state causes big changes in the outcome- often called the Butterfly Effect (2) Bifurcations, where the event comes to a fork in the road and spontaneously goes along one path rather than the other, and (3) results in a new complex or simpler state. One can also speculate about the possibility that this history is inevitable, in which case, the trajectory of this man is an "attractor". The processes involved in the rise of complexity from a simple orderly state or a disorderly state are "Non linear phenomena" and "Non Equilibrium Thermodynamics".

Most physics equations are deterministic in the sense that one can use them independent of the choice of initial or reference state. For example, the distance $D$ travelled by an object in a certain direction, from time $t=T_0$ to time $T$, with an initial velocity $v_0$ at time $T_0$ and an acceleration $a$ (all in one direction), is given by,

$$D = v_0(T - T_0) + \frac{1}{2}a(T - T_0)^2 \quad (4.1)$$

This calculation would be valid whatever time we take for $T_0$ and therefore for all initial velocities $v_0$ and positions. In other words, one can start calculating the trajectory of this object any time and follow it to later times or one can reverse the calculations and follow it back to some time and we will get the same trajectory of the object. However, the evolution of the so called "dynamical system" can be calculated only at each step in that development. This is similar to obtaining solutions to many differential equations by the method of "finite differences". In this method, a dependent variable $y$ is incrementally advanced to the next value $y+dy$ as the independent variable $x$ is advanced to $x+dx$ where the relationships between the derivatives of $y$ with respect to $x$ are given in the equation, in terms of $x$ and other known parameters. A simple example of such an equation is,

$$\frac{dy}{dx} = x^2 + 4 \quad (4.2)$$

The solution to this equation (by integration) is given by,

$$y = \frac{x^3}{3} + 4x + Constant \quad (4.3)$$

With the relation in equation 4.3, we can know the value of $y$ for all values of $x$. The constant is determined by one given boundary condition, such as when $x=1.0$, $y=5.0$, so that the value of the constant (by substitution) is $2/3$. This is a deterministic solution and would be valid for all values of $x$ and $y$. However, we can also start from the known boundary condition and increment the solution discretely from the differential equation itself, meaning, we increment $x$ by $dx$ and increment $y$ by the calculated $dy$, to obtain the new value $y+dy$ for the new value $x+dx$. When $x=1$, substitution in equation 4.2 gives, $dy/dx=5$ or $dy=5dx$. If we choose a small increment $dx=0.05$ then $dy=0.25$. The new value of $x$ and $y$ then are $x=1.05$ and $y=5.25$ (with initial value of $y$ being 5.0). We can repeat these iterative calculations by advancing $x$ by the increment $dx$ each time. The values in each step are given by,

$$x_{new} = x_{old} + dx;\ y_{new} = y_{old} + dy = y_{old} + (x_{old}^2 + 4)dx \quad (4.4)$$

This is a progressive calculation and we cannot jump to distant values, since the accurate results require that we use only small increments of $dx$.

This type of step by step approach emulates the evolution of dynamical systems. While deterministic calculations do not depend on the direction of calculation ($dx$ may be negative or positive and starting point is anywhere), the evolution of dynamical systems cannot be, in general, traced in arbitrary directions to give the same account of the history. The issue of entropy (a thermodynamic property related to increase of disorder in a system) may be involved and the so called sensitivity to initial conditions may, in real situations, make a unique omnidirectional solution, not available. While, typical scientific and engineering equations are considered workable when the parameters vary smoothly and may even be expected to have smooth gradients, dynamical systems may have abrupt changes. Therefore, dynamically varying systems, while following an equation for each step, may or may not be predictable at any given step, may vary smoothly or abruptly or may change from one solution to another for changes in the independent parameters. (An abrupt change is like a cliff on a terrain or a change of state). An example of a dynamical system is population dynamics, where change in population after a time interval can be successively calculated from the current population using known parameters of multiplication rate, emigration and immigration rates and mortality. The population dynamics would be given by the rule- $P_{new}=P_{old}+ f(P,A,B,C..)$, where $P_{new}$ on the left hand side is the population after a time interval and $P_{old}$ on the right hand side is the current population and $f$ represents a function of the population and other parameters (here symbolized by $A,B,C$) including the time interval, on which the population change depends. (For accurate calculations, the time interval has to be short, because such a calculation assumes that other parameters remain constant during this interval). One can see that this type of dynamics is applicable to a variety of systems we encounter, from weather to stock market to election forecasting. It is also easy to see that once the number of rapidly varying and strong influence parameters become large and these parameters are interdependent, the system would also become complex. But, what makes this field of inquiry interesting is that this complexity gives rise to order that cannot be seen from the inspection of the governing rules and can be observed only after the fact of carrying out the iterations or after the system has evolved.

A simple example of a chaotic system is a leaky waterwheel, which rotates when one or more buckets placed around the wheel get filled from a water fall or spout. Buckets are hung loose so that they stay upright as the wheel is rotated by weight of water and the lever arm (spoke) of the wheel. If the buckets are leaky, several things happen. (a) When the water flow from the source is small, the bucket fills slowly and drains equally or more and the bucket does not fill well enough and the wheel does not rotate. (b) When the flow is larger, the bucket gets filled adequately during its transit near the spout, even though it leaks, and so the wheel rotates, (c) as the flow gets even more, the rotation can become chaotic. As the wheel rotates with increasing speed, it may not have enough time near the spout and therefore may not fill adequately and subsequently slow down and may even come to a stop. As it slows down it would once again start filling adequately and speed back up in an oscillatory fashion. But the changes are still consequences of one parameter feeding back on another with a slight delay (inertia, filling duration etc.) and the wheel responds differently each time. When the flow gets strong, it may not only stop, but rotate in the reverse.

The mathematical theory of chaos is not only prevalent in practically every aspect of the Universe from formation of galaxies to cardiac arrhythmia, it also has immense applications in many areas. For example, in chaotic communication, pieces of waveforms are used rather than a single or combination of sinusoidal signals. Also, since the chaotic signals are unstable or not predictable, one can use them for secure communications, where the key to recover the information is separately communicated.

## Sensitivity to Initial Conditions and Butterfly Effect

In the above story, the man had a dramatic change in his life, just because he spilt coffee on his shirt. (Actually we did not go into what made him spill coffee on his shirt; even that may be a random step or be caused by a preset of steps caused by a different random condition). So on that particular day, the trajectory of life was very sensitive to the initial condition of when he left for work. This sensitivity to initial condition may not be high in many cases and may be exquisite in others. An opposite way of saying this is that, this trajectory of events has points that are arbitrarily close to points in other trajectories which develop completely differently. In mathematics, these “Dense” trajectories are observed in what is called “topological mixing”. The story of the man described above is repeated in all our lives in different forms and outcomes and similarly, the exquisite dependence on initial condition resulting in major features, is seen in nature as well. The recognition of this sensitivity in nature had its beginning more than a century earlier.

In 1570s, Tycho Brahe, a Danish Nobleman –Astronomer and a man passionate enough about mathematics that he lost his nose in a dispute over a mathematical formula, made precise measurements of planetary transits in the sky and published his results in 1572-1577 C.E. Johannes Kepler took these meticulous observations and examined fits to different shapes of orbits and determined that planets follow an elliptical orbit with specific radii and ellipticities and published his findings in 1609. The world and the churches had accepted Kepler’s laws of the revolution of the earth and other around the Sun. With the advent of calculus discovered by Leibnitz and Newton, problems of this type could be calculated in differential manner (small incremental steps) or in special cases to give an integrated effect over large intervals. Using the calculus and Newton’s law of Gravity, the actual geometric equation for the orbital path of a planet, such as the ellipse traced by a particular planet going around the Sun, could be analytically derived. But there was/is a catch. The analytical derivation can be done only by assuming that the planet feels only the gravitational pull of the Sun. In reality, of course, all the other planets and indeed the whole Universe are pulling on the planet. An analytical solution for the path of the planet cannot be obtained with the added gravitational pull of even a single additional planet, leave alone all the other planets. An approximation can be obtained in some cases where the other planet’s pull is small, but one cannot get an exact answer nor a real guarantee that the planet would definitely stay in that approximate orbit forever. In 1685, J. Flamesteed made the alarming discovery that Saturn was accelerating and moving closer to the Sun and Jupiter was slowing down, recessing away from the Sun. (As a man who believed in God unconditionally, Newton, who was aware of the problem, was less than scientific in invoking God's power and believed that God would set right any dangerous wanderings by planets.) So, the issue remained unresolved for a century. In 1785, using high order corrections, Pierre Laplace discovered the 'Law of Great Inequality' and showed why the Saturn- Jupiter change occurred and that such changes are also there in other planets. But very importantly, he showed that the Saturn -Jupiter deviations were stable oscillations with a period of 929 years. This rather brilliant achievement by Laplace gave him such excitement that he remarked to Napoleon that he had convinced himself that "he had no need" for God. Perhaps he was referring to Newton's faith. Though largely this result is correct, one can demonstrate that stable solar systems are special cases of this periodicity and in fact, the only objects (planets) that form the solar system have stable periodic orbits and oscillations. (Correctly stated, these are the planets that have stable orbits and therefore are seen to be stable. Other planets with unstable orbits have been lost). Even in cases with reasonable stability, which is the case for the planets in our solar system, the very long term stability may be in question.

Another century passed and mathematicians were somewhat frustrated that there were no general analytical solutions to demonstrate the stability of the planets in the solar system. So, on the occasion of the 60$^{th}$ birthday of King Oscar II of Sweden in 1887, a large cash prize was announced for the winning entry that provided this proof. Henri Poincaré developed a new technique of describing motion of objects in the form of trajectories in the so called "phase space" of positions and velocities of objects. In this insightful approach, now a standard and widely used method, the trajectory of an object, could be plotted starting from a position and a velocity. The stability with periodicity is assured if the trajectory returns to the starting point, since then it is clear that the rest of the trajectory would be repeated. For an orbit of a planet in a "three body problem", Poincaré demonstrated that the orbits are stable. Poincare won the Prize. But later, other mathematicians found that, like Laplace's assertion, the proof by Poincare was incomplete. Realizing this, Poincaré repeated his calculations. He found that periodic orbits were not the norm and were a special case. In examining a circular orbit of planet affected by the transit of a comet, he found a rather surprising result. While in one case he obtained a perturbation in the orbit that returned the planet to the original orbit once the comet went far away, repeating this calculation with a different starting position of the comet made a permanent change in the orbit of the planet. In essence, what this means is that such a three body system is actually very sensitive to the comet's path, in addition to the point of the comet's path at which one starts the calculation. This was the first discovery of one of the features of Chaos, namely, high sensitivity to starting condition. After this discovery, Poincare stated that one would say a situation is predictable, if an approximately correct solution is obtained for a similarly approximately correct initial condition. He added that *"this is not always so; it may happen that small differences in initial conditions produce great changes in the phenomenon"*. (Poincaré 1903).

One can see for oneself the complexity of this problem by setting up a perturbative 3 body calculation on a spreadsheet application such as Microsoft Excel or Openoffice Spreadsheet. (See figure 4.1).

(i) Consider a planet P (figure rotating in a circle of radius 1 unit, around a star A under the gravitational force of one unit. So, first, the $x$ and $y$ coordinates of a circular motion of a planet would be written down for a unit radius as $x = cos\ (angle)$ and $y = sin\ (angle)$. We also assume that the planet is rotating at 1 degree/sec. With angles as numbers in the first column, $x$ and $y$ are the corresponding numbers in the second and third columns. 3 orbits around the star would be set up with 3x360 rows with one degree increment per row, starting with say 0 deg angle. With a speed of one degree/per sec, each row is also one second. (Remember to convert degrees into radians in calculating $x$ and $y$).

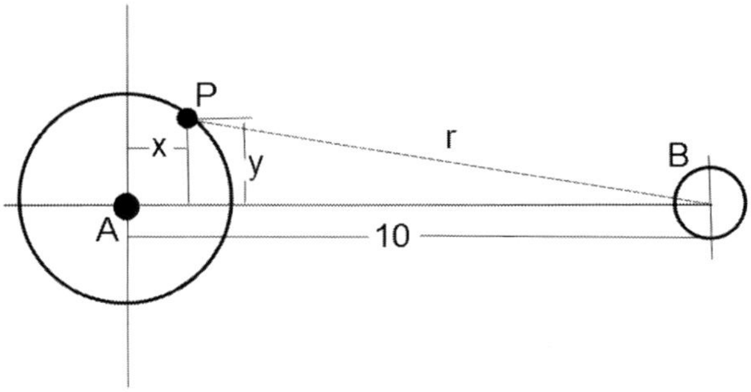


Figure 4.1. Orbit perturbation by a third body.

(ii)A small influence of a distant third body B can now be calculated as a perturbation, by calculating the acceleration towards this body along the vector connecting the instantaneous location around the circle and the location of the third body. For example, the acceleration of a second stationary star with a strength, say 30% of the first star and at a distance of 10 units along the x axis, can be written as,

$$A_x = 0.3\frac{(10 - x)}{r^3} \,; A_y = 0.3\frac{y}{r^3}$$

Where $A_x$ and $A_y$ are the acceleration of P due to the third object B, in the $x$ and $y$ direction respectively and $r = \sqrt{((10 - x)^2 + y^2)}$

(iii) The displacement in the $x$ direction due to this additional force in one second interval (using equation of motion given in 4.1 with $T\text{-}T_0\text{=}1$), would be given by,

$$\Delta x = 0.5 * A_x;\ \Delta y = 0.5 * A_y$$

Therefore each row will get a new value in the column for $x$ with $x' = x + \Delta x$ and similarly a new value with $y' = y + \Delta y$.

(iv) The new distance r' from the star, of the circling planet, is then calculated from the new values of $x$ and $y$, $r = \sqrt{((10 - x')^2 + y'^2)}$ which goes into a new column. One can repeat this calculation for further steps, advancing $x$ and $y$ with new values for $A_x$ and $A_y$ and $r$.

(v) One can plot $y$ vs. $x$ to obtain the original path of circle and $y'$ vs. $x'$ to get the new path and also plot $(x'-x)$ vs. $(y'-y)$.

The exercise usually gives stable orbits as was first found by *Poincaré*. But the next step is to include the fact that the gravitational force of the star has changed around the orbit as the path of the circling planet is not a true circle anymore. Its attraction at the local distance r', has changed from being 1 unit to $(1/r')^2$. So this must be corrected with additional columns incorporating an equation that balances the local centripetal force with the gravitational force of the star. This would change the speed locally and now with each row not corresponding to 1 degree, the problem has to be set up slightly differently. But even doing that will require another round of correction and the next and the next to potentially infinite cycles till the numbers do not change. To get a true convergence, one must also make more fine segmentation in time and repeat the calculations for say each row corresponding to 0.1 sec.

## Basic Concepts in Sensitivity to Initial Conditions

Sensitivity to initial conditions is a frequent component of chaotic systems. Under such chaotic conditions, even trajectories following the same laws and starting from close but not identical starting conditions, do not converge to a point or to periodic (events oscillating or traversing through points periodically) or quasi-periodic orbits after any length of time. The system is still "deterministic" in the sense that this result is not due to external noise or interference and can be replicated for a given deviation in starting conditions. In this sense, this is not due to quantum mechanical uncertainty. This is because of the non-linearities (see below) built into the deterministic equation. (However, it must be noted that precisely because of this sensitivity, noise or small random external influences in a chaotic system may lead to different trajectories, again, though not predictably.)

Mathematically stated, two trajectories with initial separation $\delta Z_0$ in one dimension, diverge (exponentially) to a new separation,

$$\delta Z \sim e^{\lambda t} \delta Z_0$$

where $\lambda$ is called the Lyapunov exponent and *e* is the constant equal to 2.71828. There are as many values of exponents (Lyapunov Spectrum) as there are dimensions in the problem. The largest of these exponents defines how unstable (sensitive) the trajectories are with respect to dependence on the initial condition. It is safe to say that there are few physical systems that do not benefit from this analysis. While applications to physics are numerous such as forced pendulums, particle orbits in electric and magnetic fields, fluid flow, mixing of fluids etc., examples abound even in biology and medicine, such as in molecular and cellular dynamics, movement of ions in a microplasm, blood flow, muscle contractions, human handwriting and many more. But the story of weather prediction is what really started the modern field of chaos.

## Sensitivity and Weather Prediction

Philosophically Poincaré distinguished between this sensitivity, due to inaccuracy in or lack of knowledge of initial condition, and chance happening, which he attributed to general ignorance of the conditions. (He therefore did not believe in chance). Prophetically, he used the example of weather prediction and pointed out that it is not possible to measure accurately all parameters (such as the location of formation of the cyclone to a tenth of a degree) and declared *"...but the observations were neither sufficiently comprehensive nor sufficiently precise, and that is why it all seems due to the intervention of chance."* (Poincaré 1903). One would believe that the tenth of a degree was a metaphorical number, since Poincaré himself did not know the required accuracy. For almost a century, this warning on the sensitivity of dynamical systems to initial conditions would remain a historical note. Poincaré's engagement with weather prediction is particularly poignant, because it frames and embodies a basic desire in science and technology to predict something that needs to be known to communities for daily routine as well as to prepare for major weather conditions.

This quest on weather forecasting has been there since 650 BC, when Babylonians forecasted weather from cloud formation. In fact, the vagaries of weather persuaded one meteorologist Ezekiel Wiggins, even in as late as the 1880s, that the weather was related to planetary transits. After all, tides are influenced by the moon. Mark Twain ridiculed the forecasts of this "Ottawa Prophet" in his humorous essay in Columbus Enquirer forecasting a definite meteor strike, *"As it approaches Canada it will make a majestic downward swoop in the direction of Ottawa, affording a spectacle resembling a million inverted rainbows woven together, and will take the Prophet Wiggins right in the seat of his inspiration and lift him straight up into the back yard of the planet Mars, and leave him permanently there in an inconceivably mashed and unpleasant condition. This can be depended on."* (Twain 1886). In the early 1900s, inspired by the suggestion of Norwegian meteorologist Vilhelm Bjerknes, but clearly unaware of Poincaré's insights, a Scottish meteorologist, Lewis Fry Richardson put together a computing algorithm for calculating the weather conditions over a grid of conditions, advancing over small time steps. His attempt was not necessarily to forecast the weather, but to establish a computing technique for the future, using laws of fluid mechanics coded into the steps of calculation. The scientific/computational logic would normally imply that as the computing machines became faster and bigger, the grid could be finer and the time steps could be made smaller. But, in the early 1900s such calculations would take enormous manpower and great lengths of time. In 1922, Richardson published his technique for forecasting the weather and wrote hopefully, "*Perhaps someday in the dim future, it will be possible to advance the computations faster than the weather advances at a cost less than the saving to mankind due to the information gained.*" (Richardson 1922).

The understanding of the actual import of Poincaré's conclusion had to wait for an era when a decent computer would become available, namely 1959 and for the efforts of a mild mannered but steadfast meteorologist Edward Lorenz at the Massachusetts Institute of Technology (MIT). Lorenz had come to know about the work of Jerry Namias on mid-altitude westerly winds, expressed by the so called zonal index. Namias had found a striking year-to- year repetitive periodic cycle which was later termed as Index cycle. Lorenz had discussed the observed correlations between northward and southward winds, their products and derivatives, with Victor Starr. As numerical weather forecasting became more and more popular, Lorenz was emboldened to go further in weather prediction. He realized that the correlations between wind speeds and the cyclicity of the zonal index could lead to a method to calculate and forecast the weather. Taking off from Barry Saltzmann's model involving 7 ordinary differential equations of atmospheric fluid flow, he derived the following equations:

$$\frac{dx}{dt} = \sigma(y - x) \tag{4.5}$$

$$\frac{dy}{dt} = x(r - z) - y \tag{4.6}$$

$$\frac{dz}{dt} = xy - bz \tag{4.7}$$

where $\sigma$ is a parameter called Prandtle number and $r$ is a Reynold's number ratio, all related to fluid velocities, densities and temperatures and viscosity. $x$ is proportional to the intensity of convection, $y$, the difference in temperature between the rising and descending air currents and $z$ is a parameter describing how different from linearity, the vertical temperature profile is. b is a geometric parameter. Lorenz used a desk size computer, Royal McBee LGP-30, made by the Royal typewriter company, to calculate what one might call a hypothetical or "toy" weather. Although it took a long time to make a prediction from known conditions, it was still worth it to see if weather can be predicted at all. He set up initial conditions of 12 variables such as temperature, barometer pressure, humidity and wind speed related to each other by a set of 12 non-linear equations (equations which do not only contain linear combinations of variables and have higher powers such as squares, cubes etc.). He ran certain simulations for a weather prediction and printed the numbers out in a graph as a function of time. On the fated day, he decided to extend the prediction into later future, for a certain starting condition. Instead of starting from where he left off in the previous simulation, he decided to start from some middle point and redo the calculation, this time extending it to a longer period. Since the numbers printed out by the computers were 6 digits, he decided to round the new starting condition off to the nearest 3 digits and ran the simulation. What he found was that for the overlapping duration of the two calculations, initially, the calculations started out near each other, but at a certain point the two calculations departed and then followed different curves and therefore gave a very different weather predictions. (See figure 4.2). This, he soon realized, was only because he had rounded off the starting input (the starting point was different by less than 1%) and what he was seeing was a high sensitivity to dependence on initial conditions. The implication, of course is that, weather being a non linear phenomenon, cannot be predicted far into the future. How near or how far depends on the specifics of the initial condition and how large is the system of weather one is considering. Knowing this, we can be a little kinder to our weather forecasters.

Once again, people initiated into the basic use of spreadsheet applications, can easily set up Lorenz Equations and can check the above sensitivity to initial conditions.

The previous example of a man going to work, though involved high sensitivity to initial conditions, there are other factors in a larger system and many sudden forks in the roads. This involves the broader concept of chaos. It is difficult to isolate a pure example of sensitivity to initial conditions in life, because there are so many feedbacks into the systems which tend to correct or increase the divergence. However one can see this sensitivity in certain areas. Examples are: (a) Identical twins in a family develop differently, because of small natural differences between them, which get amplified even with similar nurturing, (b) Success breeds success. So among two equally talented people, a small success of one of them can lead to a large lead over the other, through name recognition and access. This happens particularly in areas such as art, cinema, music, modeling, politics and journalism, where competition is great and talents are hard to differentiate. All the same, we know in our minds that, just like a good breakfast- the first meal of the day, a good start is very helpful in life.

We can only wonder at the sensitivity of really initial conditions, such as the conditions at the birth of the Universe and origin of life. Life, indeed, is sensitive to initial conditions.

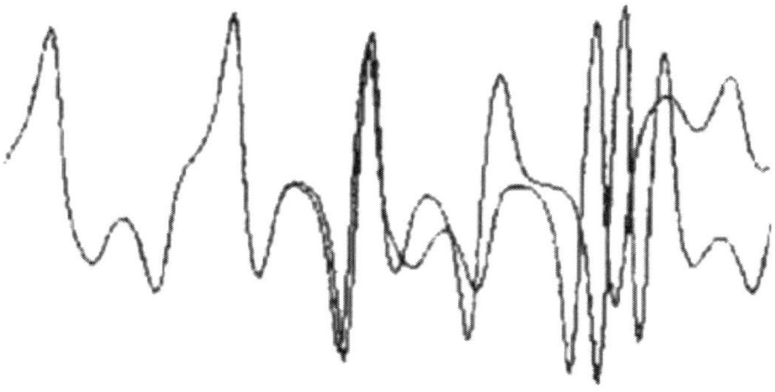

Figure 4.2. Edward Lorenz's weather forecasting calculations. Small (not discernible above) difference in initial condition (at the left) resulted in a very different forecast (two separate plots). (Source: Stewart, I.; *Does God Play Dice? The Mathematics of Chaos*, Blackwell, Oxford, U.K. (1989) p.141).

## Attractors and Strange Attractors

The term attractor is borrowed from its colloquial usage to mean that motions, orbits, events etc. are attracted to a particular point, a specific periodic motion or trajectory. A simple example of an attractor is the vertical position of a pendulum with a weightless string and a bob. Any other position would be unstable and eventually the pendulum would come to this attractor. If an object starts spirally in and then falls into a periodic circular or elliptical orbit stably, then the final orbit is an attractor. This is a case in a 2 dimensional plane. A ball rolling in 3 dimensional space from hill to hill, will come to rest in a valley and the valley is an attractor. If it were water flowing down hill, the valley would fill with water forming whirlpool patterns and therefore have additional features. The actual arrival of the object to the attractor may follow any pattern or path, but the attractor remains the same.

In a torus (for example, in a magnetic nuclear fusion device called a Tokamak, the field lines go around the torus the long way (the major circle along the toroidal direction), while winding around the short way (the minor circle, in the poloidal direction). In specific cases where the numbers of turns of winding around the minor direction is an integer number of times it winds around the major direction, the field lines would close. This is an attractor. Particles traveling along such lines would have a periodic and fixed trajectory. Such closed field lines are also obtained when the ratio of the number of transits in the two directions is a rational number. (See figure 4.3). However, when the ratio is not a rational number, the field lines wind around the torus infinite number of times and would never meet and repeat the pattern. (In these nuclear magnetic fusion machines, the "plasma" electrons and ions follow the magnetic field lines and therefore their stable or unstable behavior and how redistribute their energies is extremely dependent on the nature of these field lines). This type of configuration can lead to a chaotic regime, with very orderly behavior in some regions and unstable behavior in other regions. In particle accelerators, regions with chaotic orbits may be well contained in some regions and others may be lost.

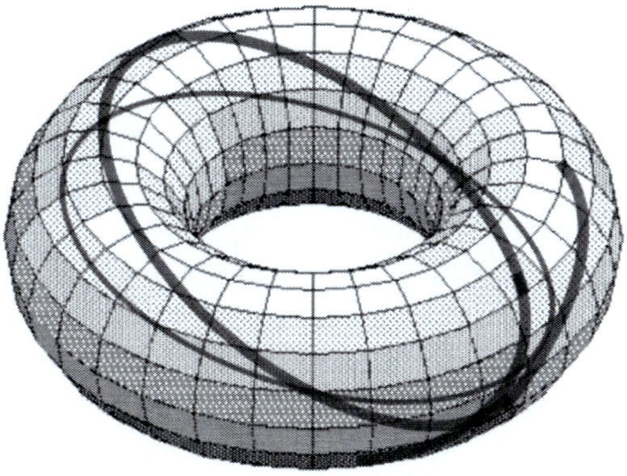

Figure 4.3. Torus, with rational and non rational orbits around minor and major radiii. The thinner line goes around once around major radius and in that transit it also goes around once in minor radius. For the thicker line, the once around the minor radius is completed before the once around the major radius is completed.

But the idea of the "Strange" Attractor owes its origin to turbulence, specifically, the description of turbulent flows. After many years of studies, turbulence is a field, which remains only partially described as overlapping flows of different kinds, cells of fluids, different periodicities etc. But David Ruelle and Floris Takens came up with another description which involved only 3 different flows (equations). One could plot the results of the dynamical development of flows using these equations, in a phase space diagram. As we saw before in the Poincaré analysis, a point (coordinates) on the curve in the multi dimensional phase space corresponds to specific values of positions and velocities in different directions at an instant. For example, in 3 dimensional space of fluid flow, there are 3 positions $(x,y,z)$ and 3 velocities $(V_x, V_y, V_z)$, a total of 6 dimensions. One can plot different sections of the trajectories from that trajectory, for example, $x$ vs. $V_x$, or $y$ vs. $V_x$ or $x$ vs. $V_y$ and so on. A Poincaré section is one in which one may choose a plane, transverse or at an angle to the plane of trajectories and mark the points at which the trajectories cross this imagined plane. (See Figure 4.4). The collection of such points would be plot in the Poincare section. A figure could develop out of these points and one could find an attractor which the trajectories finally settle down to or periodically trace.

Ruelle and Takens believed that the description of turbulent flow using the three modes proposed by them, would conform to such attractors – a flow dynamically developing into a turbulent structure in phase space diagram. This attractor, they called- the Strange Attractor, a complex attractor. The feature of such attractors is that these have no periodicity and yet they remain bounded in a certain range. Though their ideas didn't hold in the light of rigorous study of turbulence, the terminology is recognized for all trajectories arising out of complex chaotic behaviors.

Returning to the Lorenz equation, Lorenz calculated the quantities stepwise in time. The first equation for *dx/dt, dy/dt and dz/dt* would be calculated using the current values of $x$, $y$ and $z$ and fixed values of $\rho$, $\sigma$, and $\beta$ and then the new value of x would be given by,

$$x_{new}=x_{old}+(dx/dt)*\Delta t$$

where Δt is the increment in time and similarly for $y$ and $z$. When Lorenz plotted the long term history of the values of $x$, $y$ and $z$ in 3 dimensions, he found interesting features in them. First, the results tended to be stable (unchanging values for $x$, $y$ and $z$ with time increment) for low values of $r$, but as $r$ crossed the value of 24.74, the values never settled down to a constant value or even a periodically changing value. Instead, it changed within a bounded set of values. When he plotted the values in a 3 dimensional geometry, he got the plot shown in figure 4.5, which is now famously known as the Lorenz Attractor.

Figure 4.4. A crossing of a plane by a particle and Poincare plot (little circles on the plane).

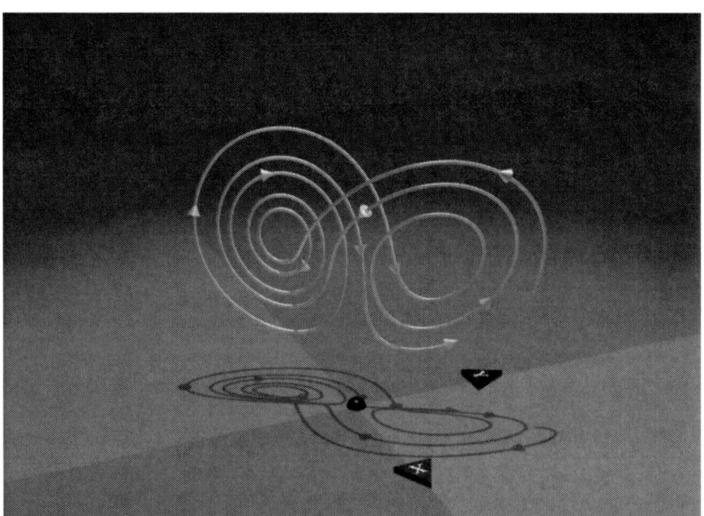

Figure 4.5. Lorenz attractor resembles a butterfly, rendered with wire to show the 3 dimension character. (Source: Wikipedia, Author: Mrubel).

Lorenz published his results in a meteorological journal and at the time, it was rarely noticed as something significant. But a mathematician, James Yorke found the paper and understood its implication and then the word spread through a meteorologist Allen Faller and the famous topologist Stephen Smale. Lorenz became a sort of a celebrity in scientific circle a few years after this discovery of the sensitivity to dependence on initial conditions. But his strange attractor was indeed a strange attractor to the broad public when a meteorologist referred to this phenomenon (quoting Lorenz), "One meteorologist remarked that if the theory were correct, one flap of a seagull's wings could change the course of weather forever." Later, he changed this reference to a butterfly. It then became a game of words and so at the 139th meeting of the American Association for the Advancement of Science in 1972, Philip Merilees gave the title for Lorenz's talk as "*Does the flap of a butterfly's wings in Brazil set off a tornado in Texas?*". It is not clear, if this reference was initiated by the resemblance of the shape of the Lorenz attractor to a butterfly, but now the terminology of the "Butterfly Effect" is irrevocably etched in the public mind and is a frequent phrase in the description of common events, similar to the terminology of "Tipping Point". (See below). Now the strange attractors are a staple of mathematics and analysis of dynamical systems. The strange attractors are also fun and artistic creations. An exploration of its potential in art can be seen in the book, "*Strange Attractors: Creating Patterns in Chaos*" by Julien Clinton Sprott. Most strange attractors are to be found in phenomena that follow non linear equations, algebraic or differential. The values of solutions trace out the strange attractors when coefficients of the equation are changed or the independent variable (such as the time in Lorenz attractor) is varied. Yet they remain bounded within a region and may also stay out of a region within that region. The Henon attractor, shown in figure 4.6, is the Poincaré section of the orbit of stars in galaxy.

Figure 4.6. Henon attractor. This is made of points on a plane (containing position and velocity) at which a star crosses in its orbit around a galaxy. (Source: Wikipedia).

The attractors are well known to us in our lives and are generally expected in straightforward tasks. We arrange things and work towards a simple goal and events would take us there like an attractor would. We study and pass an exam with a certain grade which we can more or less anticipate. But in more complex situations, the path can be only approximately expected. Often, it is never quite what we expect and can be quite different, disappointingly or exhilaratingly. Yet, except in extremely rare conditions, the outcomes remain bounded within reason. This too is similar to the strange attractors. Evolution on earth has produced many species with stark differences in appearances and yet only on very rare occasions do we see creatures with features and abilities far outside the average. So, natural phenomena and human lives themselves follow attractors or strange attractors. Jungian analysts point out (see for example, Archetypal and Strange Attractors by John R Van Eenwyki) that the psychological state oscillates between psychic balance and chaos and yet the dynamics is mostly contained within a domain of psychological behavior, indeed within the bounds dictated by life, the larger dynamical system.

## BIFURCATIONS

Referring back to the case of the man going to work, the sensitivity to the initial condition brought the man to a point in his day where he took the elevator together with a woman. At that point, his point took off irrevocably into a different direction from his daily routine. A simple example is the fork in the road of a traveler. In general, the different directions would take the traveler to different places and different experiences and different end points. We wonder at these amazing and seemingly uncommon coincidences in our lives and imagine that these might have been designed by providence of one form or another in order to direct our lives. In reality, this type of "bifurcation" is found commonly in nature. The traveler analogy is apt because in these systems too: the occurrences happen in specific schemes (routes) and when the system arrives at the critical points (forks in the road). A momentous bifurcation in an individual's life is at conception. A spontaneous selection to form a male or female zygote occurs at this point. This selection is a true coin toss (random) and yet is made only at this one critical point in the process. A baby grows into a toddler and then a young boy, gradually he grows in mind and body, becoming heavier, taller, being able to talk, then read and then write and do math, sit up, walk, run and then become athletic. But at puberty, suddenly, his voice breaks and changes into a male voice, he develops manly bodily changes, gets facial hair etc., all within a short period compared to his past life time. This bifurcation is built in and to an extent is not a choice or a random occurrence.

In physics, this process is also called "symmetry breaking" and here too, our amazing experiences are reflective of what occurs in nature- symmetries broken in general and those broken with a hidden order. In "Nature of Economies" by Jane Jacobs (Jacobs 2002), narrator Hiram describes bifurcations as *".. a system's instabilities of some sort can be so serious that for it to continue as before is not a practical option. It must make a radical change...take a fork in the road, travel into a new territory..."* and, *"Bifurcations have complex consequences. They not only embody new practices, they change the very systems that give them birth or adopt them by imitation. Accumulation of bifurcations alters the character of the civilization."* A person may start exercising regularly and eat less fat to lose weight in order to be socially attractive. But an internal change can occur from this and end up changing the person's life style, health and even the outlook on life. Many steps in the evolution of life were bifurcations directed by changes in environment.

In nature, there are several types of bifurcations. Most mathematically described bifurcations are seen in systems similar to those that have sensitivity to initial conditions, systems that follow a given equation or a given schema. While one such system is stable, another with a slightly different controlling parameter may bifurcate. A well known example is the Logistic equation, discovered by Pierre Francois Verhulst in 1845 for a model on population growth. We will come to it in steps as below:

It is obvious that the more the number of members in a population, the greater is the population, assuming that the number of members capable of mating and reproducing increases with increasing overall population. It is also obvious that for a population that has been around for periods much longer than the life expectancy of the members, the death rate is also proportional to the population. Therefore, the increase in population $P = \Delta P = RP - MP$, where $R$ is the reproduction factor and $M$ is the mortality factor. If the population change is measured in small intervals of $\Delta t$, then the rate of change of population is given by

$$\Delta P/\Delta t = (R-M)P$$

If the measuring time interval is small enough, one can write this as a differential equation for the rate of population growth,

$$dP/dt = (R-M)P \tag{4.8}$$

Integration of this equation yields the solution, $P = Constant \ x \ e^{(R-M)t}$, where $e$ is the exponential function.(Here x is the multiplication sign). If at the initial time ($t=0$) the population is $P_0$, the population grows exponentially from $P_0$ to $P$ at time $t$, given by, $(s=R-M)$,

$$P = P_0 \, e^{st}$$

But the population cannot grow indefinitely (for $s>0$), because the capacity of the environment to provide land, food, drinking water and breathable air is always limited and also because diseases increase with population and predators also increase in population. Verhulst proposed the model that uses a carrying capacity $K$ and used the expression,

$$s = r(1-P/K)$$

so that the rate of population growth is given by,

$$dP/dt = rP(1-P/K) \tag{4.9}$$

The equation has an analytical solution given by,

$$P = \frac{KP_0}{P_0 + (K-P_0)e^{-rt}} \tag{4.10}$$

This is the logistic equation for population growth. Though the equation is simplistic and rarely is population growth dependence this simple, this equation provides the first estimates. There are many applications of this equation. For example, a similar equation is obtained for rate of consumption of resources.

With $X=P/K$, where $K$ is constant, equation (4.9) can be written as

$$dX/dt = rX(1-X) \tag{4.11}$$

where $dX$ is the difference between the populations at two successive times $dt$ apart. We can rewrite this equation with $X_n$ being the population at the nth time step and $X_{n+1}$ being the population at the $n+1$th time step, the difference equation based on equation (4.11) is,

$$X_{n+1} - X_n = r X_n(1-X_n) \tag{4.12}$$

Here the time difference between adjacent units is set to one unit. This equation is slightly different and is of great interest in studying chaotic behavior. Once again, what follows can easily be tested on a spreadsheet application with $X_{n+1}$ as the value of $X$ at the $n+1$th row, calculated from the previous $n$th row according to equation 4.12, for a chosen value of $r$. For a value of $r$ well below 3.0, the values of $X$ after several iterations, settles down to a specific value, dependent on the value of $r$. (figure 4.7a). For a value of $r$ slightly less than 3.0, the iterations initially oscillate between values but gradually converge to a value. The final stable value (an attractor, which is about 0.66 as $r$ approaches 3.0), is independent of the initial value. But as soon as r crosses 3.0, the values oscillate between exactly two values, which depend upon the value of $r$. (See figures 4.7b and 4.7c for $r$=3.02 and $r$=3.2). This is a bifurcation. Another bifurcation occurs when r is increased above 3.4498. For values above this, the values oscillate between 4 exact values. (See figure 4.7d.). There is another bifurcation at a higher value of $r$=3.5441, so that the values oscillate between 8 values. The values oscillate between 16 values for $r$=3.5644 to 3.5688. The number of values keeps increasing until the value of $r$=3.56994, at which point, the values become chaotic giving a very large number of values for $X$ (Figure 4.7e), which remain bounded for $r$ less than 4 and unbounded (up to infinity) for $r$>4. (The actual values remain bounded – a stange attractor).

This behavior is displayed in the bifurcation diagram in figure 4.8 which shows the values of stable, oscillating and chaotic values of X, with increasing values of $r$ in the horizontal direction. The intersection of the plot with a vertical line corresponding to a value of $r$, gives the values of X.

Among many interesting features of this diagram, the first surprising fact is that even in the chaotic regime ($r$>3.5688), there are small islands of stable oscillating values. Figure 4.7f shows the example for $r$=3.84. This is a stark demonstration of order unexpectedly and unpredictably hiding in the middle of disorder. In the above diagram, the dark areas represent chaotic solutions and the regions that are white striped have a few stable oscillating solutions.

In population dynamics, the parameter $r$ (called the Malthusian constant) is clearly related to reproduction, mortality and other environmental effects. Therefore, the population of species can at times be extremely sensitive for critical values of $r$ and suddenly alternate between two or four or more values. The same is potentially true of trends in consumption of goods. The second important feature of the bifurcation diagram is the fact that the number of solutions (number of values) to the logistic equation doubles at every bifurcation. While this feature gives alternating values to the population, it is also observed in many situations and is called period doubling. A tap opened gradually would give one drop at a time and then two drops close to each other, then four drops etc. and then finally it would give a turbulent flow. A ball dropped along a vertically vibrating table will give similar period doubling features. All of us have experienced the doubling of the pitch of the sound when we blow hard through the "noisemaker", a common party favor.

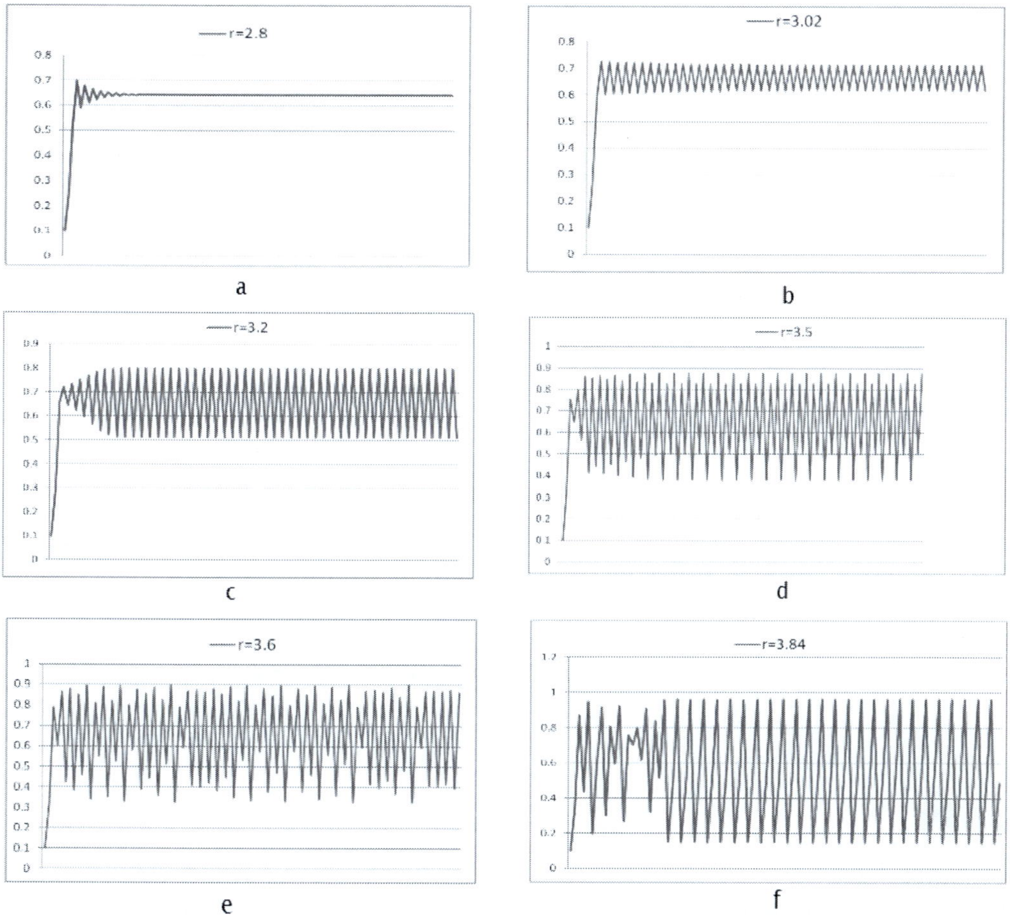


Figure 4.7. Bifurcations in values and number of values for X for different values of r as the iterations are carried out.

There is another amazing feature of this diagram, if we take a picture of the region where there are 2 solutions and 4 solutions and then compare that to one of the region where there are 4 solutions and 8 solutions and enlarge it, we would find that they look identical. The same is true of other bifurcations and so all bifurcations look identical except for the scale. In fact, even in the island stability regions in the middle of chaotic regions, such as the one we saw for $r$=3.85, the pattern is exactly repeated in an amazing detail. These are known as fractals. (See below).

Bifurcations are very common in nature. The start of every branch of a plant or a tree, budding of every flower, the meandering route of a brook or river, buckling of an overloaded column or a bridge, creation of bubbles as water is heated on a hot plate, flow of water through a tube, oscillation of a pendulum as the amplitude is increased, lightning strikes, evolution of a species, formation of life and so many others are examples to show that more often than we notice, change occurs through bifurcation and the path of further development is irrevocably changed.

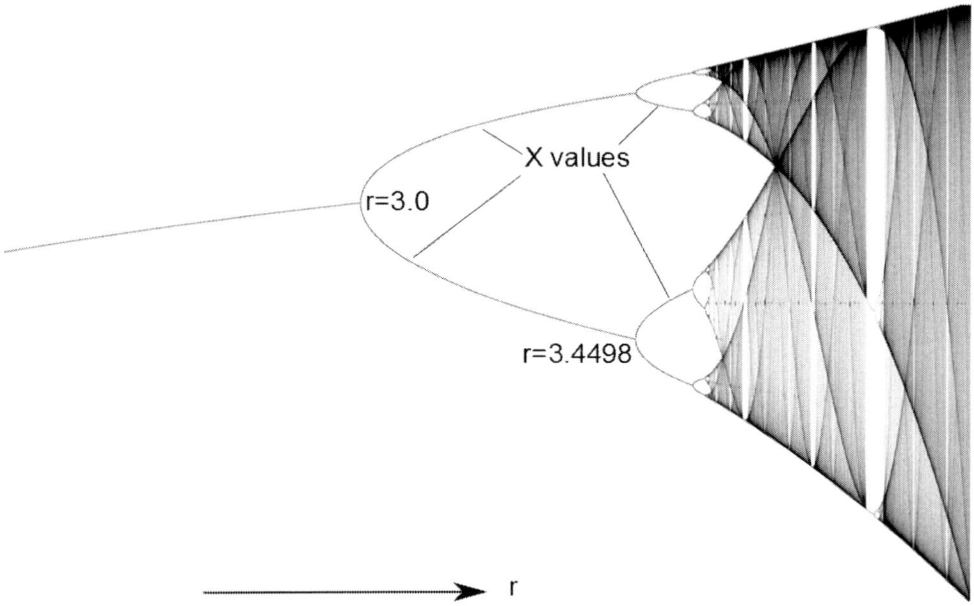


Figure 4.8. Plot of bifurcatons of values as r is changed in the Lorenz equation. (Source: Wikipedia with added number labels; Free Public Image).

## FRACTALS

The replication of the primary bifurcation in other bifurcations with a smaller scale is also one of the commonest features of nature. The basis of fractals is so powerful that most natural objects are created by this process. When we see a flower and admire its beauty and complexity, we tend to imagine that anything like that must have been designed and/or created as a whole in that exotic form. In reality it is built by repeating simple steps as in the logistics equation above. The code for that algorithm is written in the DNA of that flower and the rest of the plant of the flower provides the resources for it to build up, as needed. Many complex objects and phenomena develop through such a process. While we appreciate only those that surprise us or delight us, such sequence of repeated (iterative) steps to build something large and complicated is very common in nature and indeed, in our lives.

A well established case of fractals producing a beautiful object is the case of the "Barnsley Fern" fractal that creates life-like ferns similar to Asplenium ferns. In a two dimensional x,y plane, dots are created according to a formula for x and y with coefficients that are randomly chosen within a range.

Let us say we start with initial values $x_0=0$ and $y_0=0$. A dot is placed at $(x_0,y_0)$

1. New values of $x$ and $y$ -> $x_1=0$, $y_1=0.16y_0$ are randomly chosen 1% of the time

New values of $x$ and $y$ -> $x_1=0.85x_0+0.04y_0$; $y_1=-0.04x_0+0.85y_0+1.6$ are chosen 85% of the time

New values of $x$ and $y$ -> $x_1=0.2x_0-0.26y_0$; $y_1=0.23x_0+0.22y_0+1.6$ are chosen 7% of the time

New values of $x$ and $y$ -> $x_1=-0.15x_0+0.28y_0$; $y_1=0.26x_0+0.24y_0+0.44$ are chosen 7% of the time.

A dot is placed at $(x_1,y_1)$

2. The process is repeated now using $x_1$ and $y_1$ in place of $x_0$ and $y_0$ in the above equations for obtaining the coordinates for the next dot at $x_2$ and $y_2$. When we repeat these steps a large number of times, we get this amazing figure of a fern (figure 4.9).

Figure 4.9. Fractal of a fern using the iterative routines with equations above. (Source: Wikipedia Free Public Image).

Looking at this extraordinary rendition of a fern, one cannot deny that there is some such code in the formation of a real fern too. In fact, one might even wonder if it is the same code as above. By varying some of the parameters, one can change the figure to other forms of ferns. (One can already see the mix of order and randomness in the set up).The picture above has not been iterated sufficient number of times and also does not have the resolution to show it, but, true to a fractal each section of the figure is a scaled down version of the corresponding larger section of the figure, down to the smallest section one can draw. In fact, real ferns do not show this scaled down fractal, because of the size of the molecules and merging of cellular pieces. Figure 4.10 shows a beautiful and realistic looking fractal tree. Addition of rules for colors and transition to a different parameter or a rule gives enormous scope and artistic sensibility to the figures. The so called multifractals allow such changes in addition to stretching various features one way or another. So, artists have found a new medium, “fractal art” to create an extraordinary range of natural and original objects, sceneries with mountains, lakes and trees and abstract art. Some of these are so spectacularly realistic that it is hard to believe and others aesthetically very pleasing. Some others truly give the sense of transcending into a scientific art and are emblematic of potential development in art and design.

Fractals are also to be found in mathematical and topological iterations. The Sierpinski Triangle (figure 4.11) is a good example of this. It is made by cutting inverted equilateral triangles out of upright equilateral triangles. One can see that this figure is also self similar in detail.

Figure 4.10. A tree created by fractal application. (Source: Wikipedia, Created by Solkoll, Free Public Image).

Figure 4.11. Sierpinski triangle.

The fractal, created by the famous Polish mathematician Benoit Mandelbrot, who first demonstrated that visual complexity can be created through simple rules and coined the term fractal, is shown in figure 4.12. This is created by iterating on a complex quantity Z =x+iy, where x and y are coordinate values of a point placed on the map. (i is equal to the square root of -1). The equation for iteration is given by $Zn+1= Zn^2+C$, where C is a complex constant.

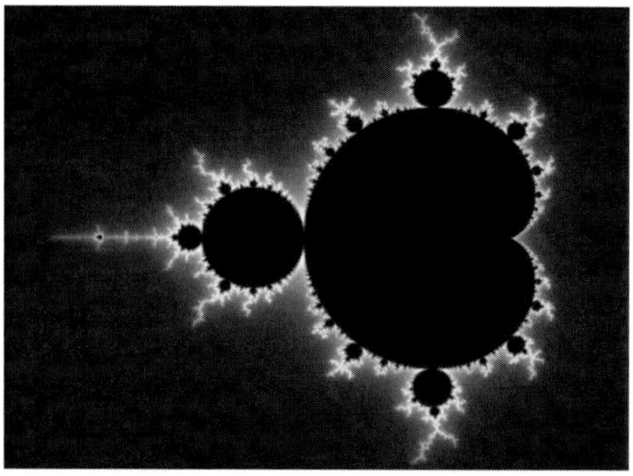

Figure 4.12. Mandelbrot Fractal (Source: Wikipedia, Free Public Image).

As we saw, the bifurcation diagram of the logistic equation is a fractal too and the Henon attractor is also a fractal (and is seen by looking at the detail of the plot). Therefore, one can see that the strange attractors, bifurcations and fractals are intimately connected and give the domain of chaotic regimes.

Fractals are the result of simple, yet powerful mathematical or process rules and are codes of nature which produce the elegant or grotesque, beautiful or ugly, magnificent or dull features that we see and experience. An avalanche on a sand hill as more sand is poured on it happens in all scales of self similar avalanches. Earthquakes happen in self similar scales and magnetic storms and many similar events have the same statistical quality to them. Within each there are these features of self similarity as well as strange attractor and bifurcation behaviors. Fractals can be generated by a more sophisticated process of interaction of sources which have different expression depending upon the environment. Variation in the distribution of these sources provides a tremendous amount of diversity, although the underlying principle would remain the same. A powerful example is the appearance of stripes or spots in animals. The pigment cells of the skins of the animals can be in two states, Differentiated (D) and Undifferentiated (U). Color is produced by the D cells, which secrete two types of molecules called morphogens, one (A) which activates the pigment and the other (I) which inhibits the pigmentation. The pigmentation then colors the U cells, turning them into D cells. The interplay, involving simple rules, between the diffusion and the presence of the two morphogens at various skin sites produces the various beautiful patterns we see in cheetahs, tigers, zebras etc. (Recent models show an additional characteristic of lateral inhibition, where selectively nearby D cells are inhibited from sending morphogens signal, is also involved). The wonderful variety and yet the universality of the rules is a powerful example of fractals in natures.

Another example of fractals is market behavior, which is supposedly human caused and so would be presumed to follow no standard rules of nature. But it is now established that a market data curve is, often, a fractal, meaning that the rarer and larger fluctuations have the same shape of ups and downs as the more frequent small scale fluctuation. As Benoit Mandelbrot wrote in Scientific American, *"The beauty of fractal geometry is that it makes possible a model general enough to reproduce the patterns that characterize portfolio theory's placid markets as well as the tumultuous trading conditions of recent months. The just described method of creating a fractal price model can be altered to show how the activity of markets speeds up and slows down—the essence of volatility."* (Mandelbrot 1999). The classical market theories do not predict or even deal with volatile markets, but fractals do explain.

To emphasize, bifurcations and fractals are clear indication that simple rules for the change of specific variable over many steps or repeated over and over in a time. can create very complex phenomena and objects, and one does not have to appeal to the divine or extremely complex design process to explain these. Simple and elegant rules repeated over many times do create complexity. As we saw in the chapter on evolution, this is the case in natural selection as well.

Fractals can be found in societies. One could see characteristics of evolution of societies as self similar to its component families. On a different aspect, the growth of a society as it matures resembles that of a child growing up. Politicians intending to convey responsible government behavior recommend adoption of the budgeting process of families. Many religions preach that God is father or mother in an adoption of self similar family structures.

One could speculate that fractals are reflected in space and Universe, objects orbiting around a central object – moons around planets, planets around stars, stars around a central black hole in a galaxy, galaxies orbiting around other galaxies. However the scale level also is accompanied by physics changes and therefore might be viewed as a form of multifractals. One also realizes that an animal body itself such as that of ourselves and our brains are structures built cell by cell, protein by protein using multifractal rules. Our bodily movements, feelings, emotions, reactions and conclusions themselves may have a fractal basis. Fractals are prevalent wherever there is complexity and wherever there is a hidden order in seeming disorder.

## Criticality, Phase Transitions and Instabilities

We see critical phenomena frequently in our daily routine, when we boil water or make ice in the refrigerator. For example, when water heats, at the specific temperature of 100 deg C, it becomes steam and not until then. It is said to reach the (critical) boiling point and have a phase transition. (The exact temperature at which this happens and even whether it happens at all would depend on other conditions, such as atmospheric pressure and solvents in it). The ubiquitous nature of such phenomena has introduced this type of interpretation in normal parlance, such as, a patient is critical or a situation has reached the critical point, implying that it is about to change to a different state. But just before that, at the critical points, the two states merge and are indistinguishable. Steam and liquid water phases are the same as a homogeneous phase. Then depending on the pressure, it continues as a liquid if the pressure is increased to above critical pressure or if the pressure remains below the critical pressure it changes into steam phase. At the boiling point when water changes to steam, additional energy does not increase the temperature, but is used up in the phase change to steam and this energy is called the latent heat.

So a normal phase is just a routine, unexciting and smooth change of associated parameter when the controlling variable is changed. Examples are, heating of water before it starts to boil or water flowing through a long pipe before the flow is increased so it becomes turbulent. This is similar to the line we see in the logistics diagram before the bifurcation, a smoothly curving single line. In case of phase transition, there are two or more forms. In the so called "first order" transition, there is a change from one state to another at the critical point as we see in boiling water at atmospheric pressure. In such cases, there is an "order parameter" which is the key characteristic that describes this transition and the reordering of the microscopic elements of the macroscopic system. For liquid-vapor transitions, it is the density. Figure 4.13 shows phase transitions including when a transition is prevented by supercritical conditions. Another example is that of iron magnetization, which is caused by all the iron atoms aligning in a particular direction. Above a critical temperature, iron is not magnetized and at the critical temperature it becomes magnetized. The order parameter is the magnetization parameter.

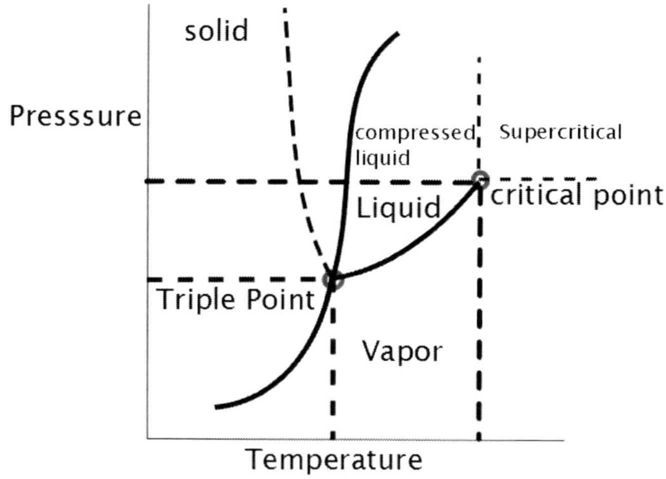


Figure 4.13. Example of critical transitions in solid, liquid and vapor.

In several transitions, there is not only a change of phase. The transition may involve precipitation of a new component which had remained hidden. Such phase transitions occurred in early development of the Universe, when different forces (governing gravitation, nuclear structure, electromagnetic phenomena and radioactivity) separated from a unified state. We can see this in forming of a cream layer when boiling milk is cooled.

A phase transition is a tussle between order and disorder. A system always tries to minimize its energy, while the entropy, which is a measure of disorder always increases. The outcome in this tussle is determined by a specific parameter, such as the temperature in the case of liquid-vapor transition. With such a pull and push regime, there is the inherent potential for chaotic behavior. Therefore, phase transitions at critical points may be of the "second order" type where there is a mixed state (order and disorder), in which some regions are in one phase and some are in the other phase. This can happen in phase transitions in magnetic materials and in superfluids.

Often, systems at the critical point defy proper description. At this point, some macroscopic quantities would become singular, meaning that they become discontinuous and cannot be handled by usual methods of mathematics. At the critical point, the correlation length, the length over which parameters are correlated (connected), tends to infinity. For example, if a sand hill has critical amount of height and width and if one adds even a small amount of sand (the proverbial straw on the camel's back), it would have a sand slide. The interesting part is not that it reacquires the critical state by the slide; it is that the whole sand pile is involved in this transition. At the onset of turbulence, every cell of a fluid "knows" and is connected to every other cell and it is as if the whole fluid becomes turbulent at the same time. Correspondingly at the transition, the transmission of information is extremely fast. Handling of critical phenomena in order to provide a description has been a great physics challenge, only solved in recent decades.

## Instabilities

Analogous to critical phenomena are development of instabilities. This is common in most systems that are in equilibrium. This equilibrium is, except in rare circumstances, not fully stable to all the modes of instability that it can incur. A ball bearing sitting at the bottom of a bowl is stable for nearly all perturbations and will return to its stable position, but a large enough displacement pulls it out of the bowl. A swinging pendulum is the same. In addition, we all know of resonances. Certain instabilities would grow because the frequency of these instabilities is equal to some natural frequency of the system. (A high quality wine glass shatters at the right sound frequency). Even in such cases, the amplitude of perturbation has to exceed a certain value, to overcome the mechanisms to damp out the instabilities, as in the case of a ball in a bowl. (See figure 4.14). The threshold then acts like a critical point, above which the instability would grow on its own. Typically, in systems like fluids, there is a threshold value of perturbation to trigger an instability. After that, the amplitude (strength) of the instability would increase exponentially. Once it reaches certain amplitude, other mechanisms may kick in and saturate the instability or decrease the amplitude. Additional processes in the non linear phase may cause the instability to change into a different mode.

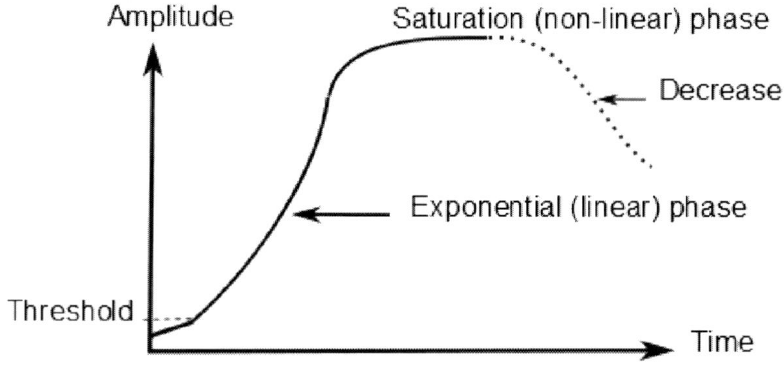


Figure 4.14. Evolution of non linear process.

A simple example of an instability is the Rayleigh-Taylor instability (a kink type instability) in fluids. This happens when a heavy fluid (say water) sits on top of a light liquid (say oil) in a fluid column. This is clearly unstable, since light fluid must float above the heavy liquid. A small perturbation causes a perturbation to grow exponentially. Once the amplitude of the instability at various wavelengths grows considerably, then these start mixing the fluids, changing the state that was. (See figure 4.15). The amplitude saturates and the heavy-light liquid situation turns over with two fluids mixed or with the lighter liquid at the top thereby relieving the driving condition. Another feature of many instabilities is the fractal nature of the perturbed state where the patterns of instabilities are self similar at different scales. A very good description of this is given in the paper "Self-similarity and internal structure of turbulence induced by Rayleigh Taylor instability by (Dalziel 1999).

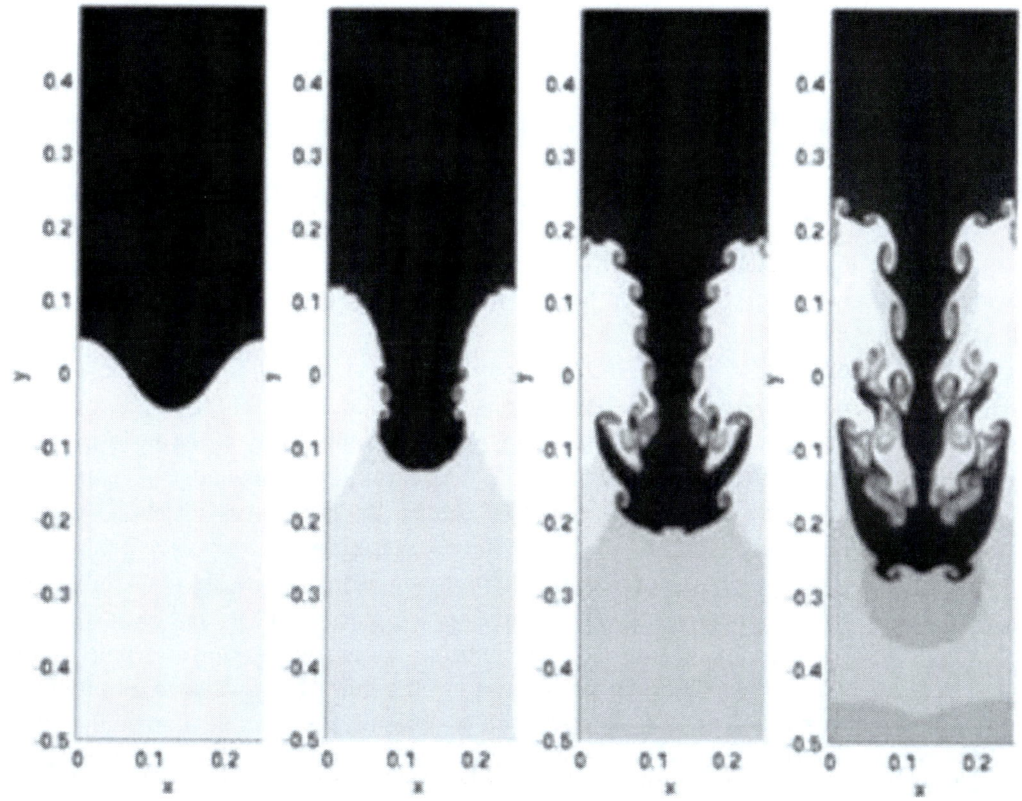

Figure 4.15. Rayleigh Taylor instability in immiscible liquids with a heavy liquid at the top. (Source: Wikimedia Commons, Author: Shengtai Li, Hui Li, Los Alamos National Laboratory, U.S.A. This image is a work of a United States Department of Energy (or predecessor organization) employee, taken or made as part of that person's official duties. As a work of the U.S. federal government, the image is in the public domain.).

## Critical Phenomena and Instabilities in Personal and Societal Experience

If critical phenomena ring a bell or are familiar, it is because one sees many similarities in other natural phenomena and in affairs of our life. Precipitation of rain, the popping of a balloon, skidding of a car on a slippery road when you reach a certain speed, the breaking of symmetry and falling of a pencil balanced on a tip, cake rising and collapsing in an oven etc. In fact, many useful objects use the principle inherently without our noticing it. A fluorescent bulb strikes lit at a specific voltage which is designed to be created in the bulb, though the operating voltage is much lower. In our personal lives too, there are many activities that embody such criticalities and transitions. When the water bill becomes overdue on a specific date and the water is cut off, a person experiences this criticality. When a person accepts employment or commits to a spouse or a partner, such a phase transition occurs. A phase transition occurs at a critical point, brought on by the inherent mechanisms in the dynamic process. These are, more often than not, internally driven and can only be avoided by an above-critical effort and feedback mechanisms. If a person has not prepared for a standardized examination well, there is certainty that the person cannot provide the correct answers and therefore would fail. The transition to passing the examination and going on to further progress is a bifurcation and a phase transition that is brought on by the internal dynamics set in motion by the prevalent rules of both personal behavior and the institutions involved. The phase transition from a homogeneous phase to separate phases is also seen, when a group is dispersed, close high school friends separate or passengers leave a train at the terminus. A phase transition in a family, when parents experience an "empty nest" as the children grow up and leave home, is also an analogy of a phase transition when a critical point in life and family is reached. Learning how to roll over as a baby and retiring from employment are sharp changes in life, are analogous, and may even have the same underpinnings as scientific phenomena. There are more stark examples like life and death, health and sickness, giving birth to a child etc.

We are both afraid of critical points and hopeful of them, depending on the context. Therefore, these are punctuation marks on different phases of our life. These also have one or more critical and order parameters and something changes dramatically in a transition. In the case of the examination, the critical parameter is the learning or knowledge and the order parameter is the grade level or in the case of staff exams, the position or wages at work. In fact, many phases in our life do have sharp transitions caused by a progressive change of a parameter, such as time or age or experience which provide us these critical points in life. The experience at the critical juncture not only changes our environment, it changes us in deep and meaningful ways. This is the same as in scientific phenomena where the whole system changes. We can even see "second order" transitions, where the change is seen in only some aspects or days of our life and these are also prevalent. When we learn and compare the above described science of the phenomenology of critical phenomena we can see that we experience what is a common and natural occurrence and we have nothing to fear and much to understand through science. However, while liquid water changing to steam at the boiling point can be reversed, many of life's critical point changes cannot be reversed or can only be reversed at a personal cost. This is because of the chaotic nature, not in the literary sense implying confusion and randomness, but in the scientific sense with hidden order and with unpredictability of changes in our multi-parametric and non-linear parameters of life, which are also observed in nature. This may be the source of the trepidation we have when critical situations arise. In a way, a good analogy is that of baking a cake with a new recipe and our trepidation about how it will look and taste. The inability to accurately predict our lives even when there is apparent prevalent order, particularly in the midst of confident predictions by pundits, causes us to be driven by impulses like hope or despair and to seek divine intervention that, we hope, brings predictable order to life.

Even more amazing is the applicability of the physics and mathematics of instabilities in personal life and society. Our experiences have thresholds for noticing or for instigating or for taking effect. There is a period of continuance of that observation, experience or incorporation, in a linear fashion. The phenomenon then takes hold and grows fast and very likely exponentially. The experience peaks and then saturates because of the effects that the phenomena brings about. A simple example is that of a person getting emotional. An event has to exceed a threshold to cause a threshold level of emotion. The emotion grows and then leads to a burst of venting or crying, expending the energy that drives the emotion, thus saturating it, or even weakening. Another parallel is the process that occurs when an important event, such as a wedding or going to college, takes place. The initial low level activity kicks into high gear and then becomes a strong activity to the maximum level limited by confusion, conflicts or availability of resources, followed by a decline when the need has been met (no driver) or the resources are spent (giving up the effort). We also see the fractal nature with self similar patterns of crisis or urgency, growth and resolution of little issues comprised in the broad non-normal events. In some cases, of course, the instability changes our life as in phase transition. But more often we see a slow change as opposed to a bifurcation or a change at a critical point. We fold this experience into our expectation and couch it in comfortable or sensational terminologies ("concerted", "frenzied", "incubating", "plateau" etc.), but the similarity with natural phenomena is too good to be coincidental and be anything but connected to science of critical phenomena.

If personal life-instabilities have similarities, instabilities at social levels are even more aligned with scientific phenomena in which a new equilibrium is perturbed and returns to its original state or changes the status of the society. One can easily relate to this, since all social movements and revolutions seem to have this nature. Let us say a simmering discontent or a lopsided or unfair or unnatural separation such as racism exists in the society. At this point, a small perturbation can cause an explosive growth of protests, rebellion or even revolution. This is like a forest dried out in the summer, where a spark can ignite a major and expanding fire. Again, the phenomena saturate and "may" burn themselves out. In a very profound similarity, like the correlation lengths becoming infinite at the point of transition in critical phenomena, a parallel feature occurs in social instability. Just around the critical social transition point, the connection between people is wide and news spreads fast and the response is quick. This, of course, feeds the explosive growth of the instability. In plasma (ionized fluid) phenomena, at the point of instability the dominant perturbations are of long wavelength and therefore include large volumes of the fluid. This is reflected in social instabilities, where disparate and distant sections seem to respond collectively with coherence at critical times. Like every grain of the sand pile is involved in preserving the critical shape, every member of the society is involved at critical point and is tuned to what is happening. When the instability grows to a certain size, this coherence may disappear and individual parts may become independent which will then cause the movement to become incoherent and saturate or decline.

## The Tipping Point

The critical point we have thus far described is similar to the points called 'The Tipping Points". The Tipping point is well known for cases of balanced arms of a scale or a see saw where just a little unbalance in weights causes the device to tip to one side or another. But the tipping point description is perhaps equally valid in transitions at critical points or a pathway of a process or an event that provides for abrupt changes. Tipping point also refers to the fact that at a certain point, the interconnectivity of various parts of a system, as in long distance correlations, will propagate change so fast that a small change in one place changes all of the system and its future state. The tipping point also occurs in a system where the rate of application of control is not fast enough to catch up with the rate of change. An example is the discussion on the potential of tipping point in climate change. (See for example, Dave Levitan in Scientific American. (Levitan 2013)). A graphical representation of the tipping point is given in figure 4.16.

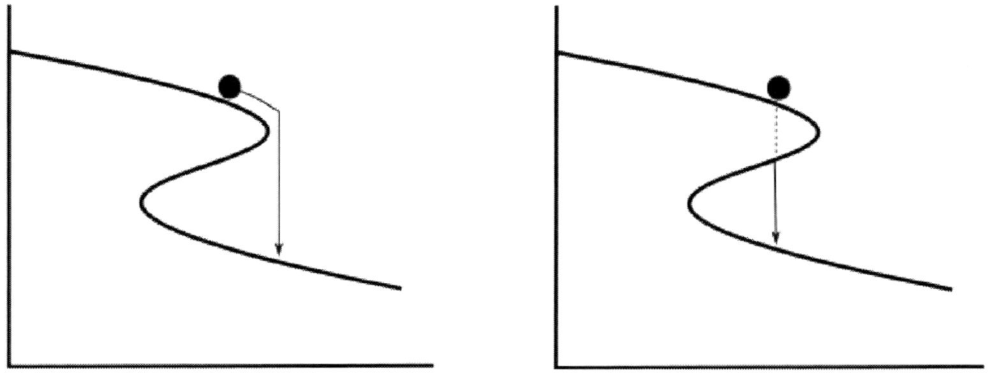

Figure 4.16. Tipping Point-The left figure shows the obvious case of the "falling off the cliff" as the tipping point. The figure on the right shows another case where a system, not close to the cliff goes down to an unstable state and then experiences a tipping point. The first part of tunneling to an unstable state happens for example in solid state devices called tunneling diodes.

In social context, a dramatic change may occur at a tipping point but it would be a tipping point only if the reactive system say a government suppressing a revolt, cannot apply controls fast enough before the revolt spreads widely. This is a well known problem in disease control as well. The breadth of the process is also important. As an example, As Elizabeth Pond states in Christian Science Monitor, (Pond 2009) *" Leipzig revolt of 1989 and in the Teheran uprising of 2009", ".. a robust civil society and middle class that habitually guarded their private sphere by eschewing politics suddenly turned political and challenged an authoritarian power structure. In both cases a mobilizing spark was the insult to citizens in apparent official falsification of formal elections that offered little genuine choice anyway. In both cases the social contract snapped; a wide range of businessmen, technocrats, and young mothers spontaneously joined the protest of elite student malcontents."* Clearly this type of tipping point has its parallel in multi component systems with different levels of interactions. In the example of liquid mixtures, the contribution to the vapor pressure from each component may be different, but the mixture boils when the sum of the vapor pressures exceeds the environmental pressure. The book "the Tipping Point" by Malcolm Gladwell (Gladwell 2002) is hardly surprising when it describes various observations as tipping points, but is surprising in its lack of mention of scientific origin and parallels and scientific description of criticality events. (As a consequence, Gladwell has been accused of oversimplification in describing the events as tipping points and thereby stating that nothing can prevent or control many social phenomena). In reality, feedback controls which sense the overall direction of the change, the current state of the systems and the rate at which the change is taking place, and apply appropriate corrective measures, do work and can prevent a "tipping" change.

## Linear and Non-linear Systems, Time Constants and Feedback

One common feature of chaotic phenomena is that the chaotic features are displayed in the so called non–linear systems. A non linear system is only described by a set of non-linear equations. Non-linear equations are equations in which the parameters that vary and characterize the system behavior, appear in algebraic powers of greater than one (or also termed as polynomials). The same is true if the system description includes powers of the derivatives. An equation such as $y=Ax+C$ where, $x$ and $y$ describe the system status and $A$ and $C$ are constants, is a linear equation, while, $y=Px^2+Qx+R$ is a non-linear equation. We saw that a linear dependence of derivative such as $dy/dx=gy$, gives a result, $y$ varying exponentially as, $e^{gx}$. Such a response is in the linear regime and does not lead to chaotic or complex behavior by itself.

Also, a second derivative (gradient of the gradient) of a parameter varying linearly with the value of the parameter, gives a periodic oscillation when the proportionality constant has the right sign. A simple harmonic motion of a simple pendulum made from a mass suspended from a rigid point with a massless string is given by the fact that the force experienced by the mass $F=ma$, where $m$ is the mass and $a$ is the acceleration. This force is balanced by a restoring force which opposes a displacement $y$ from a neutral position and is given by $-ky$. Since these are equal,

$$ma=-ky$$

Since acceleration is the second derivative of the displacement with respect to time,

$$m\, d^2y/dt^2=-ky$$

$$\text{or } d^2y/dt^2=-Sy$$

where $t$ is the time and $S$ is a constant$=k/m$. This linear equation gives the result for a periodic motion, $y=A.sin(\omega t+\phi)$, which is a (sinusoidal) periodic oscillatory motion where $A$ and $\omega$ are the amplitude and frequency of the oscillation and $\phi$ is the phase which is related to the initial displacement at $t=0$. On the other hand, the complicated non-linear equation,

$$d^2y/dt^2+q\, dy/dt+g\, sin\,(y)=-Sy+Bsin(\omega' t)$$

where $g,q$, $B$ and $\omega'$ are different constants, gives a chaotic response with the response varying depending on the values of the constants. Here other forces are acting, one a frictional force proportional to q and another, an oscillating force with a frequency of $\omega'$. This is chaotic in the sense that in seemingly changing amplitudes and frequencies, the pendulum will oscillate in different patterns, but will trace out fractal and bifurcated trajectories which are strange attractors.

## Time Constants

In physics, there is a term called “retarded time”. This refers to the fact that distribution of a quantity such as field strength, depends on propagation or emission from the source, which travels at a finite speed. Therefore, the influence of a source at a point corresponds to the condition somewhat prior to the present time at which the source condition exists, because of the time taken for the influence to travel. Similarly this time difference (either delay or advance in time) needs to be taken into account in other natural conditions. For example, the current population depends not on the current conception (fertilization) rate, but the rate which existed at the time which was just before the period of pregnancy. This period of latency or period of growth is built into all natural processes. In many cases, this is called the time constant of the process. Time constant usually refers to the duration over which, a quantity increases by a given proportion. Time constants for different phenomena affecting a single system may be quite different and this affects the evolution of the system significantly.

The most significant example of this is the climate change due to fossil fuel burning and emission of carbon dioxide by humans. The response time (time constants) of the earth's systems, such as the time constant of ocean warming or cooling, polar ice melting, land warming, absorption of greenhouse gases etc. are quite different (much slower) from the rate (much faster) at which we are burning fossil fuels emitting carbon dioxide. Therefore, even if humans stop burning fossil fuels at a given time, it would take many decades before the effects are felt and the climate will continue to change.

The time constant makes for a very significant effect. The simplest example of the extraordinary effect of delays and latencies in the system can be illustrated by taking the previously described iterative logistic equation 4.12,

$$X_{n+1} = r\, X_n(1-X_n)$$

To recap, here $X$ are values in the $n$th and $(n+1)$th step of iteration and $r$ is a constant. As stated before this iteration can be done easily on a spread sheet application where the value in $(n+1)$th row is calculated from the previous $n$th row. However, since each iteration step represents a time step, one can see what would happen if one of the terms has a greater delay than one time step. In other words, we can check what would happen if the logistic equation is written as,

$$X_{n+1} = r\, X_n(1-X_{n-2})$$

so that the value for the negative term is taken from three rows prior rather than the previous row. One finds that the characteristic behavior changes drastically. For values of $r > 1.842$, the values go seriously out of bounds and become chaotic. The figure 4.17 shows the behavior for $r=1.8$. (For the original logistic equation without a time delay, the solution is a constant – converged-value of 0.444444). This shows the population to rise and fall in a bifurcated way giving two solutions:

This particular type of behavior is actually seen in nature and is often called a predator-prey population variation. If there are rabbits and foxes in a forest and rabbit food becomes plentiful, rabbit population would increase rapidly. As a result, foxes also have greater food supply and their population grows. However, as the fox population grows, rabbit population because of being prey reduces and collapses, leading, in turn, to the collapse of fox population. With the depleted fox population, rabbit population would now increase back and so on. But, in this process, the rise and collapse are caused by the fact that there is a time delay between the rise of fox population and rabbit population. Otherwise, the population would reach equilibrium. In multi component systems with different time constants and delays, the chaotic nature can be even more pronounced and multidimensional, with each component exhibiting chaos.

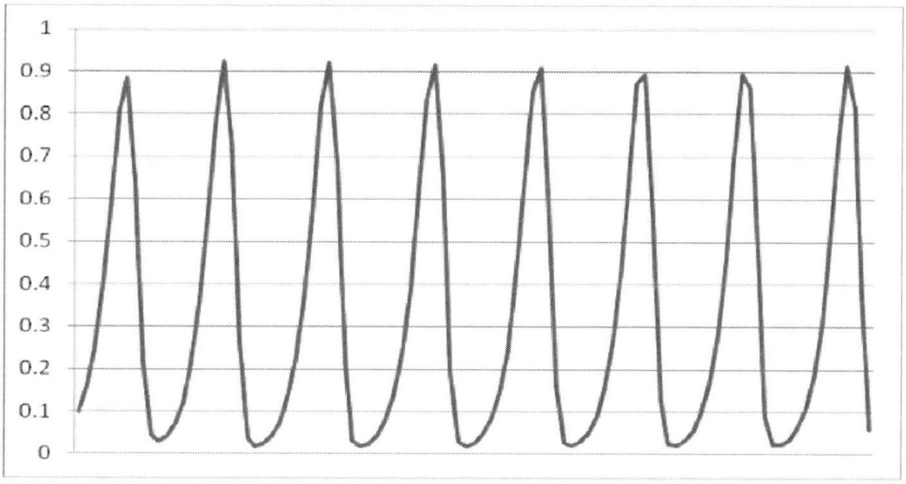


Figure 4.17. Lorenz equation with time delay of 2 time steps, r=1.8.

It should be noted that even the behavior of the system, shown in the previous logistic equation example, is related to the delay of one step. If there was no iterative calculation (instantaneously responsive system), the equation would be written as

$$X = r\,X(1-X)$$

Giving a steady state value of $X=r/(r+1)$. For example, for $r$=2.0, the solution is 0.66666, which is what we found for the iterative logistic equation too. For $r$=3.5, the instantaneous responsive system would give $X=2.5/3.5$. But the iterative procedure gave chaotic behavior. Therefore, the chaotic nature is exhibited when the coefficients exceed a threshold value AND when a delay is incorporated.

## Feedback

A simple example of feedback control is that of a rocket, which is launched towards a particular location. But, since the conditions and effects can never be perfectly calculated, sensors are placed on the rocket and relay the error from the intended path to the control station. The control may automatically be applied to correct the error. As stated before, the feedback can correct an accumulated error in the location and also correct the direction in which it is moving at present. This type of control is often used in many engineering systems and is called PID (Proportional, Integral and Differential) control. The additional Integral and Differential controls are needed because of time delays in communicating with the rocket and the correction taking effect.

Systems often have built-in feedbacks which restore equilibrium, as we saw in the case of a simple pendulum. Examples abound as in many anatomic systems controlling heart rate, body temperature, blood pressure etc. or a speed control system on a train etc. But, feedbacks can also be positive in which the displacement from equilibrium is increased causing instability or a runaway situation. An exponential growth of population (even if it is a linear process) is a positive feedback, where the rate of increase of the population is proportional to the population. On the other hand, if the rate of growth of a quantity were proportional to the negative of that quantity or alternatively stated, if the rate of decrease of a quantity is proportional to the quantity, the quantity would exponentially decay. This happens, for example, when storage of material (such as water in a dam) is emptied. The case of the chaotic pendulum is also a case of different feedbacks and additional forces on the displacement of the pendulum. If a system has many components, each providing its own feedback and impacts on itself and other components, the system would exhibit a complex behavior, which might seem undecipherable like the multi body problems, but would, all the same, have orderly and disorderly behavior and exhibit features of chaos. Systems such as our atmosphere, our fresh water supply, the ecosystem and the oceans, exhibit this multi component behavior, and are based on feedbacks between subsystems. Even more interestingly these systems themselves are subsystems of the earth system, which, in turn, is part of the solar system. In another terminology, we might call these, interdependent components. The complexity arising out of the interaction (interdependence) between each of the components of the subsystems and systems can, in certain range of parameter, be very orderly. However, once these are outside these ranges, feedbacks from and to each component can throw these systems into chaotic regime, in which the system does not behave randomly. But the pattern can be complex, sometimes too complex to be discernible. This ability to create complexity through feedbacks between various components creates extraordinary difficulties in diagnosing problems and issues and in devising solutions. The field of medicine is an example of this.

We know now that such behavior is common and approach to chaos can often be anticipated and in particular cases, additional negative feedbacks can control the system if the system is not at or beyond the critical point. A real life example is that of a feedback control of a child's behavior. One can assess and correct the child's behavior after he had done something wrong or correct the child's tendency to behave wrongly, before he does any wrong. The correction has to be proportional, not too much, not too fast and appropriate or else it could lead to other behaviors which may also be detrimental. The science of diagnosing critical points and approach to controlling the system is increasingly becoming sophisticated and effective. The ability to control spacecrafts with sophisticated instruments, to send them into complex orbits in somewhat unpredictable conditions is a case in point. A spectacular case of diagnosis of the critical situation that existed in the bronze era (around 1200 BC) is the discovery of the cause of the simultaneous and sudden collapse of Egyptian, Syrian and Greek prosperity. This collapse is now traced to a prolonged drought due to climate change. The discovery used highly honed, high resolution techniques of analysis of pollen from the period (9000 years old) at 40 year intervals, combined with archaeological and historical data. The work was done by the Tel Aviv University team which extracted about 60 feet of muddy sediment cores from the Sea of Galilee in northern Israel. At present, we are faced with a similar situation where a looming climate change is demanding urgent attention to chaotic processes and systems and potential solutions. Most agree that, in the absence of technologies of substantive and exquisite controls on greenhouse gases or the mitigation of their effects, the prudent choice is to cut back severely on the emission of greenhouse gases and stay away from any criticality point, while developing accurate models for prediction and tune our energy sources to overcome this climate change problem.

## CHAOS, BIOLOGY, QUANTUM PHYSICS AND TYCHISM

It is easy to see that fields of biology and medicine would be dominated by chaotic processes. The complex phenomenology, the multi parametric influences and a highly branched (large number of bifurcations leading to a particular outcome) make biology (and biological evolution in particular) and medicine particularly susceptible to chaotic phenomena. One sees fractals in basic building blocks of life and bifurcations and critical phenomena everywhere and yet the processes have an order in them, with built-in indeterminacy of detail. Computer models (Jorgensen 1998), that model the adaption and co-adaption of interactive ecological systems, show that such systems which are in semi equilibrium exist in the region between order and disorder at the edge of chaos like sand piles. For example, key parameters like size of change in population versus frequency of change and size of avalanches versus frequency of avalanches (here rapid loss of information on the ecosystem is called avalanche) follow a power law (Zipf's law), which is typical of critical phenomena. One also sees that with chaotic processes in the mix, sudden evolutions and punctuated equilibria (major bifurcations) are possible.

In Quark and Jaguar, Gell-Mann eloquently explains the concept of alternative histories-bifurcations at the quantum mechanics level. All such alternate histories are not realized but only have probabilities. One only experiences the history one does. This is similar to buying a lottery ticket where there are various probabilities from 0 to 100% of winning any amount from zero to the lottery pot amount. But the moment the lottery is drawn, the result is 100% known and the winning persons and the amounts are part of the history. Gell-Mann says that actual histories are realized only by coarse graining of the fine grained quantum mechanical histories, where some parameters are averaged over time or space. Also, in classical processes with Chaos, such coarse graining may be applied. As an example one can see how two trees have grown in height and width differently without seeing the details of the leaves and small branches. But the nature of fractals in chaos is unique. It would seem that phenomena well described by classical physics would not contain chaotic nature going down to the subatomic quantum mechanical level and physical bodies have a deterministic behavior in simple processes. Perhaps, quantum mechanical uncertainty and mechanisms of chaos do not work in tandem.

Chaos confirms some of the ideas of Tychism, an idea proposed by Charles S. Peirce, which states that the Universe is imbued with uncertainty and so may be a prophecy for the Heisenberg Principle. Peirce stated in his lecture at the University of Cambridge in 1898 that *"Tychism is the doctrine that absolute chance is a factor in the Universe."* (Rockwell 2012). Yet he postulated that this chance only plays a role and does not destroy orderly progression. Regularity and irregularity introduced by chance go hand in hand complementarily. The latter principle is in agreement with the theories and observations of chaotic phenomena.

## HUMAN EXPERIENCES IN ORDER, DISORDER AND CHAOS

We saw how chaos rules in real physical phenomena, where deterministic equations are supposed to provide a substantial description. As we saw, in the real world, chaos modifies and can even nullify some of the solutions and concepts in physics because the assumptions made in physics are only valid either partly or wholly, such as instantaneous balance and interactions do not occur, multi body interactions are not calculable or that multiple systems are interacting in complex ways. In addition, lack of precision and reproducibility in processes and starting points can cause major changes due to bifurcation points. In biology, particularly in evolution of species, chaos has to be included in making full assessments, because all the complex features, (non-linearity of processes, competition and synergistic co-evolution providing feedback, characteristic time delays etc.) exist in that. One sees chaotic processes in economics such as in prosperity and poverty, in stock market ups and downs, catastrophic crashes and market bubbles etc.

In religion, one sees that a religion comes into existence and becomes a major force in society purely because of circumstantial opportunities and feedback mechanisms, while other similar religions fail. An excellent example is that of Christianity. It is well known that before Christ, many teachers and perhaps messianic figures had come but were not successful in establishing a religion. But, just around the time of Christ, Romans had perfected the engineering of road construction and Christ's followers could travel to distant places and spread Christianity. The availability of roads was such a powerful non linear mechanism and a bifurcation that Christianity was established in the very Roman Empire that crucified Christ. In *"Ancient Innovators, A Modern Legacy"*, (Grossman 1998) writes "*Those roads that St. Paul and his associates traveled to spread the faith were unprecedented marvels of engineering. Without the ingeniously constructed roads that led from the streets where Jesus had walked to the cities of Syria and Greece, Christianity might have remained another obscure Judean sect like those that fill the pages in accounts by Jewish historian Josephus.*" The opposite example is that of Buddhism that never really took hold in India, its birthplace, because of the steadfast theism, but Buddhism spread in other countries very fast because the threshold for acceptance of Buddha's teachings was low. In Easter Island tall stone statues were built by an earlier civilization to appease the Gods of the seas, to bring more rain. For moving the stones and statues they cut the island timber and the deforestation reduced rains further. The islanders built even bigger statues and cut more of their forest and doomed their island's ecology. The remaining population finally abandoned the island. This is a case of an instability eventually reaching critical state.

In daily life and society there are many other examples of Chaos. Development of heart disease which involves multiple systems of the body can lead to death, a phase transition. When a good and prosperous family adopts a child from a poor orphanage, a bifurcation happens in the life of that particular child and the family. The former Soviet leader Mikhail Gorbachev instituted some political changes which suddenly transformed the political and social structure of Eastern Europe and several Asian countries. It is interesting to note that our aesthetics are closely linked with this order-disorder balance. We like a tree because of its branching in a somewhat organized fashion. But if it is too symmetric, we probably do not like it. We want leaves to be similar but not identical. In our own lives, we do not want extreme predictability and order. We do like a few surprises, excitement and spontaneity, without which we find life to be prosaic.

One can go on counting cases of transitions, bifurcations and transformations in personal lives, world events and economic and social activity. In some cases, we assign the success or failure to the efforts of humans. An example is that if a country sees economic growth, good jobs and prosperity, the government and the monetary controllers would get the credit, where as their contribution is usually only partial. In other cases where one cannot fit in a human cause, one may be tempted to title such happenings as not discernible to humans but only to the Divine. The ironic case is when the situation is created by humans themselves, but without taking into account the interdependence of component systems, the results might be wrongly assessed or designated to fate or providence. In reality, most of human processes mimic or follow the same laws that govern physical phenomena. The links between prosperity, health, art and culture, food supply and climate are very intimate and in the past whole civilizations have risen and fallen because of these linkages which create chaotic feature mixing order and disorder. Mythology might attribute the fall of civilizations to the wrath of Gods, while it is usually the link between human behavior and the supporting systems that took the civilization to a chaotic regime thus causing the collapse which could have been prevented by knowledge of the scientific/natural processes that precede it. One of the reasons why we don't experience deeply chaotic events in our lives on a very frequent basis is that we (and other living things) have evolved to control our responses and provide a feedback to the system that prevents entry of the system into the chaotic regime. Humans have also learnt to modify living conditions so that there is control from multiple sides of the system. In many cases, we experience an uncontrolled and unexpected order or disorder, only because we are not educated in the science of chaos in the naturally occurring event. This knowledge can provide the tools to understand and even control our lives better.

A spiritual person would see that chaos, like evolution, realizes neighborhood potentialities and for Christians, this might seem analogous to free-will. But broadly religions seem very strong on the divide between order and disorder, in the spirit of righteousness being black or white. Therefore, the theory of chaos might be helpful or anti-religion depending how fundamentalist the person is. This should not be confused with fuzzy logic or moral ambiguity. As stated before, chaotic phenomena are deterministic and are manifested only because of the neighborhood of potentialities. In some sense, one can argue that a certain outcome is the unpredictable destiny and can see the hand of God in abruptly causing or preventing a result (through bifurcation or an attractor or a repeller). But, the broad theme of religion is not within the realm of order hidden in disorder, except in few mythologies.

Understanding that the events that we experience, the complexity we see in nature and our own assessment of objects and phenomena can be explained by the knowledge of order, disorder and chaos, is the first step to sustainability of desirable conditions and finding solutions to issues.

## REFERENCES


Dalziel, S.B., Linden, P.F. and Youngs, D.L.; *J. Fluid Mech.*, vol.399, No. (1999), pp.1.

Gell-Mann, M.; *The Quark and the Jaguar*, W.H. Freeman and company, New York, NY(1994).

Gladwell, M.; *The Tipping Point*, Back Bay Books, New York, NY (2002).

Gleick, J.; *Chaos- Making a New Science,* Penguin Books, New York, NY(1987).

Gribbin, J.; *Deep Simplicity* Random House, New York, NY(2004).

Grossman, R.; *"Ancient Innovators, A Modern Legacy", Chicago Tribune*, (December 21, 1998).

Jung, C.G.; *Archetypes of the Collective Unconscious*, Editors: Read, H., Fordham, M., Adler, G., McGuire, W. Collected Works, 2nd Edition, Bollinger Series XX, vol. 9, part. 1. (1968)P.66, EBook, Google, Complete Digital Edition, Princeton University Press, Princeton Jacobs, J.; *"Nature of Economies"*, Chapter 7, Random House, New York, NY(2002) p.87.

Jorgensen, S.E., et.al.;"*Ecosystem as Self Organizing Critical System", Ecological Modeling*, Vol.111, No.2 (1998) pp.261-268, N.J. (2014). http://books. google.com/books?id= QVuEAgAAQBAJ&dq=mcguire+1974+jung+collected+ works&source=gbs_navlinks_s.

Levitan, D., *"Quick Change Planet: Do Global Climate Tipping Exist?", Scientific American*, London, U.K., New York, NY (March 25, 2013) EText: http://www. scientificamerican. com/article/do-global-tipping-points-exist/.

Mandelbrot, B.B.; *"How Fractals Can Explain What's Wrong with Wall Street", Scientific American*, London, U.K., New York, NY (Sept.18, 2008); Original (February 1999). EText: Scientific American, http://www.scientificamerican.com/article/multifractals-explain-wall-street/.

Poincaré, H.; *Science and Method,* translation by Maitland, F., Thomas Nelson & Sons, London, U.K., New York, NY (1903) pp.64-90. Ebook: https://archive.org/stream/ sciencemethod00poinuoft#page/n1/mode/2up.

Pond, E., "*What's the tipping point for revolution?", Christian Science Monitor* (July 1, 2009) EText: Christian Science Monitor, http://www.csmonitor.com/Commentary/ Opinion/ 2009/0701/p09s01-coop.html.

Richardson, L.F., *Weather Prediction by Numerical Process*, Reprinted in 1965 by Dover, NY (1922). Etext: Google Books, http://books.google.com/books?isbn=1439816697.

Rockwell, T.*; Mind or Mechanism, Which Came First?* In *Origins of Mind*, Editor: Liz Swan, Springer, Berlin, Germany (2012) p.244.

Sprott, J.C.; *"Strange Attractors: Creating Patterns in Chaos", M&T Books (1993).*

Twain, M.; *"Mark Twains Prophecy", New Zealand Herald*, Vol. XXIII, Issue 7800, (20 November 1886), p. 2. EText: http://paperspast.natlib.govt.nz/cgi-bin/paperspast? a=d&d= NZH18861120.2.49.24.

Van Eenwyk," *Archetypes & Strange Attractors: The Chaotic World of Symbols* (Studies in Jungian Psychology By Jungian Analysts), Inner City Books (1997).

*Chapter 5*

# MOTION, SPACE AND TIME: WHERE AND WHEN WE ARE

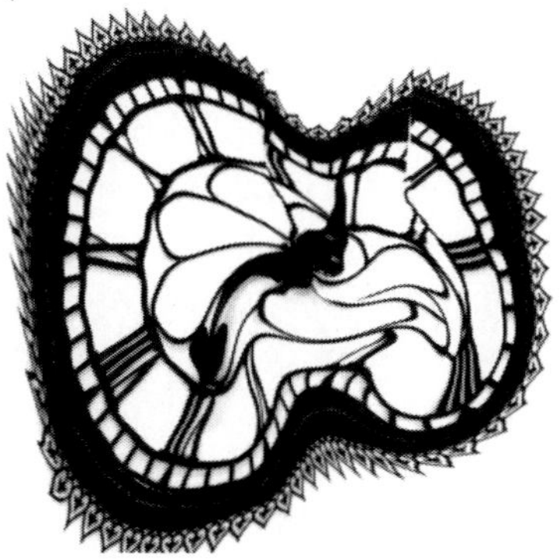

"Like as the waves make toward the pebbled shore, So do our minutes hasten to their end; Each changing place with that which goes before, In sequent toil all forward do contend." - Shakespeare Sonnet 60

Even animals understand or sense the basic relation between an object and its motion, heaviness, the force required to move it and what happens when it falls. Some animals may even be able to sense time by the circadian process. For us humans the understanding is greater, but, in day to day life, we may not deal with them much differently from animals but just as part of life to be included in our daily reckoning, without any inquiry. And so it was for many millennia. However, the philosophers, prone to worrying about the nature of things, proposed various theories about why objects move and fall. It was probably accepted without question that things had to be pushed or pulled to move them on the ground. It was accepted that the heavy objects are ones that require more effort to move. (That would be the first definition of heaviness). The Greek philosopher Aristotle theorized that all objects had four characteristics, matter, form, moving cause and final cause. (The moving cause is not to be confused with cause of motion but more like a cause for it to change form and final cause is roughly its purpose). Further he believed that matter consisted of five elements, air, water, sky, fire and aether and that motion of an object was the object’s way of bringing these elements into a balance of equilibrium. Water flows downstream, because its nature is to increase the element of earth and smoke rises because it tries to increase the sky constituent. According to his theory, motion along a straight line with constant speed was ideal on the earth while circular motion was ideal for celestial objects. Beyond this natural or ideal motion, according to him, a force was necessary to keep a body moving at a constant velocity. For Aristotle, time was a ruler to measure change and motion and therefore in terms of causality, he did not distinguish between change and time. Aristotle and his followers had to continuously tweak explanations to cover a wide range of motions, such as projectiles, collisions etc.

The $12^{th}$ century physicist and Islamic scholar, Hibat Allah Abu'l-Barakat al-Baghdaadi might have known that force caused acceleration (increase in speed), rather than constant motion because he wrote that Aristotle's physics was simplistic (Shanker 2003). He was particularly dissatisfied with the Aristotelian view of time. He concluded that time had its own existence (of being) and argued that it is recognized by us innately even without change or motion.

## FORCE IN GALILEAN AND NEWTONIAN MECHANICS

According to Aristotle's laws of motion, (a) an object is naturally at rest unless it is moving toward its natural resting place, (b) it takes a force or violent action in constant contact to move that object, (c) The speed is proportional to the force and (d) The object will keep moving until something else intervenes to stop it. Unfortunately, Aristotle's laws do not explain impulse and the later realization of force at a distance. (It is interesting to imagine how he would explain arrows and such projectiles moving through the air). In 530 C.E., the Byzantine philosopher Philoponus pointed out this major flaw and suggested that an object going through air unaided was imparted an "impetus" at the time of throwing but he believed it lost that impetus when travelling. Ibn Sina (Avicenna) agreed with this, but he believed that the impetus does not dissipate by itself and air resistance is the reason it dissipates.

In the $17^{th}$ century, Galileo started the "ball rolling" on real physics of kinematics and dynamics- motion of objects, and simultaneously on gravity, when he demonstrated with balls rolling on an inclined plane, that an "impetus" was needed to set objects moving. He also theorized that in the absence of a "retarding force" (an idea that he came up with, by watching a feather fall) an object would keep moving in a straight line. This concept of "Inertia", conceived by Aristotle, is enshrined in Newton's laws and all physics. (On a separate inquiry into celestial objects, Galileo's affirmation of the Copernican theory that Earth is not at the center of the Universe angered the church and he was in considerable difficulty).

The Newtonian theory, buttressed by the new- fangled field of calculus, co-developed by Gottfried Leibnitz, stated that objects had this inertia and accelerated only under a force. With this calculus, one could then easily obtain equations of motion for an object, to determine the displacement, speed and acceleration as a function of time. In 1350 C.E. Jean Buridan referred to Philoponus and Ibn Sina's impetus as being the product of weight and speed. Rene Descartes, the famous mathematician-philosopher of $16^{th}$ century, called this the quantity of motion. Jean Wallis and later Newton, surmised the correct nature of this quantity, now known as momentum which is the product of the (inertial) mass and the velocity. Newton showed that the rate at which the momentum of an object changes is equal to the force that is applied on it. (See figure 5.1). Newton laid out these and other fundamental concepts in his magnum opus "Mathematical Principles of Natural Philosophy". As the equations in Box 5.1 show, the velocity is the rate of change of position, the acceleration is the rate of change of velocity and the force is the rate of change of momentum, which is the product of mass and velocity. For constant mass, the resulting fact is that the force is the product of mass and acceleration. (Here rate refers to change in the quantity for a unit (a second or hour etc.) change of time. An example is that of rate of change of position, such as 100 kilometer per hour is the speed).

Figure 5.1. Force is proportional to the rate of change of momentum.

**Box 5.1. Velocity, momentum and force**

x is the displacement (distance travelled) along the x direction and t is the time taken for the displacement; all the quantities except time are vectors with direction while time is a scalar.

Velocity in x direction= rate of change of position along direction x or, $v_x = \frac{\partial x}{\partial t}$ 5.1)

Acceleration in x direction= rate of change of velocity or, $a_x = \frac{\partial v_x}{\partial t} = \frac{\partial^2 x}{\partial t^2}$ (5.2)

Momentum in x direction = mass x velocity in x direction or, $p_x = mv_x = m\frac{\partial x}{\partial t}$ (5.3)

Force in x direction = rate of change of x momentum in x direction or,

$$= \frac{\partial p_x}{\partial t} = m\frac{\partial^2 x}{\partial t^2} = ma_x \qquad (5.4)$$

Here, $\frac{\partial}{\partial t}$ is the partial derivative (rate of change) with respect to time (only) and $\frac{\partial^2}{\partial t^2}$ is the partial derivative of the partial derivative (rate of change of rate of change) with respect to time. Since we live in a 3 dimensional world with 3 directions (commonly denoted by x, y and z), one can write similar equations for each of the other two directions.

The concept of mass is discussed later, but here it is implied that it is the quantity that characterizes the inherent inertia of the body. In the 3 dimensional world with 3 spatial directions that we live in (commonly denoted by x, y and z), we write these equations as components in each direction. (When vector algebra is used, the vector quantity is a vector sum of these components). These equations would lead to other equations of motion in specific conditions. The actual derivation involves some differential and integral calculus, but among some of the well- known relations are those that relate velocity, acceleration and time, for instance, 'final velocity is initial velocity plus the acceleration multiplied by time elapsed; the distance travelled is velocity multiplied by time interval plus half the acceleration multiplied by the square of the time. (See equation 4.1). Newton's laws also imply that in the absence of an external force, momentum is conserved, just as in the absence of an external energy input or output to an external system, energy is conserved.

The everyday phenomena that we encounter can be easily explained by Newton's equations of motion and laws of conservation. For example, when an object is thrown with an impulse, the object travels horizontally with constant speed (nothing to change its speed, neglecting air friction), while vertically, because of the gravitational pull, the object decelerates going up and accelerates coming down. This means from the equations that the horizontal distance the object travels is proportional to the time it travels. But the vertical distance it travels is proportional to the square of the time taken (see above). So if we plot the vertical distance against the horizontal distance, we see that the vertical distance travelled is proportional to the square of the horizontal distance. This is the parabolic trajectory. When a billiard ball hits another at rest, it exchanges momentum and the momentum is conserved. Since momentum is proportional to the velocity and energy is proportional to the square of the velocity, the conservation of both restricts it to only the following outcome: If a very light ball hits a very heavy stationary ball, not much momentum transfer happens. The heavy ball gets a fraction of the initial momentum proportional to only the ratio of the masses, while the light ball bounces back with almost all its initial momentum. When a cannon fires, it recoils in order to conserve momentum. (See figure 5.2).

In Newtonian mechanics, the notion of relative distances and velocities were understood in a common sense way, for example the relative velocity between two moving objects with velocities **u** and **v,** as being the difference in their velocities (**u-v**) (vectorial difference, taking into account their direction). Also, mass is a constant property of a given physically stable object and since force= mass x acceleration, the force is proportional (only) to the acceleration. Time was the same for all observers irrespective of their motion. Newton's famous laws of motion are, (1) Objects remain stationary or in a state of motion in the same direction, unless acted upon by an external force. This is the law of inertia. (2) A force applied on a body of mass m results in an acceleration and the force is equal to mass x acceleration. (3) Every action has a reaction. These laws have been tested and experienced in ordinary circumstances and are common knowledge. The invention of these laws literally changed how physics was pursued. Newton's mechanical theories provided the right answers for projectiles and colliding bodies, and also explained Kepler's laws of motion of planets around the Sun. Newton is considered one of the great geniuses in human history, for coming up with the mechanics which explained so many things that were mysterious before and had a single description for the motion of celestial and terrestrial objects.

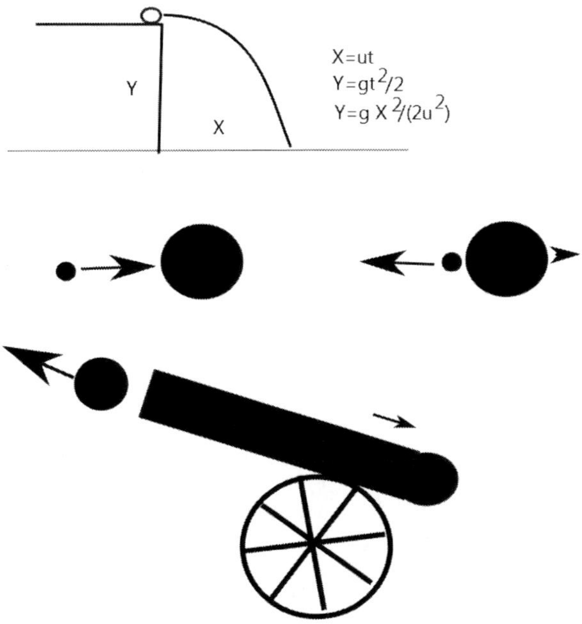


Figure 5.2. Simple examples of mechanics: (Top)-Parabolic trajectory of a projectile of an object thrown horizontally with a speed u. X and Y are distance traveled in the horizontal and vertical direction in time t and g is the acceleration due to gravity. (Middle)-A light ball colliding with a heavy ball. (Bottom) -Recoil of a cannon after firing a ball.

One could have happily gone with this hypothesis, because in almost everything we have to deal with in kinematics, including the motion of buses, trains and planes and even some classes of particles, Newtonian mechanics (and flat geometry of space), would be adequate. It is believed too that, most religions, especially Christianity, were fine with Newton's mechanical and deterministic view, because it reinforced the concept of stability, destiny and harmony of the Universe, which was held by the religions as having been endowed by God. Newton himself held off from being too rebellious of the scriptural orthodoxy, on such matters (Snobelen 1999). Even though, his discovery of the "Universal" laws and the associated mathematics of calculus was different from Biblical pronouncements, Newton's belief that the laws proved the central role of God, perhaps restrained him from an aggressive opposition to orthodoxy. So what upset the apple cart? Once again, it was the thirsty and dissatisfied intellectual physicists, trying to explain the non-normal phenomena that only irreverent and curious observers notice.

## History of the Special Theory of Relativity

The impetus for the new questioning on the extent of applicability of Newton's laws came from the new field of electromagnetism which emerged nearly a century after Newton's laws were propounded. Though electricity and magnetism were known for centuries, (Newton himself must have known about William Gilbert's work on electricity and magnetism), these phenomena were not seen as integral to the question of Universal laws in the 18th century.

**Box 5.2. Maxwell's equations of Electromagnetism**

These are given below in Cartesian coordinates (mutually orthogonal, linear, equally scaled coordinates x, y and z and with time parameter=t) as (Meter Kilogram Second (MKS) units),

$$\partial E_x/\partial x + \partial E_y/\partial y + \partial E_z/\partial z = \rho/\epsilon_0 \quad (5.5)$$

This states that the sum of the partial derivatives or gradients (or rates of variation), in each direction, of the components of the electrical field in the same direction is proportional to the density of electric charge.

$$\partial B_x/\partial x + \partial B_y/\partial y + \partial B_z/\partial z = 0 \quad (5.6)$$

This states that the sum of the partial derivatives or gradients (or rates of variation), in each direction, of the components of the magnetic field in the same direction, is zero.

$$\partial E_y/\partial x - \partial E_x/\partial y = -\partial B_z/\partial t \quad (5.7a,b,c)$$

This states that the difference between the partial derivatives or gradients, (rates of variation), in two cross directions, of the components of the electrical fields in the two cross directions, is proportional to the rate of change of magnetic field in the third direction, with time.

$$\partial B_y/\partial x - \partial B_x/\partial y = \mu_0\,(J_z + \varepsilon_0 \partial E_z/\partial t) \quad (5.8a,b,c)$$

This states that the difference between the partial derivatives or gradients (rates of variation) in two cross directions, of the components of the electrical fields in the two cross directions, is proportional to a linear sum of change of the electric field in the third direction and the electric current density in the third direction, with appropriate coefficients.

Here, $E_x$, $E_y$, and $E_z$ and $B_x$, $B_y$, and $B_z$ components of the electric and magnetic field vectors E and B, along the x,y and z coordinates, $J_z$ is the density of electric current in the z direction and ρ is the electric charge density. Here, μ and ε are called the magnetic permeability and dielectric constant of the medium in which the fields and charges are present; (The zero subscript denotes vacuum).

Equations 5.7 and 5.8 can be written in 3 ways using other coordinates and hence the notation a,b,c.

The ground for basic concepts in electromagnetism, in addition to observational data, was laid by André-Marie Ampère, Charles-Augustin de Coulomb, Michael Faraday, Hans Christian Ørsted, Carl Freidrich Gauss and others, when they described electric and magnetic phenomena in terms of the so called fields. These fields can be considered to be regions of influence. If a charge is generated by rubbing two materials together, it creates an electric field around it and in a similar manner moving charges in a wire, namely a current, create a magnetic field. These fields are measurable by the force another charge experiences in the vicinity of the charge or the wire carrying the current. The electric and magnetic fields, like velocity, are vectors in that they have magnitude as well as direction. In an amazing scientific feat, James Clarke Maxwell integrated all then known phenomena of electromagnetism into a set of compact equations and published it in 1865 in his article "A Dynamical Theory of Electromagnetic Fields" (Maxwell, 1865). These equations (see Box 5.2) give the relationship between the gradients in electric fields and magnetic fields with electric charges and electric currents respectively. The equations show that an electric charge is associated with a distribution of electric field around it and similarly, an electric current is associated with a distribution of magnetic field around it. Additionally, these equations show how a changing magnetic field gives rise to an electric field and vice versa. (There appears to be no record of religious views on the topic of electromagnetism at this time, presumably because it might have been considered a marginal finding of an exotic phenomena or a curiosity. Also, Maxwell's theory was perhaps not criticized by the church because Maxwell was a firm believer that his efforts were in the direction of Man's main endeavor to glorify God and his creation. But like Galileo's findings, Maxwell's equations too would change physics forever).

If this achievement of describing all fields in a compact form was not sufficient, Maxwell's equations form another feat. When we substitute some of these equations into each other, they give an astounding result: The electric and magnetic fields (simultaneously) follow a wave equation, meaning that in a vacuum the electric and magnetic fields oscillate and these oscillations propagate in space. The time varying electric field creates a magnetic field and the time varying magnetic field creates an electric field. Each is coupled in the same way, except that at all times the electric and magnetic fields are perpendicular to each other and perpendicular to the direction of propagation. (See Box 5.3).

**Box 5.3. Electromagnetic wave equation**

When we substitute the equations 5.7 a,b,c and 5.8 a,b,c into each other, we get the wave equations,

$$\frac{\partial^2 \boldsymbol{E}}{\partial t^2} - c_0^2\left(\frac{\partial^2 E_x}{\partial x^2} + \frac{\partial^2 E_y}{\partial y^2} + \frac{\partial^2 E_z}{\partial z^2}\right) = 0 \tag{5.9}$$

And

$$\frac{\partial^2 \boldsymbol{B}}{\partial t^2} - c_0^2\left(\frac{\partial^2 B_x}{\partial x^2} + \frac{\partial^2 B_y}{\partial y^2} + \frac{\partial^2 B_z}{\partial z^2}\right) = 0 \tag{5.10}$$

Where, $c_0 = 1/\sqrt{(\mu_0 \varepsilon_0)}$

These wave equations, shown in Box 5.3, turn out to be the equations for a wave traveling at a speed $c_0$ and are actually the set of equations for an electromagnetic wave moving in a vacuum (equations of oscillating and propagating electric and magnetic fields) with a speed $c_0$. Wonder of wonders that when $c_0$ was calculated, it matched the speed of light measured by Hippolyte Fizeau! This definitely proved that light is an electromagnetic wave. This was suspected to be the case by Michael Faraday because the so called "polarized" light responded to a magnetic field when traveling through a transparent material. With these equations, the close link between light and electromagnetism was established. So, paraphrasing the Bible one could possibly state: "God said let there be electromagnetic waves and there was light."

## Aether, Length Contraction and Time Dilation

Scientists like Christian Huygens, a contemporary of Newton and one of the first to use mathematics and geometry to express laws and concepts in physics, already knew that light behaved as a wave and had been proposing a luminiferous (light bearing) aether (or ether – here we use the spelling aether so as not to confuse with the anesthetic ether), made of fine particles for light to travel in. (After all, all known waves needed a medium). Newton had rejected this and said that light must be made of particles. Before Maxwell had come up with his theory, Thomas Young and Augustine Jean Fresnel had already demonstrated that light was a transverse wave, because the oscillating effects in the medium were perpendicular to the direction of light propagation. All this confirmed that Maxwell's equation describes the behavior of light waves and that light is constituted by oscillating electric and magnetic fields, mutually perpendicular and perpendicular also to the direction of propagation. With one set of concise equations, Maxwell had, thus, unified optics with electromagnetism. H.A. Lorentz revived the idea of aether and suggested that it was an absolutely stationary medium and light traveled through it. Then the counter question of mechanics was asked: If aether is the medium, how do solid objects move through this immobile medium and could it be the source of inertial mass of a body, or in other words, the resistance offered by immobile aether to the motion of a body? (More than a century later, this conceptual idea, although in a completely different way, would reappear as Higgs field that gives rise to mass of particles).

How the discovery of Maxwell's laws and speculation on aether led to the Special Theory of Relativity (STR) of Einstein is a legendary story and well narrated by John Stachel in his book "Einstein From B to Z" (Stachel 2002)

> "No other theory came remotely close to Lorentz's in accounting for so many electromagnetic and especially optical phenomena. This is not just my view of Lorentz's theory, it was Einstein's view. In particular, he again and again cites the aberration of starlight and the results of Fizeau's experiment on the velocity of light in flowing water as decisive evidence in favor of Lorentz's interpretation of Maxwell's equations."

A direct consequence of Lorentz's conception of the stationary aether is that the velocity of light with respect to the aether is a constant, independent of the motion of the source of light (or its frequency, amplitude, direction of propagation in the aether, etc.). Einstein adopted a slightly-but crucially-modified version of this conclusion as his second principle: There is an *inertial frame* in which the speed of light is a constant, independent of the velocity of its source. A Lorentzian aether theorist could agree at once to this statement, since it had been always tacitly assumed that the aether rest frame is an (absolute) inertial frame of reference and Einstein had "only" substituted "inertial frame" for "aether." But Einstein's omission of the aether was deliberate and crucial: by the time he formulated STR he did not believe in its existence.

Even at the age of 16, Einstein knew about aether and had thought about experiments on light. During his college days in 1897, he did not know of Maxwell's work, but had designed an experiment to study motion of light in aether, one not dissimilar to the famous Michelson-Morley Experiment. The definitive Michelson-Morley (M-M) experiment tested the idea of Lorentz's imponderous or immobile aether (also of Fresnel's, partially dragging aether), by sending two light signals, one in the direction of Earth's rotation and the other perpendicular to Earth's rotation. The Earth rotates at a speed of about 30 km/s and the speed of light is about 300,000 km/sec. Although small, the earth's motion should result in an aether wind (as seen from the earth) which would be opposite to the direction of rotation. This would affect the observed speed of light traveling in the direction of or opposite to the earth's rotation, while not affecting the speed, in the direction perpendicular to the rotation. (A small deflection in the direction of aether motion would take place). (This is similar to an object thrown from a train in the direction of the train's motion. The object would have a speed equal to the sum of the speed of the throw and the speed of the train. If it were thrown perpendicular to the direction of the train's motion, it would only have the throwing speed in the perpendicular direction). By comparing the speed perpendicular to the wind (rotation of earth) and parallel to the wind (rotation), one should be able to see the effect of aether. (This was actually accomplished by the interference between the parallel light wave and the perpendicular light wave). The details of this famous experiment, with its further subtleties, can be found from many sources such as college physics text books or on the web. (Figure 5.3). This experiment, surprisingly, proved that the difference between the speeds of light in the two directions was not greater than about one fourth to one sixth of the expected value. 20 years later, more accurate experiments have confirmed that the difference is actually negligible. This null result created a serious dilemma about the nature of waves, the medium and light. In a letter in August 1892 to Lord Rayleigh, Lorentz asked, "*Can there be some point in the theory of Mr. Michelson's experiment which has yet been overlooked?*" (Lorentz quoted in his letter of 18$^{th}$ August 1892 to Lord Rayleigh by A.J. Kox in Scientific Correspondence of H.A. Lorentz (Kox 2008)).

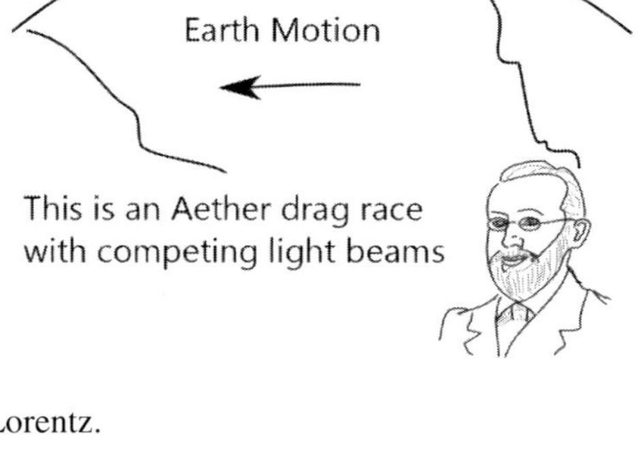


Figure 5.3. Aether drag and Lorentz.

The hint for a potential solution came, ironically, also from the field of electromagnetism, which originally created the problem. In 1888, Oliver Heaviside had discovered that as a charged particle (an electron) approached at near the speed of light, its electric field intensified by the factor $1/\sqrt{1-v^2/c^2}$, where v is the speed of the particle and c is the speed of light in the medium. This was generally noted, but no one, particularly Lorentz, seriously thought that this had any bearing on the problem at hand. But in 1889, G.F. Fitzgerald, a leading researcher in electromagnetism and the so called Maxwell stresses, did note this. He came up with the idea of length contraction to explain the M-M experiment, in which he proposed that all objects traveling close to the speed of light have contracted lengths as seen by a stationary observer. (The length contraction would intensity the electric field, because the magnitude of electric field is the gradient of the electric potential or approximately the electrical potential divided by the distance: shorter the distance, stronger is the field). Although he published a letter in the journal "Science" in 1889, somehow his ideas were not noted widely and it was only when Lorentz published his paper in 1892 that his proposition was recognized to be the same as what Lorentz proposed. This hypothesis is now called the Fitzgerald-Lorentz length contraction, which states that lengths and distances along a certain direction in a system, traveling with a speed v in the same direction, is observed by a stationary observer to be smaller by a factor, i.e. $L' = L\sqrt{1-v^2/c^2}$, where L is the length in the moving system (the aether in M-M experiment) in the direction of motion and L' is the length observed by the observer, in this case, the earth's apparatus. (Clearly, in this reckoning, only relative speeds matter and not who is actually moving). This essentially accounted for the null result of the M-M experiment, because the Fitzgerald-Lorentz contraction would appear to speed up the light in the direction it would have been dragged by the aether. (In the original Lorentz contraction, there were other multiplying coefficients which were then important. But, later these coefficients were shown to be 1.0 by Lorentz in 1904). Later, Henri Poincare', Joseph Larmor and H.A. Lorentz added the principle of time dilation (clocks moving with respect to an observer would be seen to be running slower than the observer's clocks), in which the time difference T' in a stationary frame is related to the time difference T in a moving frame by the relation, $T'=T/\sqrt{1-v^2/c^2}$. ( $\sqrt{1-v^2/c^2}$ is always less than 1.0). The Lorentz transformations are given in the Box 5.4. (The fascinating events and the science surrounding these conclusions are given in (Brown 2001)). Lorentz probably did not believe that these were basic laws of contraction and dilation and it was Poincare' who pointed out the importance of these length and time transformations as "principles of relativity", so that, (in his words), "the laws of physical phenomena should be the same for an observer at rest or for an observer carried along in uniform movement of translation, so that we do not and cannot have any means of determining whether we actually undergo a motion of this kind" (Torretti 1983). Poincare' and Einstein published a paper in 1905, showing this agreement between the theory of relativity and Lorentz Transformations. But an important fact is to be noted: both Lorentz and Poincare' thought that the length contraction was a real deformation of objects and tried to find underlying cause in electromagnetism, (which while contracting lengths, kept the speed of light constant). There is also a hidden implication in Lorentz's transformation: by comparing the lengths of the same object in different moving inertial frames, one can determine which is at absolute rest. This, as we shall see, is not true relativity.

Two important rules or laws were admitted at this point: (a) a stationary observer would see lengths of objects and rulers to be smaller on a frame which is moving at a constant speed relative to him. So a meter stick, which was 1 meter long in the observer's frame and was borrowed by the traveler, would seem shorter to the stationary observer, although it would still be 1 meter long for the traveler who is travelling with the stick. (b) The stationary observer sees that, in the frame moving at a constant speed relative to him, time is dilated. So, the stationary observer when observing clocks moving with a traveler sees those clocks slowed down (durations are longer according to those clocks), even though they were not when they were in his frame. These concepts were incorporated in the Lorentz-Fitzgerald transformation for changing time and space coordinates of an inertial frame, depending upon its motion relative to an observer in another frame. At this point, most physicists were satisfied, except a few who had different ideas. All was well, aether was the hypothetical medium for light to travel in, the cancellation of the effect of aether wind speed on light was accounted for by the fact that earth apparatus sees the distance contracted in the direction parallel to earth's motion. All this falls out readily if we agreed that aether drag and length contractions and time dilations compensated to keep the speed of light the same in all frames.

One person, who agreed with all this and yet was deeply dissatisfied, was Albert Einstein. He had to admit that there seemed to be a frame which was stationary for light to travel in. He was reluctant to name it aether. In physics, the different frames with constant speeds are called inertial frames. Newton had used this concept of the inertial frame of reference in coming up with laws of motion. But for Newton too there was only one universal frame. This stationary inertial frame and Lorentz transformation equations were clearly good for explaining constancy of speed of light, but it still violated one of the principles that Einstein believed in- all motion is relative and all inertial frames are equal and therefore all laws of nature apply equally in all inertial frames. So, there can be no justification for a universal frame.

## STR Postulates and Transformation Equations

The seed of relativity had been sown by Galileo, when he asserted that although it seemed that the Sun was moving it was the Earth that was moving and not the Sun. Einstein came up with his Special Theory of Relativity, using the Lorentz transformations as a stepping stone. The Lorentz transformation and associated aether description, though a master stroke that explained a lot, was a hypothesis. However, Einstein came up with STR with a hypothesis of his own, to overcome the conflict of declaring all inertial frames equal with the concept of one special stationary frame. He declared two postulates (as different from hypothesis):

Principle of Relativity - The laws of physics are the same in all inertial frames of reference. This means that within an inertial frame, all physical changes follow the same laws and have the same constants of nature. In simple words this means that if observers on the earth and observers on a moving ship carry out an identical experiment, each group would get identical results. However, an observer on earth, watching the experiment on the ship, say with a telescope, would not, in general, see the results on the ship to be the same as that on earth and vice versa.

The speed of light in vacuum has the same value in all inertial frames of reference. This means that the speed of light in vacuum is a constant of nature. (This could, in principle, be included in the first hypothesis as speed of light is given as a universal law). This surprising hypothesis is not derived from some property of light or some quantum mechanical treatment. This is the nature of mechanics. Light particles- photons do not have any mass and thus it can be stated that, all objects with zero mass have the same speed (speed of light) in vacuum in all inertial frames of reference. The emphasis is on speed of light in vacuum. The law does not apply to speed of light in a material medium, because material media would have properties that follow the Lorentz contraction etc.

(At this point, it becomes very important that words are stated carefully, because with these hypotheses, meaning of time and space itself changes. For example, a particular clock does not run slow or fast, it does so as seen by an observer in an inertial frame that is moving with a constant speed with respect to the inertial frame the said clock is. Similarly, a clock moving relative to an observer appears to be running slow to the observer. This type of precision is needed in dealing with physics as we go to more advanced levels of physics.)

With the second law, Einstein reversed the application of Lorentz contraction. Instead of stating as Lorentz did that objects contract in moving frames and that is the reason the speed of light appears to be constant, Einstein stated that speed of light is constant in all frames and that is why a stationary observer would see objects contract in the direction of motion in a moving frame. Using this, we need have no doubt as to the value of $c_0$ – the speed of electromagnetic wave, that we must put in Maxwell's equations in frames with different velocities- it is the same. When one applies the postulate that the speed of light is constant in all frames, the Lorentz transformations appear quite naturally and in the same way that Lorentz and Fitzgerald proposed themselves. At the same time, one does not have to find ways to justify the Universality of Maxwell's laws. (The derivation of the transformation equations in Box 5.4 is itself a very enlightening exercise and the elegance of thinking in obtaining these results can be seen in the book, "Space and Time in Special Relativity", by David Mermin (Mermin 1968). The author recommends that the reader should read this book in leisure and experience the joy of physics. A reader, who has the time and energy to go through the logical arguments using very basic algebra in this book, would be a physicist in his/her own right.)

## Lorentz Transformations

Let us consider an observer A in an inertial frame and an object PA (to identify it as in reference frame of A) in the same reference frame at a distance D' from A. Now, an observer B travelling at a speed V towards the object PA, will see the object and the observer moving towards her. Observer A passes observer B at time $T_0$=0 so that the distance of the object PA to B which we call X' is equal to X (in Newtonian physics) at time $T_0$=0. After a time T, the object would have appeared to have come closer to B and now she would see the distance to be D=D'-VT. In Newtonian physics, both observers A and B would agree. But, in STR, because of length contraction and time dilation, the Lorentz transformation multiplies this distance by a factor equal to $\sqrt{(1-v^2/c^2)}$, the inverse of which is usually denoted by the symbol $\gamma$. ($\gamma$ is always equal to greater than 1) and observer B would see this distance to be shorter equal to D=(D'-VT)/γ. The time is dilated differently in that the time elapsed in the frame of observer B as seen by observer A has additional dependences on the distance and the speed, and is divided by the above factor (multiplied by γ).

**Box 5.4. Lorentz Transformation Equations**

The equations for transforming the coordinates in an inertial frame (no prime) into a second one (with prime), moving at a velocity v with respect to the first, are, (Cartesian-x,y,z coordinates)

$$x' = \frac{x-vt}{\sqrt{(1-v^2/c^2)}};\ x = \frac{x'+vt}{\sqrt{(1-v^2/c^2)}} \quad (5.11)$$

$$y' = y;\ z' = z \quad (5.12)$$

$$t' = \frac{t-vx/c^2}{\sqrt{(1-v^2/c^2)}};\ t = \frac{t'+vx/c^2}{\sqrt{(1-v^2/c^2)}} \quad (5.13)$$

The length of a rod L' with ends at $x_1'$ and $x_2'$ can be obtained from the length L of the rod with ends at $x_1$ and $x_2$ (measured at the same time), and the contraction can be observed. In this, it is easy to be confused. The uncontracted length is in the moving frame and is equal to $x_2'$- $x_1'$ and the contracted length observed by the stationary observer = $x_2 - x_1 = (x_2' - x_1')\sqrt{(1 - v^2/c^2)}$.

The calculation of time interval is a little more involved.

The contraction of length (figure 5.4) and dilation of time are not some illusory tricks or metaphysical concepts. These have been observed and accounted for, in observations of fast particles. For example, in the laboratory, mesons, that have lifetimes much shorter than one second (microseconds to milliseconds), are observed in cosmic rays which originated from stars that are hundreds of light years away, taking hundreds of years to arrive on earth. This is because in the meson frame of reference this time is still very short, but as measured by the observer on earth, the clock for the meson runs slower compared to ours, since the meson travels at speeds close to that of light. From reading the history of STR, one would suspect that the contraction of the length of a moving object and the dilation of time in that moving frame, as seen by a stationary observer, would mean that the stationary observer should also find the cause for this difference. In other words, the stationary observer would see that forces, acting on the particles comprising a stick in moving frame, are consistent with a shorter stick. Detailed physics of forces of nature when treated relativistically, indeed show this difference.

One of the important components in the above derivation is the realization that there is no absolute simultaneity. This means that if one observer travelling at one speed sees two events occurring at the same time (simultaneous), another observer moving at a different (constant relative to the first) speed, would not see those events as simultaneous. This can be seen readily in the equations in Box 5.4, where the transformed time depends on spatial location. So at two different locations in the moving frame and same time in that frame, the time for a stationary observer at these two locations would be different. A natural question usually arises, if observer A sees that the observer B, travelling at a speed relative to A, has a shorter meter stick, then, since the length contraction is independent of direction of motion (velocity appears as square in the reduction factor), A's meter stick should appear to be shorter to B. How is this logical? The logic is acquired by realizing that the measurement of a length of a stick is obtained by taking the measurement of the position of the two ends of the stick at the same time. Since 'at the same time' means different things to A and B, (time is dilated with respect to each other) the measurements are therefore not the same. For a similar question about the time measurements as illustrated in the famed "twin paradox", see 'Introduction to Relativity' by William D. McGlinn, (McGlinn 2002). (One important point is to be noted – although the concept of time is elastic and relational, no observer can see someone's future). The lack of simultaneity and intermingling of space and time has far reaching implications to all of science and also the very foundations of philosophies.

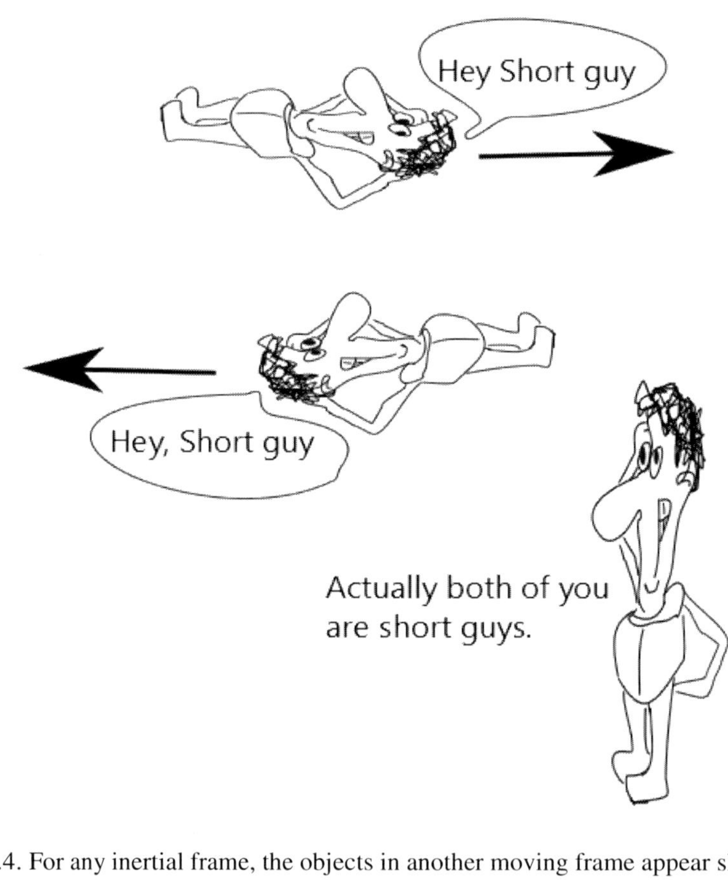


Figure 5.4. For any inertial frame, the objects in another moving frame appear shorter along the direction of motion.

John Polkinhorne (Science and Religion in Quest of Truth), points out that in the 'block universe', time and space are a block and there is no absolute past, present and future. This, he points out, is consistent with the fact that natural laws do not have a 'present moment' and this is not a 'limitation in physics', but a 'necessary denial of human experience'. (Polkinghorne 2011).

## Relative Velocity in STR

A very interesting outcome of the STR and the Lorentz transformation equations is that, while different frames see a moving object moving at different speeds (except for light) this difference is not the same as that in Newtonian physics.

Figure 5.5. Relative speed is not the sum of individual speeds.

(See figure 5.5). Let us say an observer A, who believes he is at rest sees another observer B traveling at a constant speed v in the direction x. The observer A sees a bullet fired in the x direction by the moving observer B and sees the bullet traveling at a constant speed u. In Newtonian physics, the moving observer B would see the bullet traveling at u-v. But, the velocity of the bullet u' observed by the observer B, according to the STR, is actually larger by a factor $1/(1\text{-}uv/c^2)$. If the bullet was fired at an angle so that it also had y and z velocities, even these velocities would be different for the two observers, because even though there is no Lorentz contraction in these directions perpendicular to the motion, there is the time dilation (slowing down of clocks in the frame of B, observed by A ). See Box 5.5 for the expressions of relative velocities. The derivation, given in Box 5.5, is accessible only to those who are familiar with differential calculus. (Please see the Appendix on primer to get familiar).

One can see an interesting and confirming result: If instead of a bullet, B fired a laser beam, a beam of light, then B would observe the speed of this beam to be c, the speed of light. When we substitute u'=c, we see that u is also equal to c. Both observers see the speed of the light beam to be the same. Speed of light is the same for all inertial frames, in agreement with the STR postulate. This also points to another important fact. By firing additional rockets from moving rockets one cannot keep increasing the speed of the successive objects as a linear sum each time because as it approaches the speed of light, the additional speed will be small and finally at near the speed of light, the rocket will have a vanishing increase in speed according to a stationary observer.

## MASS OF AN OBJECT AS MEASURED BY DIFFERENT OBSERVERS

Another spectacular game changing outcome of the STR is that different frames traveling relative to each other at different speeds obtain different masses for the same object. An example of this using a proper analysis is given in the Box 5.6 for ships travelling relative to each other. However, a heuristic result is given here. As stated before, all laws must be equally applicable in all reference frames. Conservation of momentum is one such law. If a collision between two objects takes place, all inertial frames irrespective of their own speeds, should see that the sum of the momenta of the two objects remains the same before and after the collision. If we check through the possibilities, we see that if the momentum of an object with a mass $m_0$ and speed v is defined, instead of the Newtonian definition as $m_0v$, but as $m_0v/\sqrt{(1 - v^2/c^2)}$, then this conservation holds. This essentially means that an object whose mass in its (rest) frame, is $m_0$, would have a mass $m_0/\sqrt{(1 - v^2/c^2)}$ or $\gamma m_0$ to the observer who sees that the object has a speed v. Since $\gamma$ increases as the speed of the object increases, the observer at rest would measure a higher mass for an object at a higher speed.

**Box 5.5. Relative velocities**

If x is the distance travelled by the bullet in time t according to the observer A, the corresponding values of the distance for the observer B, traveling at speed v, are x' and t'. The bullet is fired in the x direction and observer A and observer B measure bullet velocities to be u and u' respectively. For this, the Lorentz transformation equation gives,

$$x' = \frac{x - vt}{\sqrt{(1-v^2/c^2)}}; \; t' = \frac{t - vx/c^2}{\sqrt{(1-v^2/c^2)}} \qquad (4.14)$$

Differentiating the above (taking derivative with respect to t') with $\gamma = 1/\sqrt{1 - v^2/c^2})$

$$\frac{dx'}{dt'} = \gamma(dx - vdt)/(\gamma(dt - \frac{vdx}{c^2}) \qquad (4.15)$$

$$\frac{dx'}{dt'} = (dx/dt - v)/(1 - \frac{vdx/dt}{c^2}) \qquad (4.16)$$

Now, dx/dt=u and dx'/dt'=u', so that

$$u' = (u - v)/(1 - u.v/c^2) \qquad (4.17)$$

The velocities in the other two (y and z) directions are also affected because of time dilation.

$$\frac{dy'}{dt'} = \frac{dy}{\gamma(dt - \frac{vdx}{c^2})} \qquad (4.18)$$

With dy'/dt'=$u_y$' and dy/dt =$u_y$ ; $u'_y = \frac{u_y}{\gamma(1-uv/c^2)}$ and similarly for dz'/dt'.

In the debate between Socrates and Protagoras, the Sophist Protagoras said, "What is true for you is true for you, and what is true for me is true for me." (Sahakian 1993). The astonishing conclusion that inertial mass is relative, meaning that an observer would measure a higher inertial mass of a moving object compared to when it was stationary with respect to her and that the observed mass would be different for different observers, dismantles Newton's mechanics at various levels. The "relativistic mass", as some like to call it, is not an absolute quantity. The most evident of all is the fact that increasing force is required to accelerate an object, as its speed gets increasingly higher. Ultimately, as the object approaches the speed of light, the force required to accelerate it, tends to infinity (when v/c tends to 1, observed inertial mass tends to infinity). So this leads to the fact that, according to the postulates of STR no inertial object can exceed the speed of light. One can also note the consistency with relative speed and the previously noted fact that addition of velocities tends to the limit of speed of light. However, the rest mass $m_0$ is invariant, that is, it is the same in all frames of reference. (This is an important aspect of gravitation). (See figure 5.7).

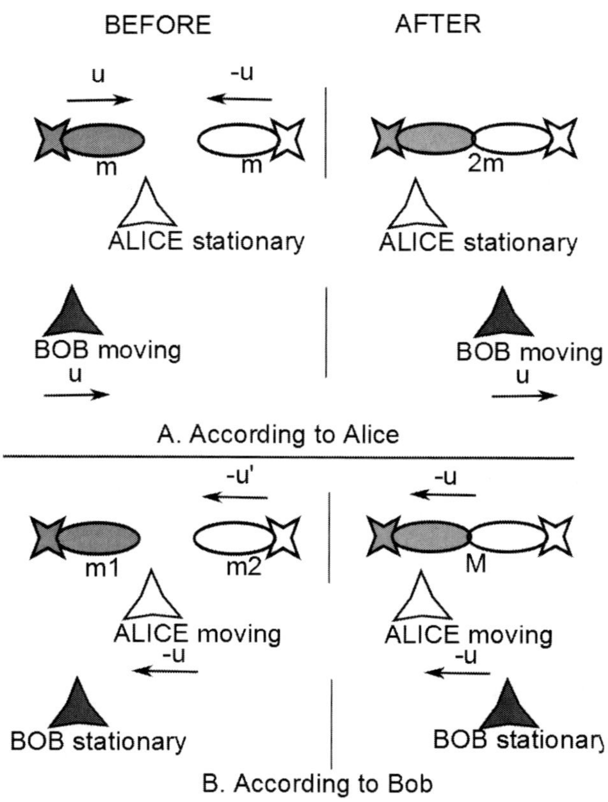


Figure 5.6. Dependence of observed mass on the speed. See Box 5.6.

**Box 5.6. Dependence of observed mass of an object on the observed speed of the mass**

Alice and Bob are observers watching space ships in their neighborhood, travelling at enormous speeds. (Figure 5.6). Alice decides to stay in one place and sees a gray ship coming towards her at a speed u. Bob goes to that ship, synchronizes his speed with the gray ship and travels alongside it. Both then see a white ship, otherwise identical to the gray ship coming directly at the gray ship to rendezvous with it. These future ships do not have to slow down, they just attach in what is called an "inelastic collision" to rendezvous, and in doing so the kinetic energy of the ships is absorbed in the process of docking and converted into stored energy.

*What Alice saw* – Alice sees that the white ship has the speed u, same as the gray ship, except that it is coming towards her from the opposite direction. Alice sees that once they rendezvous, they are joined together and they have stopped. When the ships were approaching each other (before the rendezvous), Alice measures the mass of the ships with a mass gun, and finds that they each have a mass m. She is satisfied that the total momentum of the two ships is conserved in the process, because the gray ship and the white ship each had a momentum m times u, but in opposite directions, so that their sum was zero, and since the ships stopped completely after the rendezvous, the momentum of the combined ships is also zero. But – she is puzzled. The mass gun should show that the mass of the combined ships should be 2m, but her mass gun is showing something less.

(Alice sees that Bob was travelling towards her with the gray ship before the rendezvous and is continuing to move in the same direction after the rendezvous).

*What Bob saw* – Bob, who is moving with the gray ship, sees that the gray ship is stationary, but a white ship is approaching the gray ship at some speed and then he sees that after the two ships rendezvous, the combined ship is moving at a speed of u in the former direction of the white ship and therefore Bob sees the combined ships as well as Alice receding from him at speed u. In communicating with Alice, he disagrees with her saying that the white ship did not have the same mass as the gray ship before docking. He says he is flummoxed because when he compares the momentum before and after, they don't agree. Like Alice, he is also confounded by the fact that the mass of the combined ships is not the sum of the masses of the two ships. He believes that some invisible space creature is feeding off the momentum and mass.

So there is a problem in each case. Both Alice and Bob are great experimentalists and know their instruments well and maintain them in perfect working order. So, why these problems?

## The Equations

Alice measures the velocity of the gray ship as u and white ship as –u before the rendezvous. She also measures their masses to be m each. After the rendezvous, she measures each ship's mass as $m_0$ so that the total mass of the combined ships is 2 $m_0$. The momenta of each of the ships is mu but in opposite directions, before docking. The net momentum before and after is zero. Bob has a speed u before and after the docking.

Bob measured the velocity of the gray ship to be zero and the velocity of the white ship to be –u', before the docking After the docking, he sees that the combined ship is travelling with a velocity –u. He measures the mass of gray ship to be $m_0$ and the mass of white ship to be m' and the mass of the combined ships to be M.

Alice and Bob come home and work on resolving these differences. Alice says thoughtfully, 'Bob, it looks like the mass of the ship we measure depends on the speed of the ship we measure, because I measured the mass of each of the ships to be $m_0$ after the rendezvous, when the ships were at rest and you too measured the mass of the gray ship to be $m_0$, before the rendezvous when for you the gray ship was at rest. Therefore we agree on something, namely that when the ship is at rest its mass is $m_0$." Bob nods in agreement and they try various combinations of multiplication factors to adjust the mass. Finally, Bob hits on the answer. He says 'I found it. If an object has a mass $m_0$ when it is at rest, then its apparent mass when it is travelling at a speed v is,

$$\text{m(apparent)} = \frac{m_0}{\sqrt{(1-v^2/c^2)}}.$$

Let us see if this works for Bob's case. Since Alice measured the mass of the two ships before rendezvous as m,

$$m = \frac{m_0}{\sqrt{(1-v^2/c^2)}} \tag{5.19}$$

After the rendezvous when both ships come to rest, each ship has the rest mass $m_0$. Alice is satisfied because she can account for the reduction in the mass after the rendezvous. She was already OK on the conservation of momentum.

But Bob has a complicated problem. The white ship was approaching him with a speed –u' when, according to Alice, he was travelling with the gray ship at a velocity u and since she saw the white ship also approach with the velocity –u, the relative velocity calculation (previous Box) gives the velocity of the white ship according to Bob as,

$$-u' = \frac{-u-u}{(1-u(-u)/c^2)} = \frac{-2u}{(1+u^2/c^2)} \tag{5.20}$$

Therefore the mass of the white ship before the rendezvous, according to Bob, is

$$m' = \frac{m_0}{\sqrt{(1-u'^2/c^2)}} \tag{5.21}$$

Substituting for u' and simplifying,

$$m' = m_0\{\frac{(1+u^2/c^2)}{(1-u^2/c^2)}\} \tag{5.22}$$

Therefore, the total momentum before the rendezvous = momentum of the white ship (gray ship is stationary according to Bob) = -m'u'

$$= m_0 \left\{\frac{(1+u^2/c^2)}{(1-u^2/c^2)}\right\} \cdot \frac{-2u}{(1+u^2/c^2)} \tag{5.23}$$

$$= \frac{-2m_0u}{(1-u^2/c^2)} \tag{5.24}$$

Again, according to Bob, the mass of each ship after the rendezvous (the velocity of both the ships are –u) is $\frac{m_0}{\sqrt{(1-u^2/c^2)}}$ and therefore the momenta of the two ships together and travelling at the velocity –u is $\frac{-2m_0u}{(1-u^2/c^2)}$, which is equal to the total momentum before the rendezvous.

Alice and Bob are thrilled at this revelation and go to a space bar to celebrate and tell others.

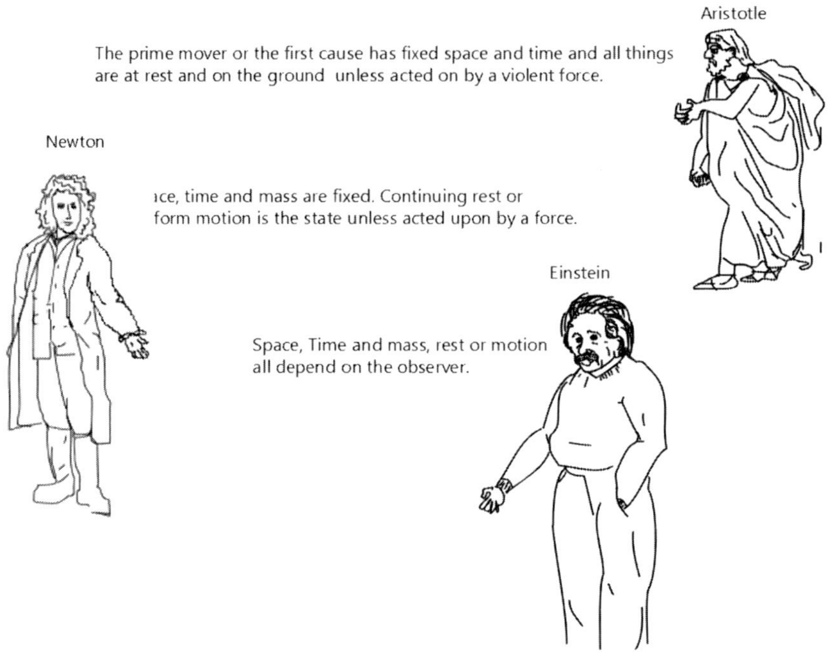


Figure 5.7. Progress in the understanding of mechanics, space and time.

## EQUIVALENCE OF MASS AND ENERGY

So, once the STR postulates are stated and the Lorentz transforms are applied, it was found that measured inertial mass is relative to the frame of reference. But an even grander result appears with relative ease, namely that mass and energy are equivalent, the law that truly made Einstein and the science of mechanics really famous even among general public. The principle of equivalence of mass and energy states that one can take account of the increase of observed mass with increasing speed and the increase of kinetic energy with increasing speed by taking an appropriate energy quantity. But, in order to sum the mass and kinetic energy in some fashion, the two terms must have the same units. Since the unit of kinetic energy is mass multiplied by the square of the velocity, the contribution to the new energy quantity from the mass requires the mass to be multiplied by the square of a standard speed, the speed of light.

The famous equation is, of course, $E = mc^2$ where E is the total energy of the object and m is the observed mass of the object and c is the speed of light. This is the sum of the energy of the rest mass, $E_0 = m_0c^2$, where now $m_0$ is the rest mass, and the kinetic energy of the object. The kinetic energy however has a different form in relativistic theory. The most general way of stating the equivalence of mass and energy is that the total energy E of an object moving with a momentum p is given by, $E^2=m_0c^2+p^2c^2$. With, $\gamma = 1/\sqrt{(1-\frac{v^2}{c^2})}$, $E=\gamma m_0c^2$ and $p=\gamma m_0v$ and the kinetic energy is $=(\gamma-1)\ m_0c^2$ where v is the speed of the object. In Newtonian mechanics, there is no knowledge of the energy associated with mass and is nearly unchanged in the Newtonian world of slow speeds, but the kinetic energy is given by $\frac{1}{2}\ m_0v^2$. In the limit of slow speeds of the Newtonian world ($\gamma\sim1$), one can show that $E -$

$E_0 = (\gamma - 1)m_0c^2 \cong \frac{1}{2}m_0v^2$. The derivation is given in Box 5.7 and is accessible only to those who are familiar with differential calculus.

**Box 5.7. Equivalence of Mass and Energy**

The momentum of an object, moving at a velocity v, is given by, p=mv, m being the mass and velocity of the object according to the observer. $m = m_0/\sqrt{1 - v^2/c^2}$, where $m_0$ is the mass when the object is at rest. Differentiating both sides we see that the change in mass dm corresponding to a change in velocity dv is given by,

$$dm = m_0\left\{-\frac{-\frac{2vdv}{c^2}}{2\left(1-\frac{v^2}{c^2}\right)^{\frac{3}{2}}}\right\} \tag{5.25}$$

$$= \left\{\frac{m_0}{\left(1-\frac{v^2}{c^2}\right)^{\frac{1}{2}}}\right\}.\left\{\frac{\frac{vdv}{c^2}}{\left(1-\frac{v^2}{c^2}\right)}\right\} \tag{5.26}$$

$$= \frac{mv.dv}{c^2-v^2} \text{ so that, } \frac{dm}{dv} = \frac{mv}{c^2-v^2} \text{ or } \frac{dv}{dm} = \frac{c^2-v^2}{mv} \tag{5.27}$$

Now, if the object is seen to be under a force F and its velocity and therefore momentum is changing and the object is observed to move a distance dx, then the work done = F.dx. But, according to Newton's law, force is the rate of change of momentum,

F= d(mv)/dt or the impulse =Fdt = d(mv)= vdm +mdv

Because of the conservation of energy, the kinetic energy dE gained by the object is equal to the work done by the force F,

dE =Fdx= Fdt.dx/dt = Fdt.v

or from the above,

$$dE = v^2dm + mvdv \tag{5.28}$$

$$= dm(v^2 + mv\frac{dv}{dm}) = dm(v^2 + mv\frac{c^2-v^2}{mv}) \text{ (using equation 5.27)}$$

Simplifying,

$$dE = dm.c^2 \tag{5.29}$$

Which essentially means that the kinetic energy dE gained can be seen also as a gain dm

$$E=mc^2 \tag{5.30}$$

With this new view, it is evident that bound up nuclear energies could be converted to masses and that release of energies into other forms could reduce mass. It is also evident how this equivalence spurred the discovery of nuclear energy and associated high energy physics. When a nucleus splits, some of the mass is converted into energy. So, when a powerful particle collider collides energetic particles, the kinetic energy (and the energy of the rest mass) of the particles becomes available for the creation of new particles heavier than the colliding particles. While in Newtonian mechanics, conservation of energy meant that the sum of kinetic and potential energy is conserved, in relativistic mechanics, the conservation has to include the energy associated with the mass in the system.

## 4 Vectors, World line and Proper Time

The Special Theory of Relativity (STR) spawned a new way of thinking about time and space and energy and momentum. In STR, space and time are united in a 4 dimensional space, for example, an event would occur at coordinates (t,x,y,z). That is, if we have a figure where the coordinates are on one set of axes and time is on another axis, each point in the figure would correspond to an event, occurring at one location in space and at a specific time. A curve through this point would represent the trajectory of events and is called the world line or worldline. The world line is a line in 4 dimensional (3 space and 1 time dimensions) space. Figure 5.8 illustrates the world lines for a 1 space dimension and 1 time dimension. A horizontal line (parallel to space coordinate axis) contains all the events at a given time and a vertical line, parallel to the time axis, represents all the events at one location. A constant velocity world line is a straight line (inverse of the slope of the line is the velocity, for axes shown in the figure) and curving world lines represent accelerating or retarding motion of objects. (Note that this is only for illustration. The accelerating frames of reference require a special treatment not available in STR, but included in the General Theory of Relativity). The speed of light has the limiting slope (minimum for the coordinate representation in the figure). An important difference between the time and the space coordinate is that while on the space coordinate one can move left (in the negative direction), one can only go vertically up (positive t) direction. (But one may be getting information from the past now and therefore the reverse direction in time is relevant). In normal usage, the units of distance are chosen so that the distance is divided by the speed of light. Therefore, with the vertical time axis being in seconds or years, the distance would be measured in light seconds or light years. In this case, the world line of light bisects the angle between the time and spatial axes. (For the case shown in the figure, it is 45 degrees). If the 2 space coordinates had been plotted on a horizontal plane and the time on the vertical axes, the light world line would be a light cone.

The proper time is the quantity related to the distance along the world line and is the same in all inertial frames (is an invariant). (Again the unit of distance is in time unit. To use usual distance units, the distance has to be divided by the speed of light).

$$\tau^2=t^2-x^2-y^2-z^2$$

One can confirm that this quantity is independent of the speed of the observer, by simple use of the Lorentz transformation equation. For example, in the above figure, the distance between two world events $(t_1,x_1)$ and $(t_2,x_2)$ is $(t_2-t_1)^2+(x_2-x_1)^2$. In relativistic description, the sign is changed and the difference in proper time $\tau$ is given by

$$(\delta\tau)^2=(t_2-t_1)^2-(x_2-x_1)^2-(y_2-y_1)^2-(z_2-z_1)^2$$

World line and proper time are key concepts describing the interaction between two observers travelling at relativistic speeds to each other. An example is that of two travelers A and B (see figure 5.9), Traveler B gets information from A's recent past but cannot get information from further past shown in dotted line.

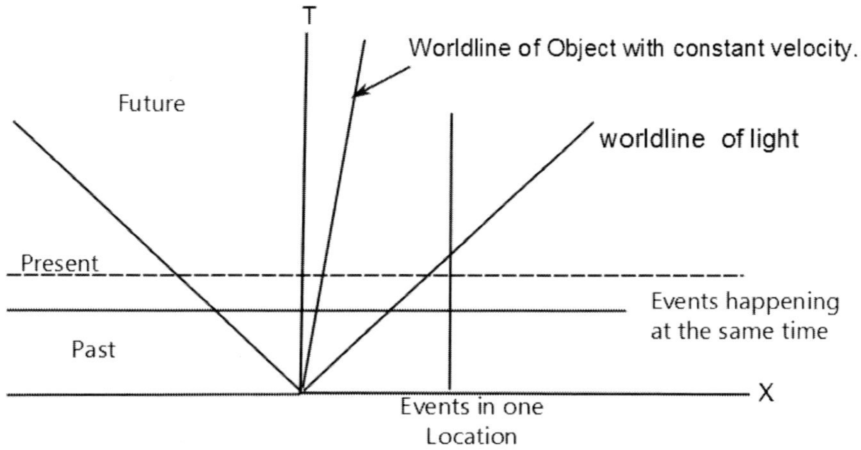


Figure 5.8. World line of objects and light beam and events in space and time.

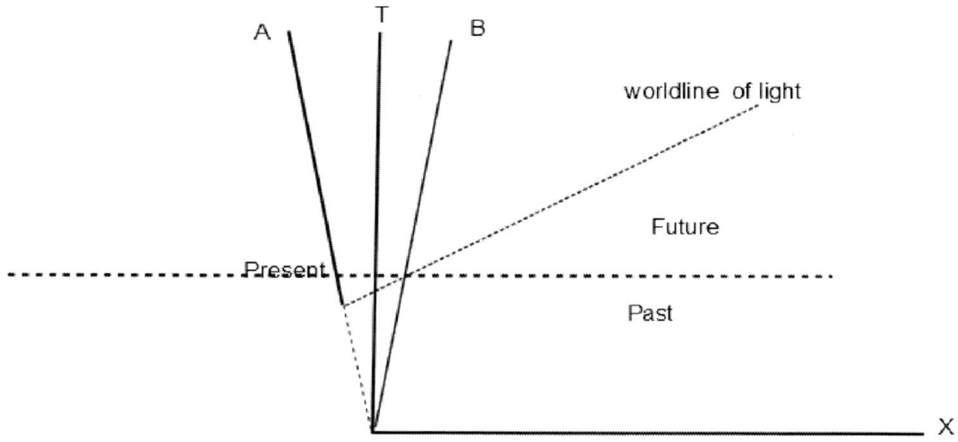


Figure 5.9. World line of two inertial frames moving at different speeds.

The world line diagram can also show two or more inertial frames (figure 5.10). The figure below shows the stationary frame (time and space axes are perpendicular). The world line of the object travelling at a speed v is tilted from the stationary frame time axis by an angle $\tan^{-1}(v/c)$ and is the time axis (T') in the travelling object's frame. The space axis of the object (X') also makes the same angle but in the opposite direction and therefore now, the time and space axes are not perpendicular. With the units remaining the same, the speed of light still bisects the angle between the time and space axes. With this diagram one can easily obtain the Lorentz transforms and see how, for example, Lorentz contraction of identical objects is seen by both frames. One can also see how two events are not simultaneous in the two frames.

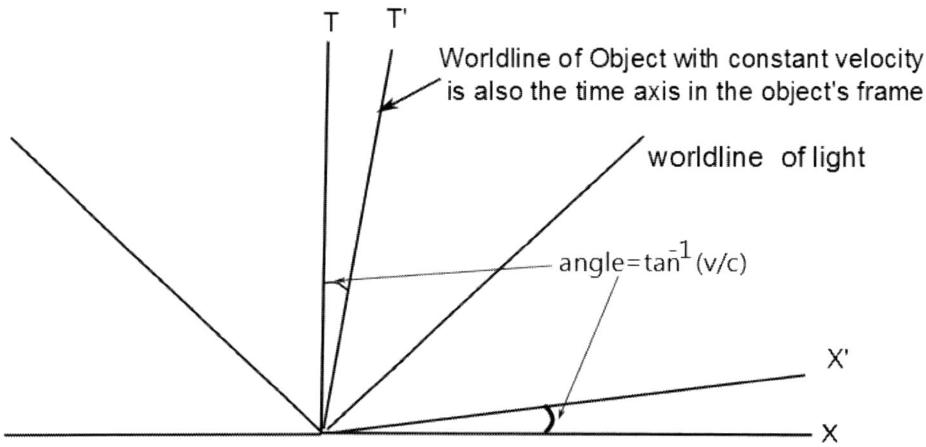


Figure 5.10. Reference axes of space and time of a second frame are tilted with respect to axes of the first frame. Even if the first frame's axes are perpendicular, the second ones axes are not.

The concept of 4 vectors of the space-time is also used for velocity or momentum. Using common units for space and time, the 4 velocity is $(1,v_x,v_y,v_z)$ where the spatial coordinates are in light time units. And the 4 momentum is $(E/c,p_x,p_y,p_z)$. These geometric representations were invented by Hermann Minkowski, (Albert Einstein's former teacher), and provide great insights into the special theory of relativity and in compact representations.

## An Assessment of the Mechanical Theories

The importance of this progression from Galileo to Einstein's STR in propelling science, thought and meaning of reality, must be evident to most readers. It is now good to take stock of all the definitions, hypotheses and postulates.

- Objects have mechanical inertia and the property assigned to this inertia is mass.
- Force is equal to the rate of change of momentum. – This Newton's law continues to be verified over several centuries. It follows that in a closed system with all forces balanced, total momentum is conserved.

- The fundamental laws (such as laws of Thermodynamics, Maxwell's equations, mechanical laws, laws of nuclear and particle interactions) are exactly valid in all inertial frames, moving or not with respect to any observer.
- Speed of light is constant in all frames of reference.
- The correspondence principle states that, for frames that are traveling at very low speeds, compared to the speed of light, with respect to an observer, the Newtonian form of the equations should be derivable from the SRT equations, with the approximation of neglecting terms with v/c and its higher powers. One can see from above that this is true.
- Mass and energy are equivalent (mass is another form of energy). Total energy is conserved.
- Mass, spatial distances and time are relative in that these depend on the speed of the observer relative to the object(s) of measurement.
- The rest mass and the proper time are invariant (same for all observers).

The Lorentzian view holds that the 3 dimensional space and 1 dimensional time (some call it ontological description), are embedded in a unique stationary reference frame and states that all reference frames are equal. But STR gave us a new basis for thinking about space and time. Beyond this, the discovery of phenomena like charges, fields etc. went on and will go on, but all those new discoveries are expected to be bound by the above postulates or hypothesis. If, a new discovery violates the above hypotheses or postulates, science will be the first to take notice. Indeed, scientists who are their own police, constantly try to subvert the hypotheses and postulates of STR, much like an expert hacker hacking into computers to explore its vulnerabilities. An example can be found even from recent news: Neutrinos are expected to have mass albeit very small and so they cannot travel at the speed of light and can only come close to it. In a series of experiments, neutrinos, streaming from the collision of protons in the Large Hadron Collider, were directed to the Gran Sasso Laboratory, which is 730km away. Originally, one team reported that the arrival time of the neutrinos at Gran Sasso showed that the neutrinos had exceeded the light speed limit. The whole world waited with bated breath and wondered about the potential of violation of STR and new breakout science. Indeed, if that were true, many systems of physics would have to be reviewed and many theories abandoned. After a few months of checking the data and the instruments, it was discovered that there was a faulty element in one of the fiberoptic time delay units and the rechecking gave a speed somewhat less than the speed of light, which is usual for all neutrino observations. Despite periodic attempts to "break" the STR and its founding principles, all evidence, accumulated over a century, points to the robust validity of STR and the validity of Newtonian approximation at low speeds (the correspondence principle).

This is also to say that no scientist would claim that STR is fully proven. For example, it is difficult to conceive of an experiment to measure the one way speed of light from differing frames, because of the problem of simultaneity (synchronizing clocks at two ends). However, there are clear indications about time dilation and Fitzgerald-Lorentz contractions and inertial mass increases in particle physics have been measured to a great accuracy. Yet, there is continued questioning on the internet about the validity of and assumptions of the STR, much of it, unfortunately, because the questioner has not done enough work and comes with a non-rational approach and agenda. But, it is safe to say that there has been a paradigm shift (The

Structure of Scientific Revolutions by Thomas S. Kuhn, (1962), University of Chicago Press) and a scientific revolution had taken place with STR. As Hermann Minkowski stated in his address to the 80th Assembly of German Natural Scientists and Physicians on September 21, 1908:

"The views of space and time which I wish to lay before you have sprung from the soil of experimental physics, and therein lie their strength. They are radical. Henceforth, space by itself, and time by itself, are doomed to fade away into mere shadows, and only a kind of union of the two will preserve an independent reality." (Lorentz 1953).

## Society's Response to the Theory of Relativity

Even after going through the details of relativistic space and time, one cannot help wonder at the unreality of it all. This is because, our minds are tuned by evolution, not to have to deal with this unnecessary detail for our survival in a Newtonian world. So, it is not surprising that we can readily "feel" that Newton's laws make sense and our minds do not readily assimilate the findings of the theory of relativity. Yet, knowing that, so far as has been demonstrated, these findings are correct, we admire the ability of these scientists to overcome the incomplete logic in our brains and find and describe these physical descriptions of nature. (Of course, as we saw, this need to describe nature which underlies our own existence is a result of our evolution too). While we can make this general statement on all of modern science's discoveries, the findings of the STR are in a category of their own, because these impact some fundamental notions and organizing of our lives. There is no question that we do not have to alter our lives, the way the Government functions or how we treat a disease etc., because of STR, but society and individuals must acknowledge the fact and function with that knowledge that how this universe came about and functions and what we do, incorporate the principles and consequences, discovered by the STR.

The perceptions of space, time and matter for people in the 21st century are very different from that for people from earlier centuries and millennia and dramatically different from prehistoric humans. This is because, the subject of space and time are, in many ways, abstractions. Because of fast airplanes, trains and cars, our perception of time and distance are much altered from the previous era. These concepts are not concrete to touch, but help us to see causalities, connections, targets and goals in our life and experience. For some, this is also exemplified in art. When we see a painting or a sculpture, we not only see matter, but we see an entangled space, time and matter. The famous painting 'Night Watch' by Rembrandt, tells us the story of a victorious army moving out of a battle field, after the Dutch asserted their independence from the Spanish Empire. When we see the painting, we see the place, the time and the event. So, the mind is capable of being an observer of other places and times that is not its own. Paul Laporte discussed this in his letter to Albert Einstein in which he talked about the relation between cubism and relativity. Though, Einstein himself saw tenuous connections with art such as Picasso's, Laporte argued that there was influence of the theory on the art, although the art creations do not have much to do with the theory itself. Einstein, himself, seemed to be unaware of the influences in art and literature around him.

The philosophy of space and time has a long history and has spawned a field of its own. From Plato's concept of time being defined by the motion of celestial bodies and space as that which contains objects, to Einstein and Minkowski's spacetime description, where they are mutable, this discipline inspires considerable discussion. Immanuel Kant held that space and time are nature's framework in which we sense things. There are those who believe that time is an illusion as in Hindu philosophy. The reality of time, its flow and the interplay between space and time (movement) is a subject of highly intellectual and often abstruse scientific-philosophical debates. This is further complicated by the findings in the General Theory of Relativity.

It has been claimed (Isaacson 2007) that Einstein's theory was responsible for a moral relativism and the appearance of modern art. But, it is hard to make the connection firmly and to conclude that the theory somehow affected people's thinking and perceptions. Most people, after all, could not understand the physics and could not connect to it. For example, the 4 dimensional descriptions where one of the dimensions is time, never reached the consciousness of people. However, higher (spatial) dimensional notions in art and literature might have been spurred on by the STR and its use of the terminology of 4 spatial dimensions. The understanding of time dilation, on the other hand, has had a significant influence. It is certainly true that a consciousness that things are not certain, may spring from a vague understanding of the STR. However, the society should be made aware that, while concepts are relative to observers, these are objective realities and are not due to uncertainties in observations. Today, although society does pay homage to the theory and may understand the theory a little better, it still does not see how the Theory of Relativity has advanced their lives, notwithstanding the fact that their own lives appear to be Newtonian. Today, the STR permeates many fields- particle physics, astrophysics, atomic physics and condensed matter physics. The technologies that depend upon these fields intrinsically, owe their development to the SRT. The Global Positioning Systems actually account for relativistic effects. As time passes, technologies advance and as we venture into deeper spaces, the role of the theories of relativity in our lives will increase.

## Impact of STR on Religion and Correspondence with Religious Beliefs

In the 5th century, St. Augustine remarked: "What, then, is time? If no one asks me, I know what time is. If I wish to explain it to him who asks me, I do not know." Clearly, considering that STR deals with observations and world view, theology and religious philosophies must have much to say about the Special Theory of Relativity and its consequences because theology deals with universal order, particularly the order of space and time and also about perception. The religious scholars have liked the "unitary" nature of the theory of relativity, namely the uniformity of its application (including General Theory of Relativity for accelerating frames), which conforms to theology. The universality of laws confirms to them the faithfulness and dependability of the Creator. The Scottish theologian Thomas Forsyth Torrance, who was a great believer in the potential for a great alliance between science and religion, has written a great deal about the intersection of Christian theology and the Theory of relativity (See for example, (Torrance1981)). According to him, the STR confirms the fact that the "invisible" explains the "visible". Torrance states that the relational nature of space is more acceptable because an absolute space would bind God, while relational space and time allows greater openness to God's creativity and gives scope for Incarnation. He also admires Einstein's concept of thought experiments and says that it validates the notion that the abstract can be understood by intuition and reflecting on experiences. An examination of interaction between relativity and religion has been carried out in the book by H.Douglas Anthony (Anthony 1927) and readers who want an extensive discussion of philosophical aspects may read this.

Much of the discussion of the theory of relativity and religion is connected to relativity of time. William Lane Craig, physicist-theologian, feels the Lorentzian picture with an aether is the more correct one, since, otherwise it would make God temporal. He maintains that there is a privileged inertial frame (Aether) from which God operates. He even quotes from Lorentz's letter to Einstein,

> "A 'World Spirit' who, not being bound to a specific place, permeated the entire system under consideration or 'in whom' this system existed and who could 'feel' immediately all events would naturally distinguish at once one of the systems *U, U', etc.* above the others." (Craig 2001).

Einstein himself said that the STR did not disprove the existence of aether, but only that it if it exists, it cannot be immobile and therefore is not the required medium with particles for light to travel. So, to paraphrase that, one does not need aether to explain physics. That is similar to the outlook of the atheists on the existence of God. However, it is somewhat disquieting to see theologians, Christians, in particular, identifying STR's application or applying its terminology to the God discussion. A very interesting example is the "Zero Thesis" of Leftow. Since God is omnipresent (and spaceless), His distance to a being is essentially zero. He also states "Local Simultaneity" in which all events are there for God. In another interpretation, theologians have stretched the concept of relativistic time to justify belief in afterlife and eternal life.

For theologians like M.W. Worthington, these theories are unsatisfying because they reduce the idea of eternity to unending space-time processes. R.J. Russell states, *"Relativity says there is no unique present. That would be inadequate to my experience of genuine openness, of novelty and of moral agency"* (Quoted in (Stannard 1996)). There is no question in any one's mind that ideas of "relativism" (a concept that points of view have no absolute truth and all scientific, social and moral ideas are only relative to these relative points of view) got a boost from the STR ( Baghramian, (2004)). There is a stronger opposition from the Roman Catholic Church to the principle of "relativism", since it would deny the existence of an absolute truth and this could lead to moral ambiguity and refusal to admit the existence of sin and God. This fear is correct, since STR does state that one cannot arrive at the "Truth" through one's personal experiences. The orthodox Christian theologian's view is that the denial of an absolute reference is a denial of God Himself and thereby an invitation to atheism. The less scholarly, more down to earth, conservative Christian religious groups view the theory of relativity suspiciously and some even (for example, the Christian Encyclopedia Conservapedia) consider it a "liberal plot". One also finds that, in order to discredit the Theories of Relativity, a shrill debate is carried on about the credit that is given to Einstein, instead of Larmor, Poincare' and Lorentz.

Hindus readily embrace this idea of varying temporal, spatial and material perception, in conformance with the idea of an illusory world embedded in the Advaita (non-duality) philosophy and mythology. There is a vague correspondence in Hindu scripture (Krishna advising the warrior king Arjuna),

> Arjuna, God abides in the heart of all creatures, causing them to revolve according to their Karma (Desires) by His illusive power (Maya) as though mounted on a machine.-*Lord Krishna in Bhagawat Gita, Ch. 18, Verse 61.*

Hindu sages variously state the following theme- The ultimate reality is absolute and undifferentiated and it is Maya (illusory or foggy perception or God's play) that creates space and time. The concept of different time for Gods and for humans is also imprinted in the Hindu mind. For example, for the creator Brahma. one day is equal to about 9 billion years for humans. (That conversion corresponds to Brahma traveling at a speed differing from the speed of light only by 1 in $10^{-24.}$). The time dilation concepts also correspond to Hindu mythological stories. In one, the King Kakudmi, goes up to Brahma in heaven taking his daughter Revati (since his daughter was so special that only a God was worthy to marry her.) He had to wait for Brahma, who was listening to a music concert. When finally he met Brahma, he told him his wish and Brahma laughed and said, while you were waiting here, many eons have passed on the earth and many generations would be gone when you return. He consoled him suggesting a worthy incarnation of a God who would be suitable for Revati when they returned. (This reminds one about the twin paradox in relativity). Such stories are a great delight to Hindus and they may even hold claim to this as the Hindu version of the theory of relativity, although there is no semblance of a scientific foundation of this in Hinduism.

The Taoist religion seems to have some thinking along the STR. From 120 B.C. onwards, the diagrams, taiji tu of the Great Ultimate, have been drawn to represent space (yu- four cardinal directions and above and below), time (chou – what is past and what is to come) and the supreme. In one part of a diagram by Chou Tun Yi, space and time are represented in the same diagram. To some extent, this represents the Chinese concept of fluidity in space and time and one can even see this mythically presented in the popular representation of eastern mysticism, particularly in Zen Buddhism.

The idea of relativism too is encapsulated in Hindu and Buddhist philosophies and the admonishments that world views and realities are subjective. The *Rg Veda* states that *"Truth is One, but sages say it differently"*. This actually means that even sages cannot glean the truth or that personal experiences are not a guide to Truth. Maya puts a veil on the absolute. This is in sharp contrast to Christian theology. The Buddhist notion of Maya is more nihilistic and does not truly correspond to the observer seeing different things, but only illusions. But, within the seemingly contradictory, yet philosophically sound teaching, Buddhism also teaches that there is no absolute either and all things are, therefore, relative. This is in total consonance with SRT. (Unlike in other religions, in Buddhism, Ultimate does not correspond with Absolute and the Ultimate is actually "emptiness").

It is claimed, (for example (Naied 2008)), by some Islamic scholars that the Quran has many elements of Special Theory of Relativity. Indeed, the Quran is also replete with how human time scales are vastly shorter than the time of heavenly beings. The passage *"But Allah will not fail in His Promise. Verily a Day in the sight of your Lord is like a thousand years of your reckoning". -Surat al-Hajj: 47- 22nd sura of Quran,* is an example of this. One statement that goes around a lot among Muslim believers is that the Quran had calculated the speed of light as 50,000 lunar years (1 day for angels), since that was how long it took angels to ascend to Him (Allah). (The angels and the Spirit ascend to Him in a day, the measure of which is fifty thousand years.- Quran 70.4) This relativistic time dilated calculation is applied to calculate the actual speed of the angels which is pretty near the speed of light! But there are no scholarly or philosophical discussions on the world view itself, perhaps because, as stated before, Islam and science have diverged as separate philosophies since the 11$^{th}$ century. However, among the ulama (teachers) of today, Einstein is held in high esteem and is considered to be a wise man.

## Concept of Time in Science and Theology

In addition to the above general observations, time is a particularly relevant quantity in both science and religion. Even with the defining of time and space as similar coordinates and plotting time from the present to the past, which in turn gives rise to notions of time travel to the past, in natural phenomena as well as in human consciousness, time flows only forward, whereas one can, by choice, go backward or forward in space. This comes about because of thermodynamics laws. Even if the universes were cyclic and the presently expanding universe shrinks back in the far future, it is extremely unlikely (even with theories that allow this) to have events recurring in the reverse just as they happened in the time forward direction. So, the saying, "Once in a life time" is true for every moment in the universe. With this in the background, it is hard to assimilate the fact that time is relative and simultaneity is not absolute. However, it is important to realize that if the clocks were synchronized in the universe at the start of the universe, each inertial frame travelling at different velocities have experienced different time intervals. But, it is also true that each of the frames, if they watched, witnessed (or will witness) all the events that another frame had watched. All memories and observations in any frame are of the past and the present, which are common to all. These just happened at different paces of time.

For the scientific theologian, this gives solidity to theologies. Theologians conclude that only God knows all times and events and the fact that different observers see different rates of happenings, is proof to them that God is the One Absolute and all worldly things are not. One of the thorny points, in which the Bible comes into conflict with science, is in Genesis which states that the world (universe) was created in six days, whereas, it is now established that the universe is about 14 billion years old. According to Gerald Schroeder, a physicist and a Jewish scholar, this can be explained by time dilation. The ratio of 14 billion years to 6 days is about 1 trillion. (A variation of this is to change 14 billion years to 6000 years which is how long ago, the Universe was supposed to have been made according to Genesis). In the book "Science of God", he claims convergence of Genesis with science using the theory of relativity (Schroeder 2009). A calculation of this kind would give a speed of the frame of God to be even closer to the speed of light, than the Quran, which only gives a ratio of 50,000 years to 1 day, compared to about 2 billion years to 1 day in Genesis. Schroeder, however, goes on reconciling each day with the scientific explanation of the history of the universe according to particle physics and cosmology, which, to this author, is a little too much. Many theologians (Cyril Domb and Alan Padgett, for example) have used time dilation in this and other contexts in describing God's existence and knowledge. (A question that does not seem to have been pondered is- why is God's frame special when speed is relative? God should find our time dilated. One possible explanation that they can give is that of the answer to the twin paradox, where if one observer stays put in one place, the other who travels will return to find that a long time has passed at the stationary observer's frame. So the implication is that God never leaves His frame, while all else travel between His and another.)

There is one concept which is dear to religion from the earliest known recorded history- Eternity. Even Plato said that God lives in eternity and yet is present at every event. The same is expressed by Thomas Aquinas, who stated that all things are present to God from eternity. This means that, to God, all events are simultaneous. Richard Swinburne (Swinburne 1977), objects to this on the basis stated above that all things do not happen simultaneously for anybody or anything. He asserts that this is only the limit of our knowledge but not of its existence. (Quoted in (Craig 2001)). Brian Leftow, a theist philosopher, has attempted to include STR in the description of the timelessness of God, by using the idea of Eternalism or Stasis theory, where all times really exist eternally and are only experienced sequentially by mortal beings:

> "An event occurs in eternity simultaneously with all other events, but this does not entail that event occurs simultaneously with all other events in any other reference-frame. Rather, in eternity, all events occur at once, and they occur in sequence in temporal reference frames. Events are present and actual all at once in eternity, but present and actual in sequence in other reference frames." (Leftow 1990).

There are several sophisticated arguments about the coexistence of three types of simultaneity, temporal, eternal- present and temporal-eternal (Morris 1987). E. Stump and N. Kretzmann give a nice analogy (Stump and Kretzmann 1987). One can see from the transformation equations that time stands still in a frame moving at the speed of light. (Incidentally, the concept of stationaryness in time is an attractive piece of physics used in description of events at the horizon of a black hole). So, if we draw a line at a certain speed and we put a dot of light on it, each position of the dot would then represent a moment. But if a second line were drawn at the speed of light, the dots in first line would be a strip of light present at every moment of the second strip (time dilation is infinite). (This is confusing, because due to length contraction, a dot of finite size would appear to shrink to nothing for observers not at the speed of light). A world line based argument to show that for a frame all events can be simultaneous is used by Ralph G. Mitchell (Mitchell 1987).

But, what is missed out by these authors eager to relate STR to eternity and God, is that the events are not causally connected and the fact that the past, the conditional present in which simultaneity can be discussed and the future are the same for all frames. So no frame is exempt from this knowledge, memory and lack of knowledge (Padgett 1992). As Max Jammer states in *"Einstein on religion"* (Jammer 2002), *'The idea of God's timeless eternity or of timeless mode of divine existence has been conceived in order to resolve the incompatibility between divine omniscience (including foreknowledge) and human free will."*

This issue of human free will is also a deeply dividing topic among philosophers in the context of STR. According to C.W. Reitdijk, there is no human free-will because, irrespective of the event, *" Each event is determined. It is already past for someone in our now."* (Reitdijk 1966). However, process theologians (the well- known mathematician-philosopher Alfred N. Whitehead is one), who profess that God just created the process and gave free will to all to determine their future according to the process, have no difficulties with determinism. Even

Aquinas stated in that God does not have foreknowledge of contingent future. *(Divi Thomae Aquinatis Summe Theologica I*, Question 14, Article 13, Senatus, Rome, 1886).

Most of these discussions have been held by Christian theologians and any discussions in other religions come nowhere close to the depth, seriousness and earnestness of those by Christian scholars. There are probably many reasons for it. One might be that science itself is a trivial worldly pursuit for some theologians and not worth engaging in. The other reason could be that these theologies are vague and the scriptures are not dogmatic or descriptive or prescriptive and are inclusive of liberal ideas (as in Hinduism). So, for these theologies, scientific reasoning is acceptable and can be accommodated without questions. There are, however, alternative viewpoints, albeit vague, where Hindu theologians take objections to science based on their own interpretations and claim that the scriptures are already cognizant of these ideas. Most Hindu philosophers and theologians have not made serious commentaries on the STR other than to draw the parallels in mythology. An example is the blog article by Gadadhara Pandit Dasa in the Huffington Post (September 20, 2011). This is also true for many Islamic scholars.

However, Hindu and Christian theologians come together in happily finding a device in the equivalence of energy and mass in order to explain Incarnations and Avatars, where God uses energy to create his own images. Even Hindu God- men (women) explain such materializations (rather flippantly). The equivalence also helps in explaining the *ex nihilo* (from nothing) creation of the universe. However, there is some confusion as to what energy the religions talk about when mentioning the spiritual nature of beings (say in creating them, providing the so called essence).

More will be said about space and time and, in fact about space-time, in another chapter. The final word on space, time and mechanics will not be written. Science promises progress, sometimes revolutionary, sometimes incremental, but always illuminating and affirming the wondrous structure of the Universe. All the above discussions prove that religions are accepting the theory of relativity and that theologians are only wondering how religious teachings and scriptures can be interpreted to conform to the scientific findings. This is indeed a victory for science as well as the coming of age of religions. For the most part, after an early false start, religious views appear to have adapted to the present scientific understanding of space and time, except among a minority of people. So, for an average person of intellect and reason, findings of the science of mechanics can be a powerful guide and religious faith will not come in the way of learning and using this knowledge.

Max Born stated,

> "I have tried to read philosophers of all ages and have found many illuminating ideas but no steady progress toward deeper knowledge and understanding. Science, however, gives me the feeling of steady progress: I am convinced that theoretical physics is actual philosophy. It has revolutionized fundamental concepts, e.g., about space and time (relativity), about causality (quantum theory), and about substance and matter (atomistics), and it has taught us new methods of thinking (complementarity) which are applicable far beyond physics". (Statement of 1963, as quoted in (Moore 1992)).

The Special Theory of Relativity is a more accurate and nuanced revision of Newton's mechanics that has been used to explain phenomena, but in the process, has raised profound philosophical questions. Indeed, as the above narrative shows, it is the new thinking that the progress from Galileo to Einstein, Lorentz, Larmor and Poincare's relativity theory enshrines and gives new perspectives on existence itself. The fact that physicists could arrive at this new thinking with no hidden agenda or dogma, with hard work of intellectual inquiry and logical and open conceptual and analytical approach, shows how science progresses for the benefit of mankind and how this knowledge has and will become part of us.

## REFERENCES


Anthony, H.D.; *Relativity and Religion,* University of London Press, London, U.K. (1927).

Baghramian, M.; *Relativism,* Routledge, London, New York, (2004).

Brown, H.; *Amer. J. of Physics*, Volume 69, Issue 10 (2001) pp. 1044.

Craig, W.L.; *God, Time, and Eternity: The Coherence of Theism II: Eternity*, Springer, Berlin, Heidelberg, Germany (2001) p.216.

Craig, W.L.; *Time and the Metaphysics of Relativity* (*TMR*), Springer, Berlin, Heidelberg, Germany (2001).

Isaacson, W.; *Einstein: His Life and Universe*, Simon and Schuster, New York, NY (2007) p.104.

Jammer, M.; *Einstein on religion*, Princeton University Press, Princeton, NJ (2002) p. 177.

Kox, A.J.; *The Scientific Correspondence of H.A. Lorentz*, Vol. I, Springer, Berlin, Heidelberg, Germany (2008) p.43.

Kuhn, T.S.; "*The Structure of Scientific Revolutions"* University of Chicago Press, Chicago (1962)p.12.

Leftow, B.; "*Aquinas on Time and Eternity", Amer. Catholic Philosophical Quarterly,* Vol 64, No.3 (1990) pp. 387-99.

Lorentz, H. A., Einstein, A., Minkowski, H. and Weyl, H.; *The Principle of Relativity: A Collection of Original Memoirs on the Special and General Theory of Relativity*, Dover, New York:, NY (1953) pp. 73.

Maxwell, J.C.; "*A Dynamical Theory of Electromagnetic Fields"*, *Phil. Trans. Royal Society of London*, Vol. 155 (1865) pp. 459-512.

McGlinn, W.D; *Introduction to Relativity*, Johns Hopkins University Press, Baltimore, MD (2002).

Mermin, D.; *Space and Time in Special Relativity*, McGraw Hill, New York, NY (1968).

Mitchell, R.G.; *Einstein and Christ*, Scottish Academic Press, Edinburgh (1987).

Moore, W.J.; *Schrödinger: Life and Thought*, Cambridge University Press, New York, NY(1992).

Morris, T.; *The Concept of God*, Oxford University Press, New York, NY (1987).

Naied, O.K.; *Relativity and Islam*, Lambert Academic Publishing, Saarbrücken, Germany (2008).

Padgett, A.G.; *God, Eternity and Nature of Time*, St. Martin's Press, N.Y. (1992)).

Polkinghorne, J.*; Science and Religion in Quest of Truth,* Yale University Press, New haven, CT (2011), pp. 62-62 & 98-99.

Reitdijk, C.W.; *"A rigorous proof of determinism derived from the Special Theory of Relativity"*, *Philosophy of Science,* Vol. 33 (1966)p.341.

Sahakian, W.S. and Mabel Lewis Sahakian, M.L.; *Ideas of the great philosophers*, Barnes & Noble Publishing, New York, NY (1993), p. 28.

Schroeder, G.L.; *Science of God*, Free Press, New York, NY(2009) p. 49-56.

Shanker, S., Marenbon, J. and Parkinson, G.H.R.; *History of Philosophy*, Routledge, London, U.K. (2003) p.76.

Snobelen, S.D.; *"Isaac Newton, heretic: the strategies of a Nicodemite",* Brit. J. History of Science Vol. 32, Issue 4 (1999) pp. 381–419.

Stachel, J.; *Einstein From B to Z*, Birkhauser, Berlin, Germany (2002).

Stannard, R.; *Science and Wonders*, Faber and Faber, London (1996) p.138.

Stump, E. and Kretzmann, N.*; "A temporal duration", Journal of Philosophy*, Vol. 84 (1987) p. 214.

Torrance, T.F.; *Christian Theology and Scientific Culture*, Vol. 1, Editor: Thomas F.Torrance, New York: Christian Journals, Belfast, Ireland (1981).

Torretti, R.; *Relativity and Geometry*, Pergamon Press, New York, NY (1983)p.83.

*Chapter 6*

# COSMOLOGY:
# HOW DID OUR UNIVERSE COME ABOUT

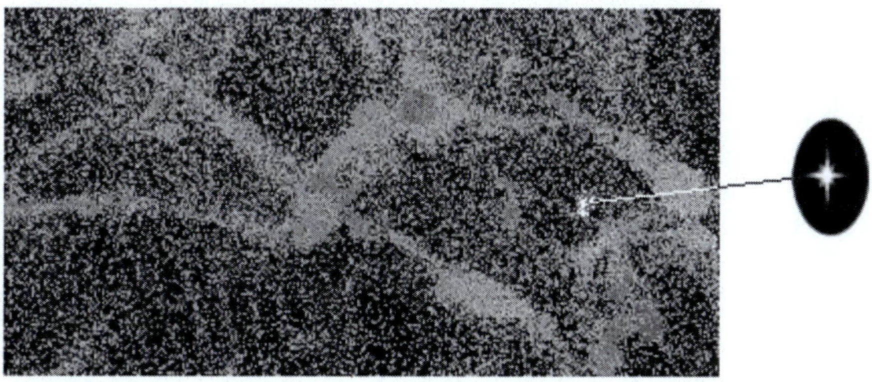

By faith alone do we hold, and by no demonstration can it be proved, that the world did not always exist.... The reason of this is that the newness of the world cannot be demonstrated on the part of the world itself.... Hence that the world began to exist is an object of faith, but not of demonstration or science. And it is useful to consider this, lest anyone, presuming to demonstrate what is of faith, should bring forward reasons that are not cogent, so as to give occasion to unbelievers to laugh, thinking that on such grounds we believe things that are of faith." (Aquinas 1952).

A passionate statement indeed! Islam, which adopts the Judeo Christian Creation mythology, states it differently –

"Do not the Unbelievers see that the heavens and the earth were joined together (as one unit of creation), before we clove them asunder? We made from water every living thing. Will they not then believe?" (Quran Sura 21, Verse 30).

In contrast, Hindu scripture *Rg Veda* states-

"The Ordainer created the sun and moon like those of previous cycles" (Rg Veda 10.190.3). The 6th century sage Sankara interprets this as, "This shows the existence of earlier cycles of creation, and hence the number of creation cycles is without beginning." (Brahma Sutra Bhashya 2.1.35)

Understandably strong and eloquent statements by the Lateran Council presided by Pope Innocent III in 1215, dissuaded any serious attempt by theologians, at a scientific exegesis of the book of Genesis. There must have been similar passionate defense of the individual creation theologies in the history of other religions. It is safe to say that cosmology is the topic if you want to have a hot debate between religions and between science and religion. Yet, it is surprising how physics, with its no nonsense approach has arrived closer to what might be the truth and how different religions are scrambling to adjust their theologies to the scientific discoveries or claiming as their own.

The creation myths are many, but major theist religions such as Christianity, Judaism, Islam and main stream Hinduism, have always postulated a perfect omnipotent God who created the Universe. (The agnostic Hindu *mimamsas* (Investigations) plead lack of knowledge on the subject of creation). Buddhism believes in a cycle of creation by a primordial wind and destruction by a great fire. But whatever these are, the creation theology or mythology is at the core of any theology, faith and worship. Even where non-religious spirituality is manifested, and even when consciousness is attributed to neuro-biological activity, the existential questions of the self and of the Universe are central to that spirituality. Cosmology is at the core of science too, which finds that the smallest particles and the largest Universe are the cause of each other - the *Ouroboros* bites its own tail. (Jayakumar 2012). The unmanifest "nothing" (vacuum) and the manifest matter interact in ways to produce myriad possibilities. Space and time become space time or space becomes time. The Universe is observable, but may also be unobservable. There may be one Universe or there may be many. This is stuff for great minds as well as lay people.

Though cosmology has been a field of scientific study from Plato's time, it was only a few decades ago that the field turned into a juggernaut with the arrival of Einstein's General Theory of Relativity (GRT) and later, observations of distant galaxies and celestial objects with precision. The GRT spurred a great debate on the nature of the Universe, whether it was static or evolving. Astronomical and astrophysical observations of the mass distribution and the expansion of Universe, the measurement of the so called cosmic microwave background and further observations on the rate of expansion have made this field explode into a mind boggling array of theories, observations and measurements. The attempts to integrate models of cosmology with particle models, and quantum mechanics with the GRT, have brought together great minds that grapple with complex theories involving spacetime geometry, quantum mechanical effects in the early Universe and thermodynamics of the Universe. One could believe that there is no greater challenge to mankind's intellect than the one to understand the scientific description of the Cosmos that is, for all practical purposes, beyond the reach of human senses. The spectacular success of physical cosmology in arriving at where it is today is a testament to the methods of science in facing this challenge. If anything illustrates that science takes the high road to necessary knowledge, it would be the discoveries in the field of cosmology. As in other sub fields of physics, the right way is found and further refinements bring us to a greater understanding, starting from the minimum necessary aspects of a problem. This chapter will give an indication of the type of approach physicists have been taking to arrive at the description of the Universe.

While the proper way to treat this introduction might be to give an account of the Gravitational theories first, it is more fun (and far less mathematics) to give the present understanding of the observable Universe, in an accessible way. (There are physics speculations and theories on the unobservable Universe too). Some concepts will have to be presented in advance in order to set the stage. We will, so to say, let the "cat out of the bag" and then try to describe the cat, as inferred from different theories and observations. Even with this advance notice, readers will clearly get the sense that the science, in this case cosmological physics, proceeds with careful integrity, exploring variations and making sure that no ground is left uncovered. No sleight of hand or assumption beyond that which is consistent is made. The description below is only illustrative and the actual cosmological theories are far more rigorous in their questions, analyses and verifications that characterize the discoveries. The following are the general properties of the Universe:

## The Cosmological Principle

The simple model that started off the description of the Universe is that the large scale Universe is remarkably simple on average, homogeneous (same energy, matter density etc. across the space and time) and isotropic (Universe looks the same in all directions). This leads to the so called Cosmological Principle that the Universe looks the same for all observers. While this was a simplifying assumption for the early theories, fortunately for the theorists the assumption has been confirmed by observations. This statement is true for distances greater than a few hundred million light years. (1 light year is slightly less than $10^{13}$ kilometers). The observed inhomogeneity and anisotropy over the observable Universe (one or more billion light years) is about 0.01%, but at smaller distances of less than 100 million light years, the variations at a given distance in different directions increases to about 30%. This is seen quite clearly from star densities and motions and so forth. But the most compelling evidence for the Cosmological Principle lies in the fact that the light that was created at the start of the Universe is highly uniformly spread in the Universe even now. Even more importantly, the temperature of this radiation is uniform with a variation of only 0.001%. (Radiation has a temperature which defines how the radiation intensity is distributed over the radiation wavelength. Hotter radiations have higher intensities at lower wavelengths than colder radiation). For cosmological models, the Principle permits less complicated models and a simpler physics understanding, yielding simple and elegant relationships, which are then debated, analyzed and verified or not, by observations.

### Universal History and Future

For the purposes of cosmological models, to the zeroth order, it is not necessary to include the sub-hundred million light year phenomena, such as star and galaxy formations and interactions. The Universe can then be treated as a single evolving object with specific features and phenomena and small variations in them. However, it should be noted that the evolution can be very different in each galaxy and region.

# EXPANSION

The Universe is expanding like a stretching rubber sheet. If we were to measure the stretching of a rubber sheet at several positions on the rubber sheet, these would be found to be proportional to the distance of the measurement positions to a chosen reference point. (See figure 6.1). The expansion carries the name Edwin Hubble on account of his publication in 1929.

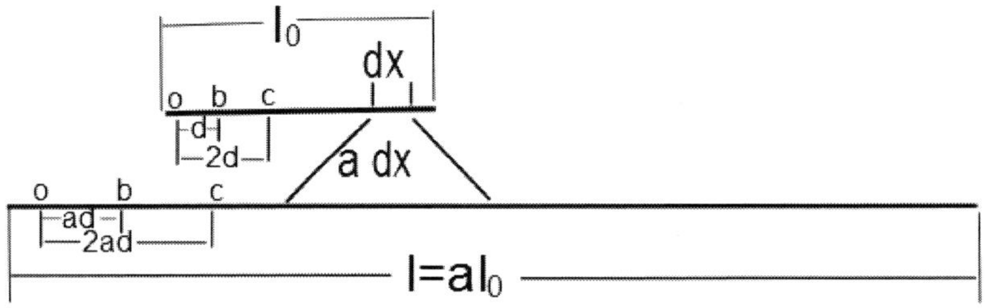


Figure 6.1. Expansion of the space (Universe) is like stretching an elastic string. The stretch is proportional to the distance from (any) reference point.

The physical observation of stars moving away at rates proportional to the distance went hand in hand with theoretical expectations. This is important to note. Where science, particularly physics is concerned, ideas rarely arise in isolation or by decree or a revelation out of the blue. (But leaps in thinking can result when a problem has been mulled over). So too is the discovery of the expansion of the Universe. In the year 1917, a decade or so before Lemaitre's and Hubble's observations, Willem de Sitter suggested solutions to Einstein's cosmological-gravitational equations that included a flat and expanding Universe with a constant expansion rate. (de Sitter's model did not specify any relation between speed and distance (expansion rates), although a solution like the Hubble expansion is included). His solution was a specific case of a vacuum dominated Universe with negligible mass and pressure. Such a model would result in an ever expanding Universe. In 1912, Vesto Slipher, at the Lowell Laboratory, Arizona, had discovered that distant galaxies exhibited spectral lines that were red shifted (wavelengths shifted to the higher -red side). A red shift of wavelength is a "Doppler shift" caused by the motion of the object away from the observer, because the light wave is stretched by the motion. (This is similar though not identical to the change in the whistle tone as it moves towards, passes and moves away from an observer. When the train approaches, the whistle pitch would be increased and after it passes, the pitch would be decreased. The change in frequency, in the case of sound, is proportional to the speed). This important discovery indicated that stars were moving away from us. Theorists such as Hermann Weyl, Ludwik Silberstein, Georges Lemaitre and Howard P. Robertson concluded that these observations could be consistent with the Cosmological Principle of all observers observing the same, which then automatically means that the expansion rate versus distance from the observer would be the same for all observers, whatever their location. That leads to the rubber sheet stretch analogy and then to a Hubble style expansion. (In theory of elasticity, it is called the Hooke's law) and hence the suggestion by Robertson that the speed of expansion would increase linearly with distance for all observers. (Before Robertson's suggestion, Carl Wilhelm Wirtz had noted that the observations probably show that the red shift and therefore the galactic velocities increase with distance. Kurt Lundmark also observed the same trend). The description of the observational confirmation of the Hubble expansion is given below. Edwin Hubble's data is used for illustration because it is better documented.

Incidentally, the confirmation that the Universe is expanding solved the problem of what is known as Olber's Paradox. In the early 19$^{th}$ century, well before the concept of expansion was proposed and when the Universe was deemed to be static and infinite, Heinrich Wilhelm Olber asked (as some before him had also asked)- if the Universe was indeed static and infinite with infinite stars, why then is the night sky dark? Light intensity drops off as the inverse square of the distance. If we were to think of the space of the Universe as consisting of a spherical shell of a given thickness, say a hundred million kilometers we would get the light from say N number of stars from the first shell, but the light from the second shell would be a quarter of the intensity. However, since the surface area increases also as square of the distance, it would have 4N stars and therefore contribute the same amount of light as the first and so on. If the distance were infinite with infinite stars, we should have infinite amount of light, day or night. Curiously, the solution, offered by Lord Kelvin, was anticipated by Edgar Alan Poe in his prose-poem "*Eureka*" (Poe 2006),

> "No astronomical fallacy is more untenable, and none has been more pertinaciously adhered to, than that of the absolute illimitation of the Universe of Stars. The reasons for limitation, as I have already assigned them, à priori, seem to me unanswerable; but, not to speak of these, observation assures us that there is, in numerous directions around us, certainly, if not in all, a positive limit—or, at the very least, affords us no basis whatever for thinking otherwise. Were the succession of stars endless, then the background of the sky would present us an uniform luminosity, like that displayed by the Galaxy—since there could be absolutely no point, in all that background, at which would not exist a star. The only mode, therefore, in which, under such a state of affairs, we could comprehend the voids which our telescopes find in innumerable directions, would be by supposing the distance of the invisible background so immense that no ray from it has yet been able to reach us at all. That this may be so, who shall venture to deny? I maintain, simply, that we have not even the shadow of a reason for believing that it is so."

Brilliant indeed for a non scientist! The Scottish mathematician- scientist did a proper analysis of this and calculated that the distance over which one has to include starlight that would give a bright night sky is 3000 trillion light years, large by any standard. This essentially means that even a pretty large but a finite Universe would not give a bright sky at night. While Poe's solution that the Universe is not infinite is correct, a greater important reason is that the Universe is expanding. The second part is necessary, because before there were stars, the Universe was filled with very bright radiation. The expansion of the Universe vastly reduced intensities and stretched the wavelength to what is now infrared (invisible). The amount of energy required to fill the sky is just too large and the Universe does not have that much energy now.

A very good athlete, a Rhodes Scholar and a law degree holder, Edwin Hubble settled down at the Mount Wilson Observatory in Pasadena, CA, after serving in World War I. He worked there till his death at the age of 63 in 1953. He had the good fortune that a 100 inch, then the largest in the world, "Hooker" telescope had just been installed at the observatory. With this telescope, the red shift (or blue shift) of the star light could be measured and more or less unequivocally showed the speed away from (or towards) the observer. On the other hand, astronomers were not very good at estimating the distance to the stars in the 1920s. So, for a long time, astronomers did not know that the blotchy and fuzzy stars were actually galaxies far away from the Milky Way. It was Hubble who recognized that. While contributing to a great many observations, he audaciously announced that there were other galaxies beyond the Milky Way. An ambitious man, Hubble first gave this story to New York Times and only then did he present it at a conference. (See "*Hubble Reveals We Are Not Alone*", New York Times, Dec. 30, 1924). This really stirred up the scientific community. One can only imagine how the religious community reacted to the news that not only was the Earth not the center of the Universe, neither was our Sun, nor even our galaxy itself. It is also revealing that humanity came to know about the existence of other galaxies, only less than a century back.

Hubble studied the so called Cepheid variable stars in detail. In 1929, he combined the data on the red shift (caused by the motion of the star away from us) and luminosity (a measure of the distance) from various sources, including those of his own from Cepheid variables, key data from Sliphers on spiral nebulae, and those of Milton L. Humason at the Mt. Wilson Observatory. (Here is a spectacular story of a physicist, which probably could not happen now. Humason was a "mule skinner" at the observatory, carrying materials to and fro from the mountain when the observatory was being constructed and later became a janitor. Seeing how hard working and astute he was, George Hale, a senior staff at the observatory recruited him to become an observatory staff member to actually carry out stellar observations. In addition to contributing to Hubble's data for his 1929 and the more convincing 1931 papers, Humason discovered a comet). Hubble's results showed that the stellar red shift increased with distance linearly. He published his findings in Proceedings of the National Academy of Sciences (Hubble 1929). (Puzzlingly or deliberately, in this paper, Hubble does not acknowledge Slipher's or Humason's contribution to the data. In a later publication by Hubble and Humason, Slipher's data were well acknowledged). Figure 6.2 shows a primary result in the paper.

Though some were skeptical of this initial data, most could see that at the low end and at the upper end, the data points to increase of speed (redshift) with distance. So, this data was a major discovery showing that the Universe is not fixed and is actually expanding. Some people just yawned and said- "So what else is new? The fastest galaxies and stars are farthest away, Big Deal!" The data has been steadily improved upon with the addition of more and more data points. In 1948, there were data for only 10 clusters and another 200 stars and galaxies, but now the data includes thousands of stellar sources.

This discovery had first been made two years earlier than Hubble in 1927, by a Belgian priest-astronomer Georges Lemaitre. He actually obtained even the rate of expansion inferred by Hubble, in his French publication. When the paper was translated in 1931 in the monthly notices of the Royal Astronomical Society, it carried the ideas of an expanding Universe but did not carry the actual calculations of the rate of expansion shown in the French paper. In his investigative paper (Livio 2011), Mario Livio cleared any suspicion of wrongdoing by the journal and Hubble himself, by finding that Lemaitre himself had asked that these be not included in the translation, because he knew the numbers had been published by Hubble and did not care if he got the credit. Many believe that Hubble gets too much credit for this discovery and the naming of the expansion should include Lemaitre. But it is to be acknowledged that Hubble's estimate of distances and his more thorough analysis of the data is what convinced the scientific community that the Universe is expanding. Science is replete with such stories, as in the case of Theory of natural Selection where Darwin gets the credit while Alfred Russell Wallace was perhaps earlier and more insightful, and the Special Theory of Relativity which is marked by very advanced thinking by Ernst Mach, Henry Poincaré and Hendrik Lorentz, ahead of Einstein. A similar story is also told of the discovery of Higgs Boson and the DNA. But these instances go to show that, even in the case of leaps in discoveries, it is a perfect storm of maturing theoretical speculations and technological advancements that enable observations and compelling questions from adjacent subfields of inquiry. Soon after, many scientists race towards the answer and the steadfast and skillful reach the answer more or less together. This explains the fact that despite centuries of lack of knowledge on a given topic, several scientists come to the same fantastic discoveries within days and months of each other. Who gets the credit or the Nobel Prize is, however, a matter of chaotic processes. The fact that the objects in space are receding from us is consistent with the theory that the Universe is expanding. What this means is that space is expanding and carrying the stars and galaxies and other objects with it.

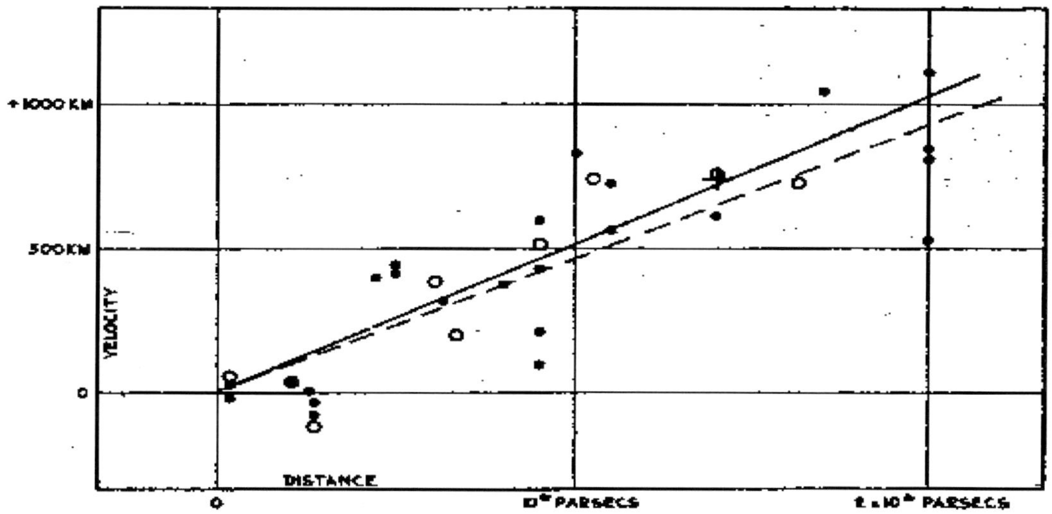


Figure 6.2. Hubble's plot of red shift versus distance, providing evidence for an expanding Universe.(Source: (Hubble 1929)).

With the sparse data that he had, of which some were incorrect, Hubble had estimated the expansion rate to be about 600 km per second per Megaparsec (speed per unit distance of the object. 1 parsec ~ 3.26 light years ~ $3x10^{13}$ km). Additional data has continuously been driving this number down over the decades. The rate of expansion has converged to the Hubble constant $H_0$ is 80±15 km per second per Megaparsec (Peebles 1993a). Recent results may be indicating somewhat lower expansion rates. (The units correspond to the fact that the speed of star movement away from us is proportional to its distance from us, so the first quantity 80 km/sec refers to the speed of a galaxy and the Megaparsec refers to the distance of the galaxy. What this means is that a galaxy or a star is moving at about 160 km/sec, if it is at a distance of 2 Megaparsec). So at a distance of *l* from the observer, the rate of expansion

$$v = dl/dt = H_0 l \qquad (6.1)$$

Here $H_0$ is called the Hubble constant. (Note that all observers are the same, since a uniform and isotropic rubber sheet would stretch the same everywhere and therefore a, the scale factor, is also same everywhere). It follows that the expansion rate is equal to the speed of light or $v=c$, at the distance of Hubble length $l=L_H= c/H_0$ and the stars and objects would be moving away from each other at the speed of light. At distances greater than Hubble length, the space expands at a speed greater than the speed of light. $L_H$ equals about 13 billion light years. The stars or galaxies beyond this distance are moving away at a speed greater than the speed of light. As noted before, the space is expanding and objects are fixed in that space and therefore they are moving with it. Expansion of space at speeds greater than the speed of light is permitted by the General Theory of Relativity. In particular, spacetime can expand faster than speed of light since this does not involve transmission of information faster than the speed of light. Space can expand at rates faster than speed of light, but objects cannot go through space at or greater than the speed of light. (See (Peebles 1993b)). Though the Hubble distance is a type of horizon, this is not the particle or event horizon that is referred to in literature. (See (Davis and Lineweaver 2004)).

## BIG BANG

The so called Standard Cosmological Model states that the Universe started, about 14 billion years ago, with an event known as the Big Bang event. The term is unfortunately the derogatory term used by Fred Hoyle and has stuck. While it seems to indicate a large explosion, there was no light and sound show. In fact, by some singular event, space and time appeared out of a small region and the Universe came into existence in a very hot very dense state. As it expanded and cooled down, light and matter appeared. At early times, all forces, the gravitational force, the strong nuclear force, the weak nuclear force that causes radioactivity and the electromagnetic force were all indistinguishable, because of the dense, high energy state. The appearance and evolution of the Universe is described by the Standard Cosmological Model which includes the Big Bang. Cosmological Model theories are applicable only below energies of the order of $10^{32}$ electron Volt (one electron Volt or eV is about 1.6 x $10^{-19}$ Joules and is equivalent to about 10,000 deg C), and densities below about $10^{94}$ gm/c.c. since below that size and above that energy density (temperature), quantum mechanical effects have to be included, the theoretical capability for which does not exist now. But, as the Universe expanded and cooled, the various forces separated and the corresponding particles with or without mass appeared and the models can be applied. The model has been verified to a fantastic accuracy by the study of the remnants of the Big Bang, particularly, the Cosmic Microwave Background (CMB). Another success of the Big Bang model is the ratios between the abundance of various light elements in the Universe that is precisely predicted by the theory, have been verified to be accurate. (The beauty of this model is that it was developed from the point of view of how particle physics observations and forces could be explained. Its success in becoming a valid cosmological theory of the Universe reflects the consistency of physics with its different branches and the rigor of the physicists).

One thing to note is that rigorously, contrary to perceptions, the Big Bang is not the starting point of the Universe. The Big Bang event did not occur at time t=0. It is more proper to say that the initial state of a non manifest Universe had existed and the Big Bang occurred after $10^{-34}$ second.

## HORIZONS

What can be observed in the Universe or what is the horizon for observation, is very complex. We discuss below what the observable Universe is or how much distance we can observe to, if we are unhindered by technology. The factors are: (1) it takes a finite amount of time for the fastest information in the form of say light, to reach us. So when we receive light from a star, it is an old light from a time that depends on how far the star was from us or how long it took for light to travel that distance. For short distances of up to a few thousand light years, this delay is approximately equal to the distance/speed of light. (2) For larger distances, as the star light travels towards us, the space between the star and us is expanding away from us and therefore the light packet is constantly falling behind. (3) For moderately large distances, the expansion of space and therefore the movement of stars away delays the arrival of the light to us but as the light packet gets closer and closer to us, the speed of expansion of space gets smaller and eventually the light can catch up to us. (4) The speed of expansion at any given distance is not constant either. In other words, Hubble's constant is really not a constant and is time dependent.

Since the distance equal to the Hubble length is where the expansion rate is equal to the speed of light, at first blush one might think that light from beyond that distance would not catch up with the expansion and the objects at and beyond that distance would not be observable by us or that there might be an "event" horizon at a distance = $L_H$, beyond which we would never see. But it is more complex. It turns out that there are several horizons.

### Particle Horizon

The first fact is that the Universe is not infinitely old. If the age of the Universe is $t_{age}$, then if the distance $l_{now}$ of the star is such that the light from the time the Universe started, is arriving now, then $l_{now} = c\ t_{age}$. Therefore, a star at this distance would be at the so called "particle horizon". The particle horizon is the edge (surface) of the sphere beyond which the Universe is not observable at this time, because particles (and any light) have not had the time since the start of the Universe. (This is the horizon Edgar Alan Poe referred to). As more time passes, the light beyond this present horizon will have had time to travel and reach us. This horizon distance is the same for all observers (assuming infinite Universe), wherever they are, but what part of the Universe they can see would be different. The reason this is called "particle" horizon, rather than just "horizon" in the meaning of seeing light, has to do with the history of the Universe. Particle horizons exist because the Universe began a finite time back. Therefore a particle horizon divides the Universe into a part which is observable now and a part that will be observable in future. (Then as we will see, there is a part that will never be observable by us. Here, "we" and "us" refer to any observer or instrument at our location).

Let us illustrate this first for a static (non-expanding) Universe. As in usual theory of relativity we draw world lines and light cones.

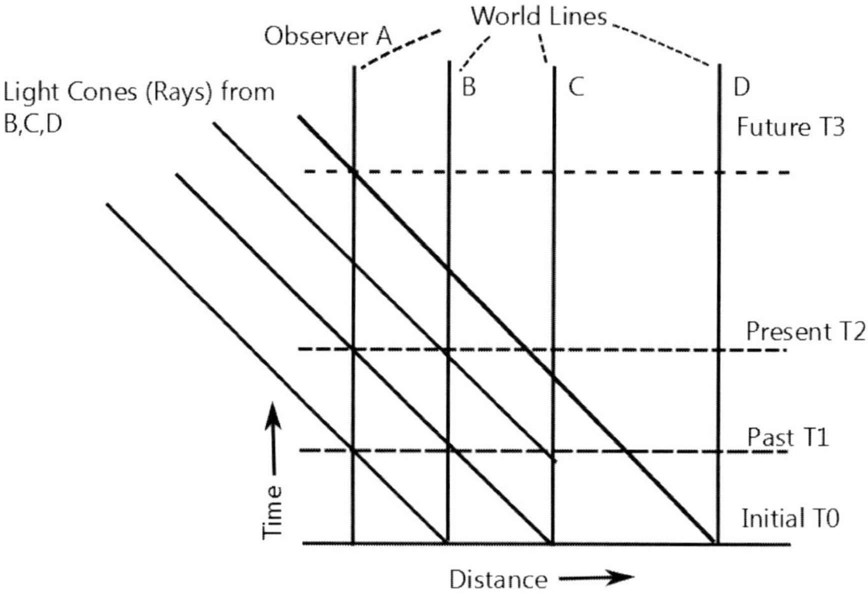


Figure 6.3. Case of static Universe: World lines (vertical) for the Observer and stars A, B, C and D and Light cones from the stars.

In Figure 6.3, which shows the case for a static Universe, the vertical world lines represent the passage of time at a given location (star) and the horizontal distance between the world lines represents the distance of the stars from the observer. The bottom horizontal line is the time when the particle or light emission took place first for a free flight. The slanted lines then represent the path of these particles (or light) as they travel towards the observer, traveling at the speed of light. (These are cones if we consider two dimensional space and are 3 dimensional cones in 3D space). We are the observer A. The light or the fast particles from star B has been reaching us since the past time $T_1$. We received the light (particle) from Star C that was emitted at initial time $T_0$ just now and light from C emitted at later time will reach us in the future and the light from D emitted at the initial time will reach us in the future at time $T_3$. So, as of just now, the star C is at the horizon and the horizon will expand to the distance of D at time $T_3$. So, in essence, since the age of our Universe is a little over 13.8 or so billion years, we can observe the light emitted within a distance of 13.8 billion light years. The particle horizon "expands" at the speed of light and so in about a billion light years, we would see stars that are 14.8 or so billion light years away. So, even in a static Universe, the observable Universe is expanding. In the static Universe then we (observers at A) will eventually receive light from all stars however far from us they are, as long as they shine. In such a case, no part of the Universe will be hidden forever. But if the Universe has an age at which it terminates or is terminated, then there is a sort of an "event" horizon, because the light switch would be turned off at the distance equal to that age in light years. Therefore such event horizons would come into existence if the Universe has an end. (By the way, the Universes horizons are unlike earthly horizons because horizons on earth have small distances and information can be passed quickly on from one observer to another, to effectively extend the horizon of an observer. Here we are talking about the extreme case of light years and light itself takes time when information is passed on from one observer to another). For an expanding Universe, there are many complexities, to the point that there are different types of horizons interacting with each other and light from the past.

## Hubble Horizon

It was shown that at the Hubble distance $L_H = c/H_0$, where $H_0$ is the Hubble constant that characterizes the speed of expansion at various distances from us, the speed of expansion is equal to the speed of light. Beyond that distance, space is receding from us (along with stars and galaxies in it) at speeds greater than the speed of light. In an expanding Universe with an expansion rate that is constant with time, the Hubble distance will remain the same at all times and indeed, the light emitted from beyond the Hubble length cannot catch up to us, because those regions are expanding faster than the speed of light. The Hubble horizon then would be the horizon beyond which we cannot ever receive light and information. (This is like walking backward on a moving walkway, the walkway being the expanding Universe and the light packet is doing the walking. If the walkway is faster than the walking speed, the walker would go opposite to the direction she wants to go and at the walking speed equal to the walkway speed; she would be stationary and not go anywhere). This would also be a type of fixed event horizon, with the additional difference that objects that are visible now will move to and beyond the Hubble distance and be lost to us.

But if the Hubble constant is not a constant and decreases with time (the walkway slows down), then this horizon will expand and light that was once outside this horizon will come inside and reach the observer. In the history of the Universe, the Hubble constant varied because it is dependent on the ratios of radiation, matter and vacuum energy in the Universe. Indeed, at the beginning of time, the Hubble constant was large and it decreased over time. (A simple model shown below would show that the Hubble constant will decrease inversely with the age of the Universe- see below). Very interestingly or coincidentally, the Hubble distance and the distance to the particle horizon are nearly equal in this era. (There may be metaphysical attributions to this and some people might find meaning in the emergence of humans at this time). We see below that the changing of the Hubble length changes has an interesting consequence.

## Event Horizon

The event horizon is the region beyond which an observer at our location can never see. (The terminology is the same as the horizon of a black hole). This horizon divides the space into all presently and future observable objects and those that are beyond the limits of observation forever. Our Universe came about from a small region of space expanding to the present size. The distances themselves are changing so that the horizontal scale keeps changing as a function of time. (As shown below, the expansion of distance is exponential for a Hubble constant. Therefore, one cannot just scale the horizontal line). This can be approximately represented by the revised figure 6.4. Now the world lines of the observer and other objects such as stars would be diverging from the initial region of expansion (shown as a point in the figure- actually it is a finite sized region). The worldlines of objects are not straight lines in a Hubble expansion (even if the Hubble constant did not vary with time). The light cones are even more complex because the space is expanding (distance to observer keeps increasing with time).

In the beginning all observers were nearly at the same place. Let us say observer B sent a light signal to A. The light cone that was supposed to come towards A, actually moves outwards, while the great speed of expansion of the early Universe is taking place. (Here the world lines are shown as straight lines for simplicity). The light ray would fall back (in analogy with the walkway above) because in the beginning the expansion was very fast and the light ray stays closer to B than coming closer to A. But as the expansion slowed, the light ray would appear to move backwards more slowly and when at the Hubble distance, where the expansion is at the speed of light, the light would appear to stop. If the Hubble distance did not change with distance, all the light would (appear to) be trapped at this distance. But then, the Hubble distance gets larger with time, because the Hubble constant got smaller with time. As a result, the Hubble horizon became larger and the light beam came to be inside it. The beam then is able to go forward towards its intended direction to the location of A at the time of Big Bang. As the light beam gets closer, the expansion of space gets smaller and therefore, the light beam gathers speed. This means that after the light beam comes within Hubble horizon, it curves back and returns to the intended place it was headed for, the location of A at the time of Big Bang.

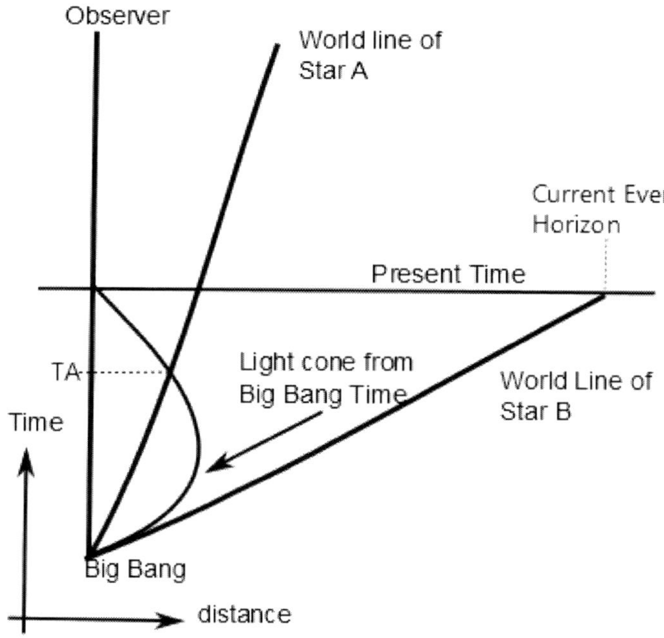


Figure 6.4. The world line and the light cone for the case of expanding Universe. The light cone from the Big Bang is being received now.

This represents a different delay in the observer receiving the light, and therefore now, we see a different epoch compared to, if it were a non expanding Universe. Only now we are receiving the light from the Big Bang, which was the originating location for us and all the stars. (This is only a coincidence). The star A remained within our horizon while, the star B is at our horizon, because it is tangential to the light cone. At the present time all stars that were outside of Star B would not be observable to us. Currently Star B's world line intersects out present time (horizontal line at 13.8 billion years) at 46.6 billion light years and this is our present "event horizon". This means that if all cosmological parameters freeze as they are now, distances beyond 46.6 billion light years would never be observable. So the Universe, for us is 46.6 billion light years in size. What this means is that after tens billion years, observers would be able to see the light from this horizon. This is not the 46.6-13.8 =33.8 billion years, because the "effective" speed of light is different in an expanding Universe. (Readers should note that the speed of light or time does not change, it is only the distance light has to cover, that changes). With an expanding Universe and with a variable Hubble constant, even the Event Horizon is not fixed and even the 46.6 billion light year distance to the horizon is only for the present. This will also change.

In our simple thinking as in static world we would receive the light, from 5 billion light year distance, that was emitted 5 billion years back. But in the expanding Universe, the light we receive was emitted when the star was closer and at a time where the light cone intersects the worldline of 5 billion light years. The light we see from star A was emitted at the time TA in Figure 6.4. If we draw a worldline for a longer distance from us now, say 10 billion light years, we see not only that light we receive from it now originated earlier, but it also was closer, when it was emitted, because the Universe was smaller. So counter intuitively, longer distances now mean closer distances at the time the light originated. The edge of the Universe we see now is that of the location at the starting point. Paradoxically, the farthest view we see today, is of things that were at our location! All world lines and backward light cones originate at Big Bang and the intersection of the world lines with different light cones arriving at different times determines the location and time of emission.

## INFLATION

There were two big problems with the Standard Big Bang theory:

### The Horizon Problem

This theory of a singular continuous expansion cannot explain the homogeneity and isotropy of the Universe (the Cosmological Principle). The homogeneity is obtained because the particles and things are causally connected and in communication with each other, but this connection is established over the transit time of interacting particles and virtual particles that travel at the speed of light, which, by physics laws, was the same as now. To see that the homogeneity we see today cannot be accounted for simple cosmological expansion, we perform the following analysis:

The earliest time the Big Bang Theory is applicable, could be taken to be when the temperature fell well below the Planck Mass (energy), let us say 100 times less ~$10^{30}$ electron Volts. The age of the Universe was about $10^{-35}$ seconds. At this point, Cosmological models give the scale factor of expansion increasing proportional to $t^{1/2}$ or $t^{2/3}$. Taking the size of the Universe to be 46 billion light years (about $4.4 \times 10^{26}$ m) (see above) and the age to be about 14 billion light years (approximately $4.4 \times 10^{17}$ sec), the size of the Universe at $10^{-35}$ sec, taking the $t^{1/2}$ ($\sqrt{t}$) dependence), was,

*Size then* = $\sqrt{(t\ (then)/t\ (now))}$ *x size now*

$\sqrt{(10^{-35}/44 \times 10^{17})} \times 4.6 \times 10^{26}$ ~ *2.2 meters.*

If we took the $t^{2/3}$ dependence, the size would be about 1.6 m. But, with a light speed of 3 x $10^8$ m/sec, it would take of the order of $10^{-9}$ sec for the farthest objects to communicate, even in this small Universe, far longer a time than their age of $10^{-35}$sec. Stated differently, within the time that was available, light would have traveled only a distance of the order of $3 \times 10^8 \times 10^{-35} \sim 10^{-26}$m. (As we saw, the principle applies even now. The light would take longer than the age of the Universe to cover the whole Universe by about a factor of three). Our Universe is very homogeneous as confirmed, with high accuracy, by the Wilkinson Microwave Anisotropy Probe (WMAP) that has measured the cosmic microwave background radiation left behind by the Big Bang. So, the question is, how was the homogeneity and isotropy that we see today, established over the Universe when light did not have the necessary time to establish the homogeneity during the early time when the particles were forming? As readers can see why from the previous section, this is called the horizon problem. An expanding Universe has a horizon beyond which objects are not causally connected. In the early Universe, there should have been a particle horizon which must have prevented homogeneity from being established. It appears our present objects beyond our present particle horizon are doing exactly what objects within the horizon are doing. How is this possible, unless this causal connection had been established long ago?

## The Flatness or the Fine Tuning Problem

The Universe started from a brief burst of energy, available in a quantum mechanical fluctuation that lasted for less than $10^{-43}$ seconds. Once the Universe is born with that appearance, there are two forces acting on it, one is the outward repulsive force due to an unknown fluid or field, which was called the Cosmological constant by Einstein, essentially an outward pressure, which expands the Universe and the other the inward gravitational force pulling it all back in very quickly in time of the order of $10^{-40}$ sec. (Once Universe has sufficient volume established, additional repulsive forces such as that due to vacuum energy would come into play). (The term “Cosmological constant” is strange for a fluid or the power or force of the fluid or the energy associated with this force to push the Universe out. This type of nomenclature becomes common once we start working in coordinates in which the speed of light is taken to be 1 and every other unit is adjusted to that). When the Cosmological Constant and gravitational forces are exquisitely balanced then the Universe is flat. (See critical density below). The relative forces are determined by the energy density (mass density vs. thermal energy density). Given the early temperature of the Universe, if the density is off by just 2 parts in $10^{21}$ (a really really small difference), the Universe would be flown apart or be pulled in very quickly. For the Universe to have lasted for this 10 plus billion years and have had the time to create stars and galaxies with nucleosynthesis and people, this extremely improbable fine tuning had to have happened. This also implies that the Universe has to be very flat. (In fact, the early Universe had to be extraordinarily flat, because as the Universe evolves, curvature starts developing). As always in science, these types of requirements cast doubt on the model and additional explanations are needed or the model is deemed to be wrong.

Alan Guth came to the rescue with a great insight. He suggested that the Universe could smooth and flatten itself out with a very fast expansion which he called "Inflation". In his paper entitled, *"The Inflationary Universe-A possible solution to the Horizon and Flatness Problems"*, (Guth 1980), he proposed that these problems are solved, if the Universe expanded very rapidly, exponentially (see above for Hubble expansion with a fixed Hubble constant) by about 100 doublings (an enormous factor of about $10^{30}$), in a matter of $10^{-34}$ sec, soon after coming into existence. After this inflation, the early Universe, which would be well packed with hot, dense matter, causally connected, and therefore smoothed like a smoothie in a blender because of the extremely small size, would then expand very quickly and many fold due to the appearance of the pressure, with the homogeneity intact. Inflation also flattens the Universe. Inflation stretches a round object $10^{30}$ times. So, the radius of curvature would increase $10^{30}$ times which is essentially flat to satisfy the above shown requirement. Therefore, the inflation is like an explosive blowing of a balloon and this expansion smoothes out any wrinkles (inhomogeneity) on the balloon, increases the radius and thus flattens the Universe. However, we pay a price for this inflation- one thing the inflation also does, in creating the homogeneity and flatness, is that it destroys any details of pre-history and gives the Universe, a sort of amnesia. Therefore, it is not possible to know what the Universe was like prior to the inflation, except for quantum fluctuations that survived inflation. The development of theory of inflation and its refinements have correspondingly expanded the physics of cosmology along a steep path. (Figure 6.5).

## PHYSICS OF THE EXPANDING UNIVERSE AFTER BIG BANG

A bounded Universe would violate the Cosmological principle of homogeneity and isotropy. An observer, not at the center of even a uniform sphere, will see that the number of stars in one direction is less than in the opposite direction. But observers living on the surface of a sphere and who are creatures of the surface (the direction away from the surface is not available to them) may see an isotropic and homogeneous Universe. So the Universe may be closed in this case. Matter and everything exists on this surface Universe, which is almost flat, but may have a small curvature. Figure 6.6 shows an illustration of an expanding or contracting Universe, represented as a 2D circle. In this Universe, one dimensional beings live on the circumference of a circle on these cones. We live in a 3 dimensional space version of this illustration.

At the bottom is the start of the Universe which we can designate as Big Bang. So as one can see, taking the 2D analogy to our 4 dimensional spacetime, there was physically nothing before the birth of the Universe and the Big Bang happened "everywhere" which was a very small place at that time. There was no "place" outside that. Now, (now in figure 6.6), there is no place outside our "small circle" or our 3D surface of sphere.

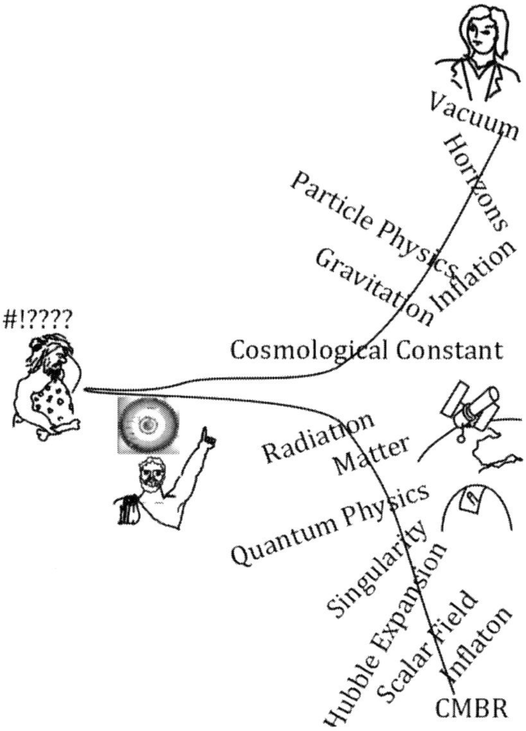


Figure 6.5. Exponential expansion of physics of Cosmology.

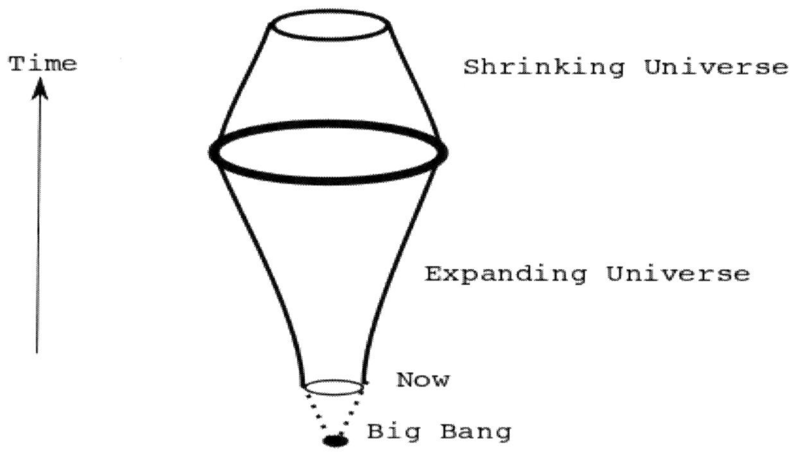


Figure 6.6. History of an expanding and then shrinking Universe (only for illustration). Our Universe, in present understanding, is not showing signs of shrinking.

Once the expansion started, the Universe also started cooling down, because by thermodynamical laws, an expanding system, with no external energy, reduces in temperature. This is very similar to expanding gases cooling down. The cooling down caused the manifestation of different types of fields, forces, radiation and matter. First there were the fundamental particles – radiation, quarks and leptons (such as electrons), and then the simplest of the atoms, hydrogen, were formed. The emergence of each of the fields, with associated forces, force carriers and matter particles, as a separate manifestation occurred as the Universe cooled down below specific temperature. Figure 6.7 shows the different epochs. In several ways, the expansion of our Universe is quite marvelous and while Cosmological physics is interesting for its own sake, our Universe is, in many ways, special. Its origin and its expansion have had the particular advantage of making the Universe look the way it does, last as long as it has and to have the characteristics that it does so that we, the human beings who have evolved may look and marvel at all of it. Yet, some aspects are charmingly simple. (1) The relatively gentle expansion is the result of the fine tuning of the Universe, between two opposing forces- the gravitational attractive force pulling everything into themselves and the Cosmological repelling force that pushes everything out,. This is not unlike throwing an object up from earth. There is the force that threw it up and then the gravitational force pulling it down, (2) The continuing expansion, causes the space to expand and this expansion is very simple, as noted below, that of a 3 dimensional elastic sheet.

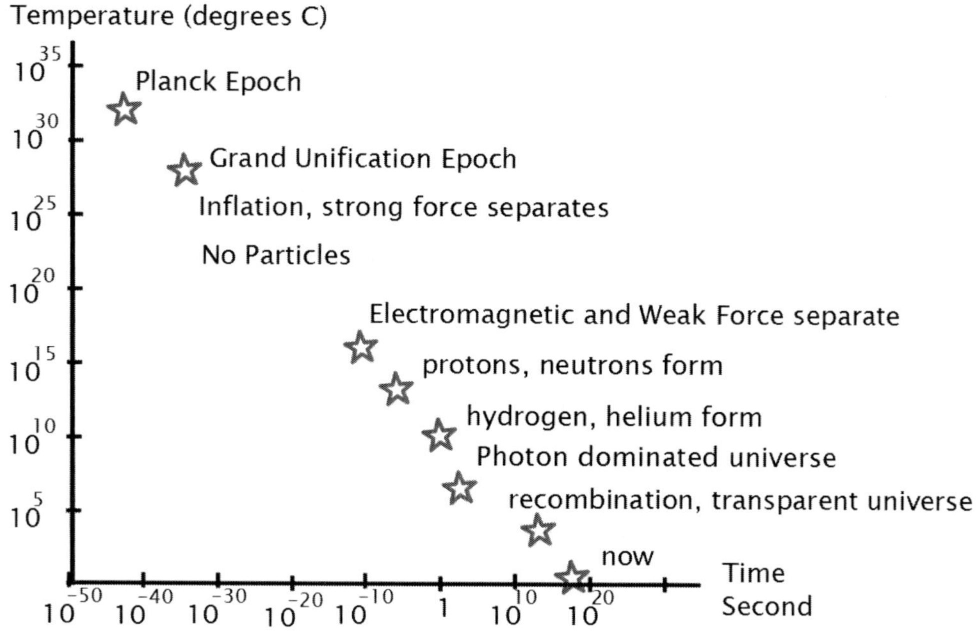


Figure 6.7. History of the Universe.

## Physics of Hubble Expansion

It is important to understand that this expansion is not tearing anything apart but is just the grid of coordinates expanding. The space is expanding taking all the objects with it. This is again like a balloon expanding with dots placed on it moving apart without them moving. The Sun and the Earth may be expanding away, but the gravitational forces between them are much greater than this stretching force. So, the solar system or galaxies are not stretching significantly. Note also that the scale factor $a< 1$ corresponds to a shrinking Universe and $a=0$ corresponds to the Big Crunch to a singular state of infinite density. $a=1$obviously corresponds to a stationary Universe.

**Box 6.1. Expressions for Hubble Expansion**

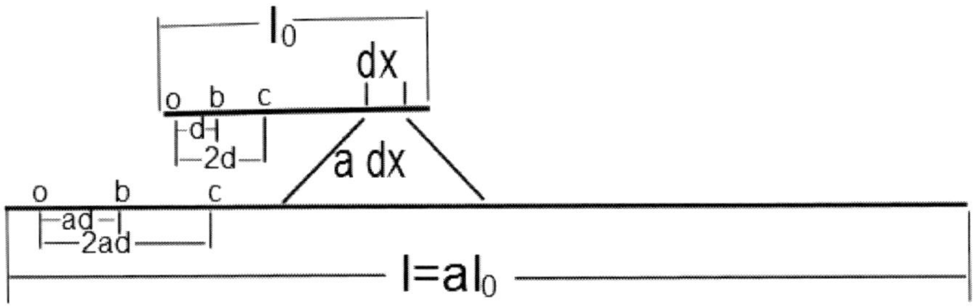


Figure 6.1. Shown again. A length of space expands by a factor a.

Writing this in general form,
$l = al_0$
v= velocity of expansion =dl/dt = (da/dt) $l_0$
Eliminating the (arbitrary) initial length,

$$v = \frac{1}{a}\frac{da}{dt}l = \frac{\dot{a}}{a}l = Hl \quad (6.2)$$

Where, $\dot{a} = \frac{da}{dt}$.

$$H = \frac{\dot{a}}{a} \quad (6.3)$$

is the Hubble Constant (note that in our Universe, H is only a constant for all distances, but is not a constant in time).

But, if H were independent of time and equal to $H_0$, then

$$v = \frac{dl}{dt} = H_0 l \text{ or } \frac{dl}{l} = H_0\, dt$$

Integrating this equation and setting $l=l_0$ at t=0,

$$l = l_0 e^{H_0 t} \quad (6.4)$$

This gives an exponential expansion for a Hubble constant that is unchanging with time.

Now as to how the mechanism of expansion works, there is a simple description, as noted above. With very simple Newtonian physics, we can illustrate the way the Universe expands and arrive at the Hubble expansion. Consider an object of mass *m*, the force of gravity is *mg*, where *g* is the acceleration due to gravity (Newton's law – force=mass x acceleration). If we move an object to the top of a pole of height *h*, we perform work against gravity. The work done is force multiplied by the distance or equal to *mgh*. Since work is energy and energy is conserved, the work done is stored in the object as potential energy. It is released and converts to kinetic energy as it is dropped from the pole and reaches the bottom with speed. If we designate kinetic energy as positive, then the potential energy is negative, since at the bottom, the sum of kinetic energy and potential energy must be zero for this object which was at rest before the work was done. The same situation applies when we launch an object into the sky. Initially we impart energy at the bottom, which is all kinetic energy and as it gains height, the kinetic energy converts to (negative) potential energy, as the object slows down in gaining height.

With this background, one needs to add the fact that $g$, the acceleration due to gravity, depends upon the height itself. It decreases as the inverse-square of the height. Therefore, the product, $gh$ actually decreases inversely as the height increases. This, essentially, means that the gravitational attraction decreases with height. So, if the object is launched with a high speed, high enough that it goes to a great height where gravity becomes too small to pull it back, the object would overcome the gravitational pull and escape gravitation. (Figure 6.8). The speed of launch at which this happens is called the escape velocity. Below this velocity, the object would fall back and beyond that the object will keep going away from the earth. So, for launch speeds greater than the escape velocity, the kinetic energy remains higher than the potential energy and the object keeps on moving and as the object moves away, the potential energy decreases further and the object regains speed and approaches the launch speed. This is how rockets work and this is similar to what happened to the Universe. The initial push away by the Cosmological Constant, though very close to the pull of gravitation, still overcame the gravitation and the Universe started expanding. Beyond this point, whether the Universe falls back or goes to a steady state or expands (escapes) forever, depends on what is contained (energy densities) in the Universe. At present, the measurements indicate that the Universe has been and is accelerating away and it is unlikely to fall back. But the acceleration is gentle and the fact remains that the Universe is fine tuned.

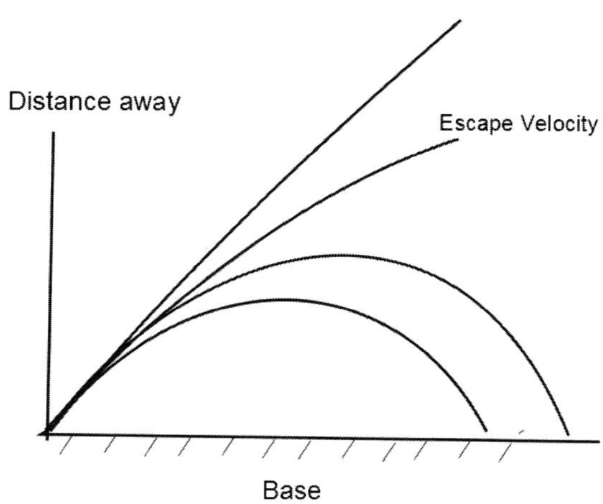


Figure 6.8. Escape velocity.

There is a very convenient further principle arising out of the inverse square dependence of force with distance. When we consider a collection of galaxies in a sphere, a particle that is leaving from somewhere inside this sphere is only affected by the gravity of those masses that are inside the surface of the sphere on which the itinerant particle is, and the masses outside this sphere do not influence this escape velocity. The average of the forces, due to the masses outside this inner sphere, cancels out. Therefore the above situation holds even if the object m and the patch M are inside a larger distribution of objects. (See figure 6.9). Therefore an object m at the surface of the smaller sphere and trying to escape the larger sphere has to contend only with the gravitational pull of the objects inside the small sphere.

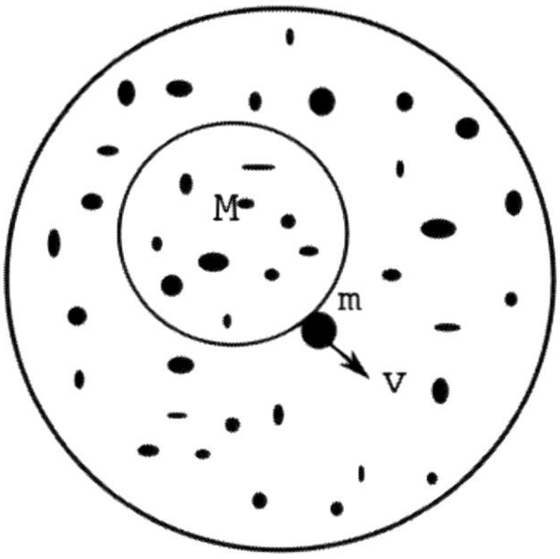


Figure 6.9. Object m's velocity to escape the large sphere only depends on mass of objects in the small sphere.

Closely related to this escape velocity and the balancing of the pushing out and pulling in of forces, is the flatness of the Universe. As stated before, the Universe is fine tuned to be flat or very close to being flat. This fact can be demonstrated using just Newtonian physics. With such an analysis, one obtains the famous Freidman Robertson Walker (FRW) equation that gives the Hubble constant for various energy densities in the Universe. In such an analysis it is seen that the energy density $\rho$ of the expanding Universe has a critical value, $\rho_{cr} = \frac{3H^2}{8\pi G}$, where $H$ is the Hubble constant and $G$ is the Gravitational Constant =6.6738 x $10^{-11}$ $m^3$ per kg per $sec^2$. NASA's WMAP probe measurements show that the average energy (mass) density of the Universe is 9.9 x $10^{-30}$ gms/c.c. which is within 0.5% of the critical density corresponding to the observed Hubble speed of about 70 km/sec/Megaparsec. The ratio of density to critical density is denoted by the symbol Ω. As stated above, for our present Universe, it appears to be very close to 1.

More than 95% of this energy density has not yet been detected. We can only account for about 5% of the matter in the form of atoms, neutron stars and black holes. Observations of the behavior of galaxies indicate that 24% of the matter is what is referred to as dark matter which is undetectable because they do not radiate energy or only weakly or do not interact with our sensors.71% of the energy density is in the form of dark energy which, too, is unknown at this time.

**Box 6.2. Critical Energy Density for a Flat Universe**

Consider a large patch of mass M (a collection of galaxies) and a particle (galaxy) of mass m that starts with a velocity v from this mass M. (Figure 6.9). (Same situation as an object launched into the sky). The kinetic energy of this moving object is given by $mv^2/2$ (outward). The potential energy is proportional to the distance D it has gone from M (the height it has reached) and is given by -mgD, where, again, g is the acceleration with which m would fall back down towards M (similar to the acceleration due to gravity on earth). The gravitational attraction force between m and M at this distance is given by Newton's law of gravitation as,

$$F = \frac{mMG}{D^2} \tag{6.5}$$

where G is the gravitational constant.

Since this force is equal to mg,

$$g = \frac{MG}{D^2} \tag{6.6}$$

The potential energy is,

$$-mgD = -\frac{mMG}{D} \tag{6.7}$$

The energy E of the particle when $D=D_0$, $v=v_0$

$$E = \frac{mv_0{}^2}{2} - \frac{mMG}{D_0} = \text{constant (energy conservation)} \tag{6.8}$$

where $v_0$ is the particle speed when it left M. With a given M and G, the mass m will return to the patch of mass M or escape from it, depending upon the velocity $v_0$ of mass m. For $\frac{v_0{}^2}{2} > \frac{MG}{D_0}$, the object will escape the patch and travel beyond and for $\frac{v_0{}^2}{2} < \frac{MG}{D_0}$, the object will return (fall back to) to the patch. $v_0$, corresponding to $\frac{v_0{}^2}{2} = \frac{MG}{D_0}$, is called the escape velocity. (For an object thrown vertically from the earth, the escape velocity is a little over 11 km/sec).

Rewriting the energy conservation equation for an expanding space with a scale factor a, or

$$D=aD_0, \text{ where } D_0 \text{ was the distance (radius of the sphere, at time t=0).} \tag{6.9}$$

$$v^2 - \frac{2MG}{D} = \text{constant (mass m absorbed in the constant on the RHS)} \tag{6.10}$$

Now since,

$$v = \frac{dD}{dt} = \frac{da}{dt} D_0 = \dot{a} D_0 \tag{6.11}$$

Substituting the value of v from equation (6.11) in equation (6.10),

$$(\dot{a} D_0)^2 - \frac{2MG}{D} = \text{constant.} \tag{6.12}$$

Now, the mass M (density x volume) inside the sphere of radius D $=4\pi\rho D^3/3$, where ρ is the mass density inside the sphere. Therefore, rewriting equation (6.12)

$$(\dot{a}D_0)^2 - \frac{2(\frac{4\pi\rho D^3}{3})G}{D} = \text{constant}$$

Since, D=a.$D_0$

$$(\dot{a}D_0)^2 - \frac{8\pi\rho G}{3}a^2{D_0}^2 = \text{constant}$$

which gives,

$$(\frac{\dot{a}}{a})^2 - \frac{8\pi\rho G}{3} = \frac{k}{a^2} \quad (6.13)$$

where k is a constant.

This equation is often written as

$$(\frac{\dot{a}}{a})^2 - \frac{8\pi\rho}{3m_{pl}^2} - \frac{k}{a^2} = 0 \quad (6.14)$$

where $m_{pl}$ is the Planck mass $=\sqrt{\frac{hc}{2\pi G}}$ and h being the Planck constant. In these units for Planck mass, h is equal to 2π and c is taken to be equal to 1.

Now since the Hubble Constant $H=\frac{\dot{a}}{a}$, this equation can be written as,

$$H^2 = \frac{8\pi\rho G}{3} + \frac{k}{a^2} \quad (6.15)$$

This equation (without the Cosmological constant which we will explain later), is the Freidman Robertson Walker (FRW) equation for Cosmology and we have derived one of the most important cosmological equations while using only simple Newtonian physics and without using Einstein's General Theory of Relativity!

A flat Universe has k=0. With a specific value of H, the density has to be accurately held to the critical value given by,

$$\rho_{cr} = \frac{3H^2}{8\pi G} \quad (6.16)$$

Equation (6.15) can be rewritten as (using equation (6.16),

$$\Omega - 1 = \frac{k}{a^2H^2} \quad (6.17)$$

where $\Omega = \rho/\rho_{cr.}$

Figure 6.10. Flat earth humor.

## Variation of the Hubble Constant and the Scale Factor with Time

How the Hubble constant and the scale factor varied with time is an important history of the Universe. As we saw in the introduction to Inflation, the reason we need the inflation is because, we can trace back to the origin of the Universe and see a discrepancy in the homogeneity issue. If the Hubble constant is not only a constant in space at a given time, but were a constant in time too, the space would expand exponentially with time. This happened only during the inflationary period. The detailed Cosmological theory traces back how Hubble constant varied with time after the inflationary period. Since early time, the Hubble constant has been decreasing. The scale factor would still increase exponentially but would have a more complicated exponential dependence.

As stated before, Newtonian physics goes a long way in describing models of Big Bang and the evolution of the Universe. Though the original FRW equations were derived using the General Theory of Relativity, the fact that Newtonian physics can obtain it easily, demonstrates the equivalence principle. It is not too inaccurate to describe the evolution of Universe without defining and describing the constant $k$. The parameter $k$ represents the curvature of the space and can, in principle, be included in the Newtonian derivation, but the better way to do it is through the General Theory of Relativity (GTR). Use of the GTR requires considerable learning and a presentation here would be beyond the scope of this book. Still, a lot can be gleaned from the application of Newtonian physics. For example, one critical piece of information is to know how the Hubble constant and the scale factor change with time. This would help us in tracing back the history of the Universe, in particular ideas of horizons and relating present day observations of stars and galaxies that are far away. This history of expansion is the critical piece of information that shines light on how everything that we see today has come about. This is given in Box 6.3. (For this discussion, it is reasonably accurate to use $k=0$, since the Universe is pretty flat. –see figure 6.10).

**Box 6.3. Variation of parameters of the Universe with time**

If $t_{age}$ is the age of the Universe and a galaxy at a distance d had expanded to a distance D (D>>d) in that time, then the distance (neglecting the small initial distance at the time of Big Bang),

$$D = t_{age}v$$

Since, v=Hd,

$$D = t_{age}Hd$$

and therefore, D/d = scale factor= a = $t_{age}$.H.

Since a is a constant in all of space at a given time of $t_{age}$, H is proportional to 1/ $t_{age}$ or the Hubble constant is inversely proportional to the age of the Universe.

The scale factor is, however, a function of time. If a varied as a polynomial of time, that is,

$$a = Ct^p$$

Then,

$$\dot{a} = \frac{da}{dt} = Cpt^{p-1}$$

Therefore, $H = \frac{\dot{a}}{a} = \frac{p}{t}$ (6.18)

or the Hubble constant still decreases inversely with time. For the Hubble constant to increase with time, the scale factor would have to increase with an exponential of time with power greater than 1.

Now, we can actually determine the value of p for a specific scenario, for example, when a flat Universe is made of only gravitating energy (mass and similar form of energy). We return to mass M and see what happens in the expanding volume of radius D= a.$D_0$. Then, as we saw at any time t in an expanding universe, the density of a Universe with mass M and radius D is,

$$\rho = \frac{M}{4\pi D^3/3} = \frac{3M}{4\pi a^3 {D_0}^3} \quad (6.19)$$

But from equation (6.15) with k=0 and using equation (6.18),

$$H^2 = \frac{8\pi\rho G}{3} = (\frac{p}{t})^2 \quad (6.20)$$

Substituting for ρ from equation (6.19)

$$\frac{2GM}{a^3} = (\frac{p}{t})^2$$

Since a=C t $^p$, $\frac{2GM}{C^3t^{3p}} = (\frac{p}{t})^2$ (6.21)

This identity, with this time exponent on each side of the equation, is possible for all values of M only if 3p=2 or p=2/3. Substitution of p=2/3 in equation (6.21) gives C=9GM/2 and therefore

The variations of the scale factor and Hubble Constant with time are given by,

$$a = \frac{9GM}{2} t^{2/3};\ H = \frac{2}{3t} \quad (6.22)$$

For a radiation dominated Universe, the energy density $\rho = C'/a^4$. (C' is a constant)

Substituting this in equation (3.16), $\frac{8\pi G}{3}\frac{C'}{a^4} = (\frac{p}{t})^2$

Or $\frac{8\pi G}{3}\frac{C'}{(Ct^p)^4} = (\frac{p}{t})^2$

Or $\frac{8\pi G}{3}\frac{C''}{t^{4p}} = (\frac{p}{t})^2$. C'' being another constant.

In order for this identity to be valid for all conditions, 4p=2 or p=1/2. Then,

$$H = \frac{1}{2t};\ a = C\, t^{1/2} \quad (6.23)$$

## Influence of Matter, Radiation and Vacuum on the Expansion

The science behind what follows is a carefully put together scenario by the use of astronomical and astrophysical observations, thermodynamics of the Universe and particle physics theories and observations. This is clearly a voluminous work by the scientific community and even a proper scientific summary and conclusions would take a separate book. But a broad timeline of major changes can be narrated (see figure 6.7): In the early Universe after the inflationary period, the Big Bang started and after the electroweak forces became distinct from nuclear (the strong) force at the age of $10^{-11}$ seconds, the Universe was more or less filled with photons (light). So, this is the moment, according to the Old Testament, God said, "Let there be light." The Universe was small and bright at this time.

When two high energy photons collide and if the photon energies are adequate, these produce a pair of matter -antimatter particles, such as quarks-antiquarks, electron-positron etc., but if the temperature is too high, the pairs themselves would recombine to go back to the photons with no net production of particles. But at $10^{-11}$ sec, the Universe had cooled enough to allow quark-antiquark particles to materialize and stabilize and, 1 microsecond later, electron-positrons to stabilize. So, a microsecond after Big Bang, photons started being converted to particles and matter started forming in the fog of light. As a result, the amount and energy density of radiation decreased and the amount and energy density of matter increased as a function of time). Before the Universe was 1 second old, quarks bound together protons, neutrons and mesons.

At about 4 minutes after the birth of the Universe, the Universe was too cold to create new quarks and leptons. The amount of normal matter more or less became fixed at this time. The expansion of the Universe diluted the energy density of both radiation and matter, but as we saw, the energy density of radiation decreases faster with expansion *($1/a^4$)* than that of matter with *($1/a^3$)* dependence. The Universe had cooled down to about a billion degrees to form helium and hydrogen nuclei, but was still too hot to form atoms. (The helium and hydrogen stayed ionized). After about 50000 years, the power densities of radiation and matter equaled. All this time, an observer would have seen that everything was foggy dark since the light was pretty much scattered by the hot energetic electrons and positrons. (We can see an example of absorption of radiation by matter in the Sun's so called Fraunhofer lines in its spectrum which shows up as dark because hydrogen atoms in our atmosphere absorb those wavelengths. These absorptions are due to excitation of electrons bound to atoms, but the scattering of radiation in this epoch was due to free energetic electrons

The last change happened around 400,000 years after the birth of the Universe, when the Universe cooled to about 3000 deg C and hydrogen and helium atoms formed, but were still ionized. Following this, the electrons were captured by the hydrogen and helium ions. This is called recombination. By now the radiation energy had gone down considerably and there were fewer free electrons to scatter the radiation. The time this happened is called the recombination era. For about 150 million to 400 million years the only new light that was emitted was the 21 cm line of neutral hydrogen and this is known as the Dark ages.

After about 500 million years, matter started condensing into structures. When these structures combined they reheated and reionized hydrogen and helium. These objects started becoming bright by themselves and protostars started forming. This is the era of Reionization. 1 billion years after the Big Bang, stars and galaxies formed and 9 billion years after the Big Bang, our Solar System was born and we see pretty much the heavens that existed then. At the age of 10 billion years, life evolved on earth.

What this means is that if we look now at a faraway distance (near the edge of the visible Universe) we would see the hot and opaque Universe that existed at early times. At this time, the Universe was bathed in light. Again, this means that we would not be able to see beyond this "surface of last scattering", even if it is within the horizon and light from that region had enough time to reach us. (Now it is slightly inside the horizon). We see only the region that lies inside the recombination era. This is another type of horizon. One can still see neutrinos from beyond that surface, but that source also has a surface of last scattering. All the light that was emitted in the earliest epoch has been expanded in wavelength to the low energy (2.7 deg K) microwave radiation. This Cosmic Microwave Radiation, its distribution and variations in intensity and energy tell us a lot about the history of the Universe.

## Scale Factor as a Function of Time

We have to remember again that all this comes about because of expansion of the Universe and therefore, one can relate this history with how the scale length changed with time and how the Hubble speed changed with time.

In the simple derivation shown in Box 6.3, we see that if we take the Universe to be populated with matter that feels the normal gravitation force that we feel on earth, we see that the scale factor changed proportional to the age of the Universe ($t_{age}$) to the power of 2/3 and a Hubble constant that is equal to 2/(3 $t_{age}$). Since mass and energy, such as that contained in the motion of the masses are equivalent, this time dependence of scale factor is applicable to the era when matter energy dominated the Universe. Let us check the numbers, and compare with the present observations:

$t_{age}$= present age of the Universe = 14 billion years. Since, 1 year = $3.15\times10^{7}$seconds,
$t_{age}$ = 14 x $10^{9}$ x $3.15\times10^{7}$seconds= 4.4 x $10^{17}$ seconds.

This gives a Hubble Constant $H=2/(3 \, x \, 4.4 \, x \, 10^{17}) = 1.5 \, x \, 10^{-18}$ per second. Let us compare this with the observed Hubble constant of 70 km/ sec/ Megaparsec. 1 Megaparsec ~3.1 x $10^{22}$ meters, so that the observed Hubble constant = $80 \, x \, 10^{3}/(3.1 \, x \, 10^{22}) \sim 2.5 \, x \, 10^{-18}$ per second- less than a factor of two different, but not bad.

We now know that in the early days of the Universe, the Universe was radiation dominated. There were $10^{9}$ photons for every proton. So, the Universe was dominated by the energy of these photons. We now see the remnants of those photons as the Cosmic Microwave Background. The one difference between the gravitating mass and gravitating photon is that the photon energy depends on the cavity (the box) that the light is in, in other words, the energy spectrum depends on the size of the Universe. (Light wavelength stretches to fill the box). From physics of photons, we know that the energy of a single photon is given by $E_\nu = h\nu$ where $\nu$ is the frequency of the light wave=$c/\lambda$ where $\lambda$ is the wavelength or $E_\nu = hc/\lambda$. But as the Universe expands the wavelengths also stretch and that is the reason, why we now see the high energy (high temperature era) photons with short wavelength, X ray energies, stretched out to the long wavelength light in the microwave energy now. Therefore, the energy of the photon $E_\nu$ varies as *1/a,* (*a* being the expansion scale factor) or $E_\nu = C/a$, where *C* is some constant.

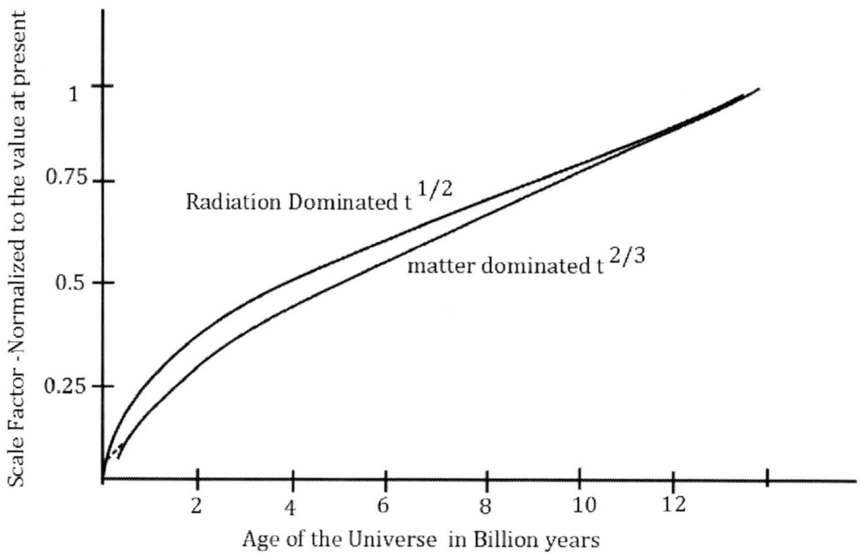


Figure 6.11. Scale factor *a* as a function of time for the radiation dominated (upper) Universe and matter dominated (lower) Universe. The transition from the first to the second is shown by the dotted line (near the left bottom).

The energy density $\rho$ in a cube of dimension $a$ =energy/volume $=(1/a^3)\ (C/a)\ =(C/a^4)$, since volume is proportional to the cube of the size of the box. When we go through the process shown in the box 3.4, under this condition we get a scale factor variation with time of $t^{1/2}$ or $\sqrt{t}$ (See equation (6.23) in box 6.3). This corresponds to the energy density itself varying as the inverse square of the time. Therefore, while the scale factor varies as time to the power ½, for the radiation dominated Universe, it varies faster with time as time to the power 2/3, for the matter dominated Universe. Since the early Universe was dominated by radiation the scale factor earlier varied as $t^{1/2}$ and as the matter started forming, it slowly switched over more to the $t^{2/3}$ dependence. Figure 6.11 shows the dependence of scale factor on time, for the radiation dominated scenario and the matter dominated scenario. The dotted line shows the switch from the former to the latter which happened at about 50,000 years after the Big Bang.

## Variation of Pressure and Energy Density with Time

As we saw before, the Universe was born with an initial repulsive force, which imparted an expanding character to space and time. But, as the radiation and matter came into existence, the energy density of these entities applied a gravitational pull trying to retract the spacetime back. While the energy density of the matter is the same as mass density ($E=mc^2$), the energy of the radiation can also be shown to be a type of pressure (opposite in comparison with the conventional gas pressure that pushes out, because it is not particles moving through space, but space that is being expanded). While most definitions would give pressure as force per unit area, it is also energy per unit volume, which is also termed as energy density. In the case of gas pressure, it is kinetic energy of gas molecules which is related to transfer of momentum as they hit the walls of the container. In the same way, consider a photon, with energy $E$, momentum $p$ traveling across a box of length $L$. When it starts from one end and bounces back from the opposite wall, it reverses its motion and also its momentum. The net change in momentum $=p-(-p)=2p$. This is also the impulse. The time taken to return to its starting position = $2L/c$, where c is the speed of the light photon. The force experienced by the wall = Impulse/time = $2p*L/2c= pc/L$. Now, for photons, (following Einstein's energy relation, with photon mass being zero), energy is related to the momentum by the relation $E^2=p^2c^2$ or $E=pc$ and hence the force on the wall $=E/L$. (See Chapter 5). Then the pressure is force on the wall/area of the wall, $P=E/(L.A)=E/V$, where $A$ is the area of the wall of the box and $V$ is the volume of the box. So one can see that the pressure of the radiation is energy per unit volume or the energy density of the radiation. Now since the pressure acts on 3 faces of the cube box, the energy density of radiation

$$\rho_R=3P_R \tag{6.24}$$

This type of relation between density and pressure is called the equation of state.

We have seen two kinds of pressures, one that is due to radiation, which give the gravitational-like pulling in pressure (matter energy density also does the same, without any pressure) and the other one is the force of the Cosmological Constant that pushes out everything and therefore has a pressure sign opposite to that of radiation. The energy density of the Cosmological Constant force is the symbol denoted by $\Lambda$ and its pressure is just the negative of the Cosmological Constant energy density or $P_C=-\Lambda$.

The so called equation of state relates the pressure with the density. If we designate this with a proportionality constant with the expression,

$$P=w\rho \tag{6.25}$$

then,

$w=1/3$ -for radiation -from equation (6.24) above.
$w=0$ -for objects and particles with mass- because these do not apply pressure
$w=-1$ for Cosmological Constant.

(For extremely light particles, such as neutrinos, $w<1/3$). As one would expect, the energy density gets diluted as the volume of the Universe gets larger. The knowledge of this is important because the balance between the different energy densities is what drives the expansion or contraction of the Universe. When we work out how this happens, as shown in Box 6.4, equation (6.26), we see that the energy density of the Universe changes with expansion scale factor $a$ as, $\rho = \rho_0 a^{-3(w+1)}$, where $\rho_0$ is the initial energy density.

One can then see that for matter dominated Universe w=0 and therefore, the density falls off as the cube of the scale factor $(\rho = \rho_0/a^3)$, just as particle densities would change with the increase in the size of the box. If the scale factor itself changes with time $a\sim t^p$, this would mean that the energy density changes with time as, $\rho \sim t^{-3p(w+1)}$ (equation 6.27 in Box 6.4). Since p=2/3 for the matter dominated Universe (see above), we see that the energy density in the matter dominated Universe changes with time as (p=2/3, w=0), $1/t^2$, that is the energy density of the matter dominated Universe decreases as the inverse square of the time, becoming one quarter for double the interval.

For the radiation dominated Universe, $w=1/3$, $p=1/2$ and therefore, the energy density decreases with scale factor as $\rho = \rho_0 a^{-4}$, as already noted. The energy density of the radiation dominated Universe therefore decreases with time as $1/t^2$, which is coincidentally the same as for the matter dominated Universe.

However, for the Cosmological Constant which is the expanding force in the Universe, an interesting situation develops. In this case, $w=-1$ and this means that the energy density decreases with scale factor as $\rho = \rho_0 a^0$ or $\rho = \rho_0$. In other words, the energy density of the Cosmological Constant remains the same as the Universe expands. This can mean only one thing, as the space expands, the energy of Cosmological Constant keeps getting generated at the same rate. (It is not surprising then that there is a desire to relate this mysterious Cosmological Constant with the energy of the vacuum. But the two energies are vastly different). The Cosmological Constant and the source of this energy generation is one of the big unresolved cosmological questions.

As shown in Box 6.5 equation (6.28), the rate of energy density reduction with time is proportional to the Hubble constant. There are two contributions to this rate. One contribution (or take away) equal to $3H.P$ is the flow of density due to reduction of pressure because of the Hubble expansion and the other is the contribution $3H\rho$ due to outflow of the density itself.

One can once again see that for the Cosmological Constant, $P=-\rho$ and therefore, the energy density remains constant!

**Box 6.4. Variation of energy density with scale factor and time**

We can determine the dependence of the energy density on the volume. Taking a box with sides of area A, as a sample volume, the force on the wall of the box due to pressure =pressure x area =PA. The work done to move this wall inwards by a small distance dx=dW= PAdx. With dV=change in the volume due to this displacement =A.dx or dW =PdV.

This is the work done in the expansion or contraction of the Universe. This work is stored as energy in the Universe and therefore dW=-dE or dE= -PdV.

But we also know that

E=energy density x volume= $\rho.V$

Differentiating,

$$dE = V.d\rho + \rho.dV = -PdV$$

Or $V.d\rho = -(P+\rho)dV$

Using equation (3.21),

$$V.d\rho = -(w+1)\rho dV$$

Or $\frac{d\rho}{\rho} = -(w+1)\frac{dV}{V}$

Integration gives,

$$\ln\rho = -(w+1)\ln V + \text{Constant}$$

This can be written as, $\rho = \rho_0 V^{-(w+1)}$

Where the integration constant is selected to give the density at the initial time to be $\rho_0$Since, the Universe volume increases as the cube of scale factor $V \sim a^3$. Therefore

$$\rho = \rho_0 a^{-3(w+1)}, \quad (6.26)$$

With the scale factor a varying as $\sim t^{p}$,

$$\rho = \rho_0 t^{-3p(w+1)} \quad (6.27)$$

**Box 6.5. Variation of Energy with Hubble Constant**

Now returning to the definition of the Hubble constant $H = \frac{\dot{a}}{a}$ and substituting the equation (6.26) for expansion in equation (6.15),

$$(\frac{\dot{a}}{a})^2 - \frac{8\pi\rho_0 G}{3}\frac{1}{a^{3(1+w)}} = -\frac{k}{a^2}$$

$$H^2 - \frac{8\pi\rho_0 G}{3}\frac{1}{a^{3(1+w)}} = -\frac{k}{a^2}$$

Now, in a well- known concept of conservation law, also called the continuity equation, the rate of increase of a conserved quantity such as density with time in a given volume, is equal to the net inflow of that quantity (- divergence of that quantity times the speed v of flow of the quantity), or,

$$\frac{\partial\rho}{\partial t} + \frac{\partial(\rho v)}{\partial t} = 0$$

where the partial derivatives are used. We apply this to the change of energy density in the Universe. If the distance of expansion is l, then the speed of expansion is dl/dt.

But, $l = a.l_0$ and therefore, $v = \frac{dl}{dt} = \left(\frac{da}{dt}\right)l_0 = \dot{a}l_0$

From equation (6.26) $\rho = \rho_0 a^{-3(w+1)}$

Differentiating, $\frac{\partial\rho}{\partial t} = -3(w+1)\rho_0 a^{(-3w-4)}.\frac{da}{dt}$

$$= -3(w\rho_0 + \rho_0)a^{-3(w+1)}\frac{\dot{a}}{a} = -3(w\rho + \rho)\frac{\dot{a}}{a}$$

But, $w\rho = P$ and $H = \frac{\dot{a}}{a}$,

Therefore, the fluid conservation equation is given by

$$\frac{\partial\rho}{\partial t} = -3(P+\rho)H \quad (6.28)$$

## Physics of the Origin of the Universe, and Inflation Theory

Before we go into the conditions and Inflation prior to Big Bang, let us look at the details of the need for invoking Inflation as the process for creating flatness as a starting point for the Big Bang.

We saw that during the radiation dominated epoch, the scale factor increased as $t^{1/2}$ and switched to the $t^{2/3}$ variation for the matter dominated epoch. In either case the Hubble constant $H = \frac{\dot{a}}{a}$ varies as $1/t$. This then means that the product of scale factor and the Hubble constant $aH$ varies as $1/t^{1/2}$ for the radiation dominated epoch and $1/t^{1/3}$ for the matter dominated epoch. Essentially the product $aH$ decreases with time. However, the important parameter that determines whether the Universe expands or contracts is given by the quantity $\Omega$ which is the ratio of the density to the critical density (to balance the pushing and pulling forces so that a flat space exists), which is given in Box 6.2. Equation (6.17) gives,

$$\Omega - 1 = \frac{k}{a^2 H^2}$$

This means that $\Omega$-$1$ keeps increasing proportionally with $t$ or proportional to $t^{2/3}$, with time. Since, $\Omega$-$1$ keeps increasing with time and $\Omega$-$1$should be and remain zero for a flat Universe, the Big Bang model (without the Cosmological Constant) does not lead to flatness over time and actually leads the Universe away from it. Therefore, one needs to have the inflation to have taken place to lead to flatness that we see today. This, in turn, means that one needs a pushing out force both to explain the origin of the Universe which overcame the Gravitational force in the beginning to launch the Universal expansion and an early phase which produced flatness and homogeneity as said before.

## THE COSMOLOGICAL CONSTANT

The Cosmological Constant was an ad-hoc term included in the equations by Einstein to make for a static Universe (because he thought, like Newton, that the Universe was static). Without this term, the Universe would contract. When the Cosmological Constant $\Lambda$ is included, the above equations include the continuing force that pushes the Universe out. This is like a space vehicle that is launched from the earth with rockets and rockets continue to burn along the way up and out.

Equation (6.14) can be rewritten as

$$\left(\frac{\dot{a}}{a}\right)^2 - \frac{8\pi\rho G}{3} - \frac{\Lambda}{3} = -\frac{k}{a^2} \qquad (6.29)$$

and as,

$$H^2 = \frac{8\pi\rho G}{3} + \frac{\Lambda}{3} - \frac{k}{a^2} \qquad (6.30)$$

If there were only the Cosmological Constant ($\rho$(mass density)=$0$ and $k$=$0$), then $H = \sqrt{(\frac{\Lambda}{3})}$. This gives the case of a Hubble constant, which does not change with time and corresponds to the case shown in Box 6.1. The scale factor then increases exponentially as, $a = e^{\frac{\Lambda t}{3}}$. In the Inflationary period such a Cosmological Constant term exponentially expanded the Universe about 60 e-fold or 100 doublings or by a factor of $10^{30}$.

The argument for including the Cosmological Constant is the same argument that we used for invoking fine tuning by Inflation to make for a flat Universe (see above). We saw that without the Cosmological Constant, the Big Bang theory takes the Universe away from flatness (k=0). It can also be applied in the context of the dynamics of evolution (or time dependence and not the initial state) with an ad-hoc Cosmological Constant. But, if a small perturbation expands the Universe, it would create a vacuum creating a negative force that, in turn, makes the Universe expand further. In other words, the situation would be unstable. When this was pointed out to Einstein, soon after he had suggested this term, he admitted that it was a big mistake. However, he would prove to be prescient. At present, the Cosmological Constant is revived again, since it is now confirmed that the expansion of the Universe is accelerating and only a Cosmological Constant can do this.

The Cosmological Constant is neither a constant with time nor is it identified with a specific phenomenon or source. At the beginning of the Universe, when there was nothingness, this Cosmological Constant would have been a very large negative energy density associated with that vacuum state and this is what gave the push to expand and give birth to the Universe. The same Cosmological Constant would also have been the driver for Inflation. These Cosmological Constants are far greater than the energy density of the vacuum. The present Cosmological Constant that has been accelerating the expansion of the Universe since about 9 billion years of age may be of a different kind and is now identified with the so called Dark Energy.

## Inflation in More Detail

Alan Guth is one of the people who solved the primary problem of fine tuning of the Universe to yield a homogeneous flat Universe. Guth was influenced by physicist Robert Dicke who, in 1961, had pointed out the flatness problem. He said that we are extremely fortunate to be in a highly balanced situation of a flat Universe (now called the Dicke Coincidence) or something was missing from the theory of Big bang.

There are very few articles where one can find the idea of inflation in a way that is understandable. Let us revisit the problem. The Universe that we see today is very homogeneous and very flat (no curvature of space). As time passes and we see that as new objects and new parts of the Universe come into our horizon of observation they look no different from our past observed Universe. This means that the whole Universe that is observable by us and not only what we see today, must be homogeneous. Since, in our present era, the new regions come into the horizon only as the time passes, we were not causally connected to these regions in the past. This also means that, the homogeneity we see with these regions could not have been established in the past. Could it have been created at the time of the birth of the Universe? But, as stated before, the Universe was formed in very short time scales of the order of $10^{-35}$ seconds because that is how long the energy to form the Universe was available. The Universe, even that small at that time, (of the order of 1 meter) had to have all regions, with different energy densities, to be causally connected to smooth out inhomogeneity. On the other hand, all information is passed on and causal connections are made only at the speed of light. For a 1 meter size Universe to be in communication, it would need about $3x10^{-9}$ seconds, which is enormously longer than the age of the Universe. So there was no time to establish the homogeneity. Without a process to homogenize the Universe soon after it formed, the Universe would have had a blotchy non uniform appearance. The homogeneity would be limited to very small regions and in the present Universe, even these formerly small uniform regions would have created non uniformity because of gravitational effects. This is the problem. (See figure 6.12).

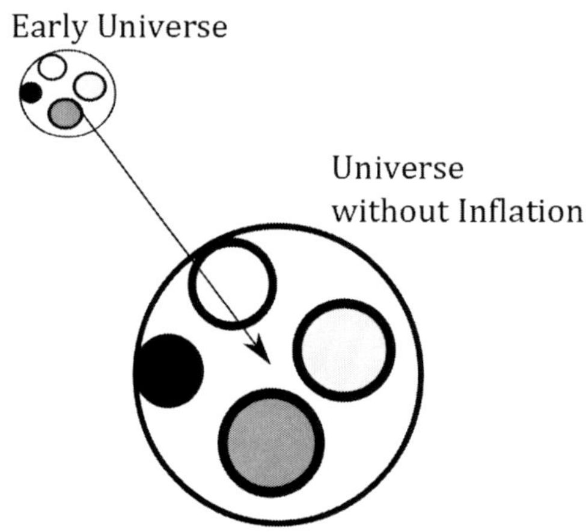


Figure 6.12. Without Inflation, the expanding Universe would be increasingly inhomogeneous, blotchy and non uniform.

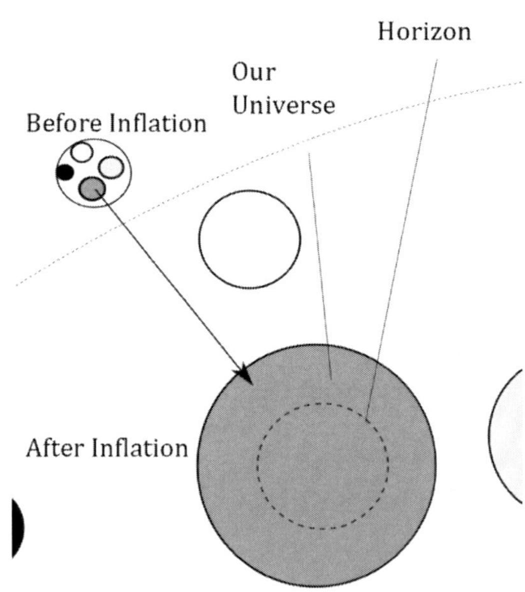


Figure 6.13. Universe with inflation.

But there was a very rapid expansion or inflation soon after the Universe came into existence. In an expansion which is much faster than the speed of light, the space would expand superluminally, much too fast for light to keep in touch with farther regions. In such rapid inflation, the horizon moves very slowly (only at the speed of light) while the early

Universe expanded enormously and therefore, our small part of the Universe kept in causal touch and remained flat and homogeneous. (See figure 6.13). As the horizon moves out at the speed of light during the slower Hubble expansion, our Universe remains homogeneous and flat with us. The conceptual understanding of this is difficult to obtain and the above explanation is inadequate. Further explanation would require more mathematical and conceptual steps involving the Hubble horizon, the product $aH$ of scale factor and Hubble constant and the idea of co-moving frames. Interested readers are urged to read R.J. Peeble's Introduction to Cosmology, cited above.

The early impulse for the rapid expansion or inflation was given by a sort of energy that is opposite of gravitational energy. These forces have been brought to an exquisite equilibrium (fine tuning) of slow expansion of a homogeneous and flat Universe, by Inflation. A cosmological Constant type (negative) energy density of $10^{71}$ gms/c.c appears to have driven the inflation. So what is the difference between Inflation and the expansion after the Big Bang? The Big Bang model takes the energy density away from the critical energy ($\Omega$= 1) for flatness, and Inflation brings the energy density to the critical value for flatness. This is because, in presence of the Cosmological Constant, the departure from critical density is given by the equation,

$$\Omega - 1 = \frac{k}{a^2H^2} - \frac{\Lambda}{3H^2}$$

The Cosmological Constant term has an opposite sign to the first term and therefore reduces $\Omega$–1. Interestingly, the acceleration of expansion or deceleration can simply be obtained from the Hubble expansion, as an equation of motion by differentiating the FRW equation and this is the equation (6.31) in Box 6.6. (The second derivative of the scale factor, $\ddot{a}$ is the acceleration). This equation is important and shows more clearly that, in presence of mass and gravitating pressure due to radiation, there is deceleration of expansion ($\ddot{a}$ is negative) and the Cosmological constant causes acceleration ($\ddot{a}$ is positive) of the expansion (motion) of the Universe. Therefore, we see that the introduction of the Cosmological Constant to keep the Universe flat might not have been wise, but it is necessary to make it and keep it flat. (As noted before, this is also the term related to the Dark energy that is presently accelerating the Universal expansion).

One of the tenets of modern physics is that quantum mechanics is at work even when there is no matter or energy or anything. Nothingness or a quantum vacuum is actually filled with potentialities and these potentialities arise out of the fact that this vacuum is seething with ghostly particles and waves which are highly transient. A pair of high energy matter and anti-matter particle can be created out of a virtual photon which in turn arises from a fleeting availability of energy out of this low energy state. An electromagnetic wave can also be created. The famed quantum mechanical principle states that the product of uncertainty in time and uncertainty in energy is a constant equal to $h/2\pi$, where $h$ is the Planck constant= $6.62\text{x}10^{-34}$ Joule-sec. Energy conservation is (allowed to be) violated by making a particle or a wave with energy $E$ for a duration $h/(2\pi E)$. So, in the time of $10^{-42}$ second, an energy equal to 2 x π x $6.62\text{x}10^{-34}$Joule-Sec/ $10^{-42}$ second, equal to $3\text{x}10^{9}$ Joules is available. This is of the order of the Planck Mass and is created out of the nothingness. This is the simplest picture of how the Universe was born.

Figure 6.14. Economic Inflation and Cosmic Inflation.

We do not have a clear picture of the primordial state and have only theories, with few or no observational data. The current theories point to the primordial Universe state to be a singularity meaning like an entity where there was no property and only potentially emergent feature of quantum fluctuations. Physicist-Mathematician and originator of the concept of the "Omega Point", Frank J. Tipler states in *'The Physics of Christianity"* (Tipler 2007), *"It is essential to realize that, although the laws of physics require the initial singularity to exist, the laws of physics cannot apply to, cannot constrain, the initial singularity. This is because the laws of physics are equations that are defined only for finite entities."* (This argument is very similar to that of Hindu Upanishads and exegeses where the cruelty or justness of the Universal God is discussed. See below). It is also useless to talk of sizes at this point because there was no space or time, but the conditions described above correspond to a wavelength (attributable as a dimension of the original size of the Universe) of $10^{-35}$ meter. From the limits of energy forms, (as also from the energy and the dimensions mentioned here), some calculate that the energy density of the primordial state of the University to be $10^{97}$ to $10^{116}$ kg/m$^3$, others estimate a much lower value (as low as $10^{17}$ kg/m$^3$, still quite large). Among the many potential Universes that could come out of this seething quantum singularity, our present Universe emerged with an initial energy density. This much energy was packed into the singular state. Once it comes into existence, it can either "big crunch" back, because of the sharp gravitational forces and go back out into nothingness or expand within a fleeting moment out again to a void filled continuously expanding Universe, like a ball thrown by an impulse. The unknown outward driver, the Cosmological constant may be the energy associated with vacuum then or the so called dark energy or something else.

**Box 6.6. Equation of motion for the expansion of the Universe**

From Equation (6.29) with a flat universe (k=0)

$$(\frac{\dot{a}}{a})^2 - \frac{8\pi\rho}{3m_{pl}^2} - \frac{\Lambda}{3} = 0$$

Differentiating, we get

$$\frac{2\dot{a}\ddot{a}}{a^2} - \dot{a}^2\frac{2\dot{a}}{a^3} = \frac{8\pi}{3m_{pl}^2}\frac{\partial\rho}{\partial t}$$

Or $\frac{\dot{a}}{a}\left\{\frac{\ddot{a}}{a} - (\frac{\dot{a}}{a})^2\right\} = \frac{4\pi}{3m_{pl}^2}\frac{\partial\rho}{\partial t}$

However, from equation (6.28)

$$\frac{\partial\rho}{\partial t} = -3(P+\rho)H$$

Substituting in the above equation and with H=$\frac{\dot{a}}{a}$

$$\frac{\dot{a}}{a}\left\{\frac{\ddot{a}}{a} - (\frac{\dot{a}}{a})^2\right\} = \frac{4\pi}{3m_{pl}^2}\{-3(P+\rho)H\} = \frac{4\pi}{3m_{pl}^2}\left\{-3(P+\rho)\frac{\dot{a}}{a}\right\}$$

Rearranging the terms,

$$\frac{\ddot{a}}{a} = \frac{4\pi}{3m_{pl}^2}\{-3(P+\rho)\} + (\frac{\dot{a}}{a})^2$$

Substituting from equation (6.29)

$$\frac{\ddot{a}}{a} = \frac{4\pi}{3m_{pl}^2}\{-3(P+\rho)\} + \frac{8\pi\rho}{3m_{pl}^2} + \frac{\Lambda}{3}$$

Rearranging the terms,

$$\frac{\ddot{a}}{a} = -\frac{4\pi}{3m_{pl}^2}(3P+\rho) + \frac{\Lambda}{3} \qquad (6.31)$$

Equation (6.31) gives the equation of motion for the Universe. One can see that the Cosmological Constant gives a positive value for the acceleration, while the pressure and energy density cause slowing down of expansion.

What possible causes are there for the Cosmological Constant? Quantum mechanics shows that the lowest energy state of vacuum has the so called "zero point" energy. As space expands, vacuum fluctuations appear and particle physics shows that this requires energy and therefore a deficit in energy (negative energy) is created when space expands, but this in turn opposes the gravitational pull and therefore this is like a self organizing phenomenon. (This is similar to a piston being pulled out, creating space and results in the reduction of the energy density of the gas inside). Detailed physics of Cosmological Constant involves many different aspects because the physics changes depending on the epoch. Even though the Cosmological Constant energy density remains constant in expansion, phase transitions from one epoch to another (due to the so called symmetry breaking) change the Cosmological Constant energy densities. The question is whether the vacuum can account for the Cosmological Constant.

One can look at the Cosmological Constant at present, although, this may not give much direct information for the early Universe. Measurements of the vacuum energy density have been carried out from careful observations of the deviation, from Kepler's laws, of planetary orbits. This gives a density of the order of $10^{-16}$ kg/m$^3$. Similar measurements of Milky Way galaxy and the geometry of the Universe give $10^{-24}$ kg/m$^3$ and $10^{-29}$ kg/m$^3$ respectively to account for the inflation that provided the background for the Big Bang. One can calculate the vacuum energy using particle physics theories, but the "Cosmological Constant Problem" is that the observed vacuum energy density is way too small (by 40 or higher orders of magnitude) from the calculations from the Quantum Field Theory. This is one of the foremost physics problems of today. It had to be far greater in the early Universe. We know there is a "dark energy" in our present expanding Universe, which not only counters the gravitational pull of the masses and some amount of radiation energy, but overcomes it and the Universe is accelerating its expansion by a little bit (74% of the magnitude of energy density comes from this in the present Universe). The presently termed dark energy may or may not include the description of the nature of the negative pressure of early Universe. So other candidates for dark energy are also being looked at and other alternative theories are being proposed too. There is also a problem for the future, because in ten billion years from now, the dark energy will be 96% of the energy density. Is the Dicke Coincidence question still in play?

## The Inflaton, the Scalar Field of the Early Universe

In the primordial, extremely dense, extremely hot state of matter, there were no distinguishable forces and there was no matter or light. The starting point of the Universe was probably a singularity of space and time and this packed a very large energy density. For this size and this energy, both the gravitation forces and quantum mechanical (termed quantum gravity) descriptions apply. This theory (sometimes called the Theory of Everything) is still in development and is in its infancy. What might describe the Universe at this point? We know that the electromagnetic forces, the weak interaction which causes radioactive decay, the strong interaction force of quarks and nuclei and the gravitation forces were all merged (unified) into one. There is a Grand Unified Theory (GUT) which attempts to describe this state. (Alan Guth came to the concept of inflation through this GUT theory applied to particle physics. He was investigating, why monopoles, which would be expected to be aplenty from the GUT, were not yet found in the Universe. The theory of inflation explained the freezing out of the formation of monopoles, because the Universe cooled down quickly due to Inflation).

The fields, that we are generally familiar with and that are associated with forces and particles, are normally vector fields, although their source may be ascribed to a scalar potential. (A vector field has a strength and a direction and its components in different directions are proportional to the rate of change of the scalar potential in that direction). Often, such a field influences particles of specific types. The well- known example is that of an electric field arising out of a distribution of electric potential and the electric field exerts a force in a particular direction, on an electrically charged particle like electron. But, as we know, the Universe is isotropic, meaning there is no preferred direction. So, the GUT theory uses a scalar field for the description, because vector fields, such as electric fields would select a direction at each point of the Universe for description. The famous Higgs Boson that confers mass (a non -directional property) to particles, is associated with a scalar field. While, previously there were no examples of scalar field observations, the discovery of the Higgs Boson confirms the existence of such fields. The scalar field that provides a description of Inflation and the later evolution of the Universe is called an inflaton denoted by the Greek letter $\phi$. Since the only character the Universe has is this scalar field, the state of the Universe is represented by the scalar field itself. So there are no particles associated with the inflaton. In other words, the distribution of $\phi(x,t)$ is the state of the Universe, where x is the position and t is the time. The description of the behavior of the Universe is then is given by how the magnitude and distribution of the inflaton changes with time.

The Inflaton has an associated kinetic energy and a potential energy. The Inflaton then behaves like a particle. Equation (6.37) in Box 6.7 is the equation of motion for the inflaton and is like the equation of motion of a particle (this is only an analogy), with an acceleration (second derivative of $\phi$) due to a gradient in the potential (like gathering speed on a down slope) and a friction term (like gravel on the sloping ground), that is proportional to the velocity (first derivative of $\phi$ ) . So, if the Inflaton is sitting at the top of a down slope of a potential, initially it has a high potential energy and no kinetic energy. It will roll down the slope and gather speed and the friction impedes the force of the potential gradient.

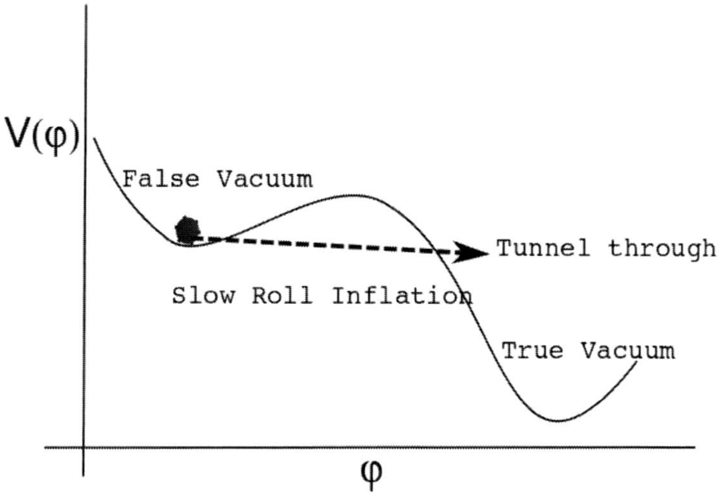


Figure 6.15. Inflaton rolling down a potential gradient.

**Box 6.7. Dynamics of the Inflaton, the scalar field potential**

The energy density associated with Inflaton has components similar to a particle in a potential. In analogy with kinetic energy of a particle and its potential energy (say, a particle rolling down a hill), the inflaton energy density is given by

$$\rho_\phi = \frac{\dot{\phi}^2}{2} + V(\phi) \tag{6.32}$$

Where, the dot represents time derivative of $\phi$ and V the potential energy of the field is the function of the inflaton. (So, in this equation, $\phi$ is like the position variable x and the first term is like the kinetic energy $mv^2/2$, where $v=\dot{x}$. Unlike matter and radiation, the energy density and pressure of a scalar field are independent and the pressure of the inflaton is given by

$$P_\phi = \frac{\dot{\phi}^2}{2} - V(\phi) \tag{6.33}$$

Since for inflation, the inflaton field has to dominate the matter and radiation energy density and pressure, we can neglect the collapsing energy and pressure of the masses and radiation. We can use equation (6.31) with equivalent $\rho_\phi$ and $P_\phi$ and drop the Cosmological Constant.

$$\frac{\ddot{a}}{a} = -\frac{4\pi}{3m_{pl}^2}\left(3P_\phi + \rho_\phi\right) \tag{6.34}$$

Rewriting the mass (energy) conservation law in equation (6.28) for the inflaton,

$$\frac{\partial \rho}{\partial t} + 3\left(P_\phi + \rho_\phi\right)H = 0 \tag{6.35}$$

Substituting for the scalar field pressure and density in equations (6.32) and (6.33) in (6.35),

$$\frac{d\rho_\phi}{dt} = 3\dot{\phi}^2 H \tag{6.36}$$

Since $\phi$ and $\dot{\phi}$ are independent, (also from the continuity equation given in Box 6.5), the total derivative of energy density with time can be written in terms of partial differentials,

$$\frac{d\rho_\phi}{dt} = \frac{\partial \rho_\phi}{\partial \phi}.\frac{\partial \phi}{\partial t} + \frac{\partial \rho_\phi}{\partial \dot{\phi}}.\frac{\partial \dot{\phi}}{\partial t}$$

$$\frac{d\rho_\phi}{dt} = \frac{\partial V_\phi}{\partial \phi}.\dot{\phi} + \dot{\phi}.\ddot{\phi}$$

where the partial derivatives of $\rho_\phi$ are obtained by partial differentiation of equation (6.32). The double dot represents second derivative with respect to time and is the acceleration. So, using equation (6.36) and rearranging the terms, the acceleration of the inflaton down the slope is given by,

$$\ddot{\phi} = -3\dot{\phi}H - \frac{\partial V_\phi}{\partial \phi} \tag{6.37}$$

This rolling down a slope is the system's way of reaching a minimum energy state, as is the tendency of the system according to thermodynamic principles. At the beginning, the inflaton is in a metastable state of a false minimum energy (See figure 6.15), a false vacuum. There is, however, a true vacuum, a true minimum energy state past the immediate hill of the false vacuum. Because quantum mechanics gives finite probability of tunneling through a hill, the inflaton, in a given Universe, tunnels through this hill then "falls down" (following the equation of motion) the large gradient in energy density rapidly, maintaining a positive acceleration. This was Alan Guth's model (using quantum mechanical particle theory). It explains the fine tuning to give flatness and the homogeneity. However, there is one problem, why would not the inflation be a runaway phenomenon leading to an almost instantly formed cold Universe? This is very similar to expanding a gas too rapidly, which leads to supercooling, which means that the gas does not nucleate to create liquid drops (opposite of superheated water) but stays as very cold gas. The formation of bubbles is called a first order transition, in which the bubbles, thus formed, interact and coalesce to form liquids and similar new states. (One can see this in the bubbles of boiling water, but here the space is fixed, whereas the Universe expands). Similarly, if the Universe expanded too fast and supercooled, there is no time for bubbles of "real vacuum" to interact. Since the bubbles had to interact to create new forms of energy including matter and the manifest Universe, the Universe would not come into the existence the way we see it. In summary, Alan Guth had imagined that the Universe would inflate and then undergo a phase transition to the new Big Bang state. This was called the "Graceful Exit", but it had the above stated problem.

## The Slow-Roll Inflation

Guth tried to fix these problems, but the new approach required its own fine tuning. A solution was almost simultaneously and independently proposed by Andrei Linde in Russia and Andreas Albrecht and Paul Steinhardt in the United States, which used the concept that if the inflaton rolled very slowly down the potential gradient and for a sufficient time, these issues would not exist. Currently, a version of this, called the Chaotic Inflation, is what most cosmological models now adopt. The description is given in Box 6.8. The slow roll condition requires that speed of rolling and the acceleration be small. This, in turn requires that the gradient of the potential should be small and that the gradient should also not change much (or the gradient of the gradient should be small).

**Box 6.8. The Chaotic Inflation Model**

We again restate all the previous results for the inflaton in the Universe without mass or radiation densities.

And Equation (6.34) gives $\frac{\ddot{a}}{a} = -\frac{4\pi}{3m_{pl}^2}\left(3P_\phi + \rho_\phi\right)$ (6.38)

Since, from equation (6.32) and (6.33) from Box 6.7,

$\rho_\phi = \frac{\dot{\varphi}^2}{2} + V(\phi)$ and $P_\phi = \frac{\dot{\varphi}^2}{2} - V(\phi)$

For the slow roll approximation, Inflation only requires $\ddot{a}$ to be positive (in equation (6.38), so that $P_\phi$ <- $\rho_\phi$ /3. If, for slow roll inflation, the inflaton rolls down slowly, then the kinetic energy is much smaller than the potential energy and can be neglected in the above equations for the pressure and density. In that case, $P_\phi$ =–V(ϕ)=– $\rho_\phi$. This is just like the Cosmological Constant described before. Now, if the "velocity" $\dot{\phi}$ has to remain small, the acceleration $\ddot{\phi}$ should also be negligible.

Equation (6.37) gives, $\ddot{\phi} = -3\dot{\phi}H - \frac{\partial V}{\partial \phi}$

For slow roll we need $\ddot{\phi} << -3\dot{\phi}H$ and $-\frac{\partial V}{\partial \phi}$ or (6.39)

The two terms on the RHS should be nearly equal, so that,

$\frac{dV_\phi}{d\phi} \sim -3\dot{\phi}H$ or $\dot{\phi} = -\frac{1}{3H}\frac{dV}{d\phi}$ (6.40)

For kinetic energy $\frac{\dot{\phi}^2}{2}$ to be small compared to V, one condition for the slow roll inflation is that

$(\frac{1}{3H}\frac{dV}{d\phi})^2 \ll V$ (6.41)

Differentiating equation (6.40),

$$\ddot{\phi} = -\frac{d}{dt}\left(\frac{dV}{d\phi}\right).\frac{1}{3H} = -\frac{d}{d\phi}\left(\frac{dV}{d\phi}\right)(\frac{\partial \phi}{\partial t}).\frac{1}{3H} = -\frac{d^2V}{d\phi^2}\frac{\dot{\phi}}{3H} \ll -3\dot{\phi}H$$

Or the condition is, $\frac{d^2V}{d\phi^2} \ll 9H^2$ (6.42)

The equation (6.15), with k=0 for a flat Universe and the early Universe without matter or radiation and only the scalar field is, (as above, V(ϕ)= $\rho_\phi$, when we neglect $\dot{\phi}$).

$$H^2 = \frac{8\pi\rho_\phi G}{3} = \frac{8\pi V G}{3} \tag{6.43}$$

So that the slow roll condition for a small $\ddot{\phi}$ becomes,

$$\eta = \frac{1}{24\pi V G}\frac{d^2V}{d\phi^2} \ll 1 \tag{6.44}$$

The other condition for slow roll in 6.41 is also written the same way so that

$$\epsilon = \frac{1}{24\pi V G}\left(\frac{dV}{d\phi}\right)^2 \ll 1 \tag{6.45}$$

## Reheating

An inflated Universe would be in a highly cooled state having expanded $10^{30}$ times. So the temperature must have dropped by about $10^{27}$ deg K to $10^{22}$ deg K (the drop in temperature depends on the model). In this cold state no matter or force carrier particles can be created and that is reason monopoles which could come into existence in the earlier hot state, were not produced. In order to create the particles that we see today, there has to be reheating to pre-inflation temperatures. The large, scalar field potential energy was large, but since the gradient was very slow to maintain inflation, there was plenty of potential energy left for the inflaton to use. So, most of the balance of the Scalar field potential energy was thermalized by falling off the potential (see figure 6.16) and reheating. This thermalization was accompanied by oscillations in the scalar field. The post inflation period is rich in oscillations in the scalar field. After inflation, the scalar field decayed into another scalar field- the Higgs field and also the fermion field. This changed the friction term in equation (6.37) from being proportional to the Hubble constant to a more complex, the daughter scalar and fermionic (Dirac), field strength dependent, term. In a recent theory, this is shown to be a resonant process and therefore the decay of the original scalar field into new particle associated fields, seems to have happened very quickly. The friction term during this decay process became the source of reheating. With a hot mix of the two new scalar fields, the stage for the Big Bang was set and the particles and associated fields started appearing. But this reheating was insufficient to create substantial monopoles.

So, in the early Universe, the energy of the scalar field would have been more or less constant and then after the reheating, much of this energy was converted to other fields and in reheating of the Universe, so that the present level of this scalar energy is extremely small. As the Inflaton rolled down gently, it rode along with a level of quantum fluctuations, but as it fell down the steep part during reheating, the fluctuations were localized and locally amplified and farther ones were separated. This is like a wavy line of tins going over a water fall. There will be a few who would fall first and then during the fall would accelerate to go farther away from the later set of falling tins. This formed the separated Universal objects such as black holes, which then gathered the small scale inhomogeneity like galaxies.

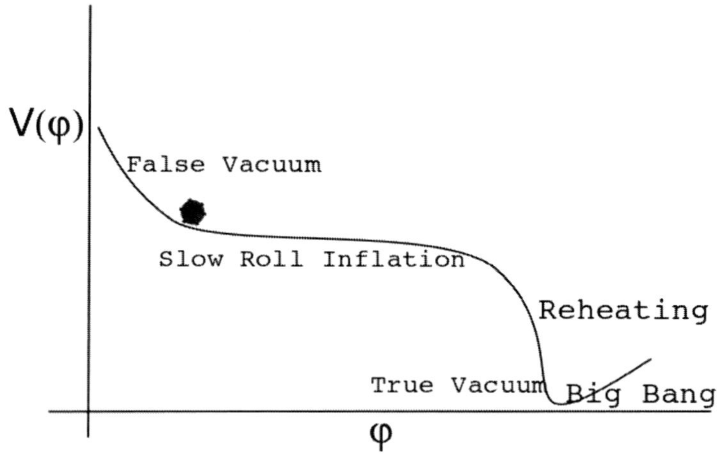


Figure 6.16. The Slow Roll and Reheating of the Universe due to the steep gradient at the end of roll and conversion of the potential energy into thermal energy.

In the Chaotic Inflation theory or Eternal inflation theory, all of the regions of the original Universe do not go through inflation and reheating at the same time. If there are regions of fast inflation, such regions (Universes) would dominate and there could be other regions that are still inflating. Some have proposed that there could be many bubbles of false vacuum with different laws and different constants of nature.

***

The pursuit of the physics of the Cosmos has been rewarding particularly during the past decades. The highlights of this continuing voyage of discovery are:

- Early wonderment about the Cosmos and attempts at a description of what could be seen.
- The discovery that we are not at the center of the Solar System.
- The stars are like our Sun, except they are far away.
- These stars are in a group and the vague milky line at the zenith is the Milky Way galaxy where our own solar system is.
- There are other galaxies just like our own.
- There are billions of such galaxies with billions of stars in them.
- These work under the same principles of mechanics and Gravitation as is evidenced on earth.
- This gravitational mechanics can be applied to the whole Universe.
- The Universe is expanding, as evidenced by observations and this expansion follows the Hubble law, where the speed of expansion is proportional to the distance.
- The Universe came from a meter size space at times smaller than $10^{-33}$ sec, through Big Bang and has since been expanding.
- Per our current horizon, we will only see over 46 or so billion light years at any time in the future and this horizon is increasing in size.

- Currently our Universe is 13.8 billion years old and we can only see about 13.3 billion light years, as of now and therefore, at the farthest distance we are seeing the light from the Big Bang.
- At the early time, the Universe was dominated by radiation and 50,000 years later matter came about and the matter density has been determining the expansion of space. The energy densities of the Universe were different in each of the eras.
- Earlier than the present expansion, there was a period of Inflation from a singular source of Universe and that homogenized the Universe and created a flat Universe.
- There is a Cosmological Constant or a Dark energy that caused and causes the Universe to expand first quickly and then slowly to allow for matter to evolve and produce life against highly improbable odds.
- Sophisticated instruments and ultra high technology space and physics tools such as instruments mounted on satellites, space probes and telescopes, like the Hubble Telescope have provided a vast range of unimaginable information. High precision observations of remnant radiation from the Big Bang (the Cosmic Microwave Background radiation) over the whole Universe to confirm the homogeneity and flatness of the Universe, the particle physics experiments that enable the investigation into early Cosmology are examples of giant steps in experimental science that accelerated the development of modern Cosmology.

## Cosmology and Society

One of the statements, common and yet profound, made by many people, is that when they are out and they see the night sky shining with stars, they realize how vast the Universe must be and how overwhelmingly large the number of stars and galaxies are. At once they realize how small our own Earth is, how petty their own problems are and how, indeed, these events that affect their lives are just part of the Universal functioning. Then come the thirst for seeing the Cosmos, to wander in the stars and vastness of space and bathe in the colorful lights of the stars and nebula. And then, there is an aching desire to know how the Universe and we came about and how the two are connected. Even without religion or science, this spiritual epiphany and joining with the Universe in a harmonious embrace are the hallmarks of human intelligence, consciousness and spirituality. This common wonder when brought inside the home, the scientific academic halls, the church, the temple, the mosque or governmental buildings, changes into multi various responses and drives. Sometimes, in the midst of worldly concerns, the deep seated drive thins out into static images of objects and then, often, gets lost as mist in the blinding heat of the immediate.

The scientific perception of the Cosmos by individuals varies between a vision of a planetary system to a vague acknowledge of the existence of stars, galaxies and black holes to the persistent curiosity about space and the enterprise associated with it. Cosmology takes a back seat to general inquiry and comes up only in conversations relating to spirituality and religion. Therefore, now as in history, religion deserves a lot of credit for maintaining the interest of people in Cosmology.

During the Neolithic age, Cosmology that people could investigate, were related to weather, Sun and the Moon. Most things were magic and much was attributed to the supernatural. This made the societies use considerable effort and resources to please gods and over the years, the prayers and offerings brought their own lessons. Once people started aggregating into cultural groups, the belief system translated into mythology and the creation myth was a prime part of it. Many myths were rooted in ancestral knowledge of events, such as major floods and earthquakes and some were based on encounters with strange animals and geological features. Some of these remain with us as fairy tales or legends. In later organized, agricultural and developed cultures such as the Sumerian, Babylonian, Egyptian, Greek, Chinese and Indian cultures, the planets, stars and constellations were gods wandering the heavens. These gods with their movement and their relative brightness were deemed to control human history and were duly worshipped to please them. The knowledge of the movements of these objects became a necessity to predict events and to have some control over the future. This cultural-religious basis of Cosmology advanced astronomy far more than many other non-utilitarian fields of science. The Sumerian and Egyptian calendars were developed partly in response to this and partly to plan economic activity and festivals.

With the rise of Judaism and Christianity, and later Islam, the Universe became Earth (human) centric. The teleological interpretations implied that God always had Man in mind when He created the Universe and the life in it. This limited and enhanced scientific research, the scientific temperament and ideas were finally accepted by the society only through the lens of religion. Perceptions changed quite a lot in the Western world after Galileo and Newton's discoveries, when suddenly humans and the chaotic earth were no more at the center of a placid Universe. The feeling of how this affected the populace is echoed by Blaise Pascal (original 1670) (Pascal 1995):

> "... engulfed in the infinite immensity of spaces whereof I know nothing, and which know nothing of me, I am terrified. The eternal silence of these infinite spaces alarms me."

However, in the rest of the world, older religions persisted with notions of a God who was not parochial to earth or humans. The pagan religions all over the world, Hindu religion and Chinese Daoist religion engendered cultural and intellectual explanations for Cosmology. As a result, these societies have experienced a long historical continuity in Cosmological thought and how it affected these cultures. However, in all civilizations, a link between worldly and personal events and Cosmological events is commonly accepted and even becomes a part of its culture. People undertook the passing of creation mythology either through scriptures or through stories as part of a heritage. While spiritual in origin, the subconscious knowledge that the Universe and everything including us are of the same origin, we are all made of star dust and that our destiny is linked to that of the Universe in the very long term, expresses itself in this form.

As material progress takes place and as people spend more time inside their air conditioned or heated homes, offices, factories and school buildings, the mental separation between the self and the heavens increases and the basic human yearning to know and be part of the Universe gets muted or hidden. Nancy Ellen Abrams and Joel R. Primack point to this departure from early human awareness, (Abrams 2001)," *"The universe" is irrelevant to most people in the West, except as a fantasy outlet. The universe plays no part in mainstream religions, except perhaps to demonstrate the glory of a creator. How many people recognize the possibility of a sacred relationship between the way the expanding universe operates and the way human beings ought to behave? What religion teaches that this could be a source of harmony among humans?"* But Cosmological discoveries resonate in society in subtle ways. The increasing sophistication and coherence of the understanding of Cosmology has reflected on to the society in the sense of what is called "critical realism". People see with the corner of their eye that there is a larger and much bigger Universe, even if they don't know much about it. Large sections of society can and do understand fundamental science more through Cosmological discoveries (which, the less informed or connected, may attribute to space research), than any other discoveries. Through this understanding, more advanced societies support research in Cosmology generously, although it is perhaps most removed from any immediate benefit in the form of technology or resource generation.

As to whether the new discoveries make any impact on societal advancement, only the future will tell. We must remember that it was Kepler's laws on heavenly bodies, Galileo's peering at the stars and planets and Newton's description of Gravitation that ushered in the modern era of a materially and technologically advanced world that must have been magical to the people who lived in that previous era. So, it is certain that present Cosmological theories and discoveries that are as far reaching, if not more so, would change the world in centuries to come. This change can come through both scientific advancement and a better appreciation of the science by the public. This will lead to better harmony of the human spirit with the Universe if the societies nurture this knowledge as sacred. But it can go sour if this is deemed as an arrogant acquisition of knowledge for exploitation and conquests.

## RELIGION'S RESPONSE TO MODERN COSMOLOGY

The much celebrated and contested confrontation between Galileo and the Catholic Church is only part of the story of Cosmology. In India, as in Islamic countries, there has never been opposition to scientific discussions and scientific conclusions on Cosmology. Hinduism has much in common with modern Cosmological discoveries because this flexible religion has always been and continues to be open to new ideas. Islam concerns itself less with fundamental Cosmology and its heritage of putting legitimate inquiry ahead of theology, allowed it to permit progress, at least until the 11th century.

The real story is how far Christian societies have come from the days of Galileo. This transition of theology to be consistent with Cosmological findings has been brought by a series of efforts by scientists and scientist- priests from Newton to modern day philosophers. At first, the Aristotelian idea of a Universe that had no beginning or has no end was changed by the Catholic Church to include Creation. By the time Newton had come along, the Biblical idea of waters above and waters below had been abandoned by the church and water, in that context, was taken to be a symbol of ether. But with his Universal Gravitational theory, Newton also quietly challenged the bible statement, *"There are celestial bodies and there are terrestrial bodies; but the glory of the celestial is one, and the glory of the terrestrial is another, There is one glory of the sun, and another glory of the moon, and another glory of the stars; for star differs from star in glory."*- (1 Corinthian 15:40). But Newton maintained much as modern day theologians would, that while his theories explain the planetary motion, God governs all things and He set the planets in motion. By the 19th century, Christian churches had accepted most of the Modern Cosmological ideas. In the 20th century many scientist- Christian believers like the physicist- priest Georges Lemaitre', Edward Arthur

Milne, Robert Boyd and Agnes Giberne engaged in Cosmology, Astronomy and Astrophysics feeling no pressure from religion in their pursuit of scientific discoveries. This remains true to this day.

The allowance by Christian religious institutions, particularly Catholic ones, is not a passive one. If the theologians and theists have a quarrel with Cosmologists, it is not on what happened, but who made it happen. Volumes and volumes have been written by modern theologians as they bend backwards to understand the relevance and implication of Modern Cosmology to theism and their faith. To get a sample of these writings see, (Halvorson 2013) and all the references therein, (Barbour 1997), (McMullin1981) and (Craig 1993). The field of Cosmology in Christian theology has become very rich and various approaches have been taken to explain Christian Bible in the context of Modern Cosmology. The discussions are vigorous ranging from stating that science is still far away from the truths, to how Christian faith should change to incorporate the scientific findings, while not compromising on the existence of God. It is not even possible to understand fully and summarize these highly intellectual writings adequately. But the main theological points are stated below.

When there was no theory of the Big Bang, the infinite Universe with no end or beginning was a problem for the Church. But present day theologians like Frank Tipler as well as some early Christian scholars like St. Thomas Aquinas, can accommodate such a Universe either by redefining time or by separating God's experience from human experience and allow for such a Universe to be created in infinite times. But admittedly, the proof of Big Bang theory in cosmic background radiation and its acceptance by scientists is a shot in the arm for the Biblical creation theology. This gives a clear beginning to the Universe. But, above all, the Biblical statement of God creating the Universe *ex nihilo* (from nothingness), is a major point of agreement with science. The quantum singularity from which the Big Bang resulted using the large energy available in a brief instant, is very much in the line of Biblical teachings- a causal point, a true creation from nothing. But, according to these modern theologians, the First Cause is still there (namely God). In the so called Kalam Cosmological Argument by the Islamic scholar Al-Kindi, the causal nature of the Universe implies a cause and this, in Old Testament and Christianity, is the First Cause. But Al Gazzhali ((see (Rafiabadi 2005) for this philosopher's views), is said to have protested this interpretation. He says, *"According to the hypothesis under consideration, it has been established that all the beings in the world have a cause. Now, let the cause itself have a cause, and the cause of the cause have yet another cause, and so on ad infinitum. It does not behove you to say that an infinite regress of causes is impossible."* (Kamali 1963). Tipler answers this by stating that if we keep going backwards for causes and forward for events, then the whole series is like a geometric series. A cause causes an event which, in turn, causes the next event and so on. Tipler, however, assumes that the successive events are smaller (in scope or impact) and therefore this is like a geometric series with a common ratio of magnitude less than 1. In this case, the sum of the series is a finite number and similarly, Tipler argues that, the causes sum up to a first cause. The atheistic (for example, Stephen Hawking's) argument that there is no first cause is based on two points. One is that there is no merit to the question of what was before the Singularity, because time did not exist till the First Event occurred and the second is that such events have always been happening, creating different Universes with different times and different properties (the Multiverse theory). Indeed, these two points mark the sharp contrast between theist creation and an atheist existential explanation.

The second Cosmology finding that is dear to the theistic philosophy, is that the Universe is "fine tuned". The initial process of inflation leading to a flat Universe is one aspect of the fine tuning Also, the Cosmological Constant and its energy density, after the period of Inflation (and such as the value at the present), became so negligibly small compared to other forces and their energy densities that the Universe expanded slowly allowing for the evolution to attain bigger and better and long lasting things. This, as we said before, is the Cosmological Constant Problem in physics, one of the most fundamental issues that are unresolved. Theologies state that this fine tuning (and the extraordinary fine tuning of all the celestial objects and earthly events to allow for humans to evolve) is the Doing of the Divine. The fine tuning of the expansion, the specific set of physical constants such as the speed of light, Planck's Constant, Boltzmann Constant etc., the precise ratios between the strengths of different forces in nature, the precise nature of the unfolding of the events such as the development of the Solar System, the right distance of the Earth to the Sun and the Moon, the tectonic plate movements and the creation of Himalayas had to be precise for humans to evolve. Such a coincidence is, indeed, mind boggling and drives one towards a supernatural and teleological explanation. Polkinghorne states, (Polkinghorne 2011) *"The religious believer will see Cosmic fine tuning as an endowment of potentiality given by the Creator to creation in order to bring about a fruitful history that fulfills the divine purpose."* In fact, even a Universe with more or less than the dominant 3 spatial dimensions would either be too unstable or too simple to have brought about life. Andy Fletcher lists the processes that have created the favorable conditions of life in *"Life, the Universe and Everything: Investigating God and the New Physics"* (Fletcher 2006). Indeed, as Roger Penrose has done, if we calculate the probability of the early and the emergent Universe to have the right amount of order, then it would come out to be extremely small.

The so called Strong Anthropic Principle states that the Universe has (must have) conditions compatible with life that would observe it and the proponents of this principle believe that it is then unremarkable that the conditions are so tuned for life. Many scientists who do not even believe in this principle often make statements that sound like they do. However, the weak Anthropic Principle states that we discuss this uniqueness because we are here and the Universe has indeed brought us about. Or if the Universe had been different and the earth not being the way it is, there would be no such discussion. The reason we see that conditions are just right for life to exist, because otherwise we would not be here and now but somewhere else at some other time and discussing the right conditions there. The Universe appears to be fine tuned for conscious life, but it is only a matter of consistency.

There is the possibility that Universes are popping up everywhere all the time. An estimate of the possible number of Universes is $10^{500}$, a very large number and yet not adequate to explain the tuning of the Universe for life, though many atheists believe that this may give the possibility of our Universe and the weak Anthropic Principle to function. (See (Stoeger 2007) for a good discussion of the Anthropic Principle and multiverse). However we slice the cake, the fact remains that we humans are in a special place and special time. We are amazingly fortunate to be here and experience it, however miserable and pointless any one person may imagine his or her life to be. The wonder and humility are expressed in "*Contact*", the book by Carl Sagan (Sagan 1985) and its film adaptation.

The vigorous debate on Cosmology and Christian religion's response show that the science of Cosmology is widely accepted among all circles. Christian scholars have respectfully accepted the verified and verifiable history of the Cosmos found by physics.

They do not question the scientific discoveries on matters of stars, galaxies, black holes, hypernova etc., but their eyes are fastened on the only one area – whether these findings are consistent with theology of the existence of God who is the Creator. These scholars deserve the credit for keeping their eyes and mind open and reexamining and reinterpreting the tenets of their religion. For the average Christian, this should give further strength in his or her trust in science.

The foundations of Hindu scriptures, the Vedas, are said to contain all the knowledge. In particular the *Rg Veda* explores and makes statements on the origin of the Universe itself. Creation is discussed without the context of time and space in an agnostic vein in the Hymn of Creation in *Rg Veda:*

There was neither non-existence nor existence then.
There was neither the realm of space nor the sky which is beyond.
What stirred?
Where?
In whose protection?
Was there water, bottomlessly deep?
There was neither death nor immortality then.
There was no distinguishing sign of night nor of day.
That One breathed, windless, by its own impulse.
Other than that there was nothing beyond.
Darkness was hidden by darkness in the beginning,
with no distinguishing sign, all this was water.
The life force that was covered with emptiness,
that One arose through the power of heat.
Desire came upon that One in the beginning,
that was the first seed of mind.
Poets seeking in their heart with wisdom
found the bond of existence and non-existence
Their cord was extended across.
Was there below?
Was there above?
There were seed-placers, there were powers.
There was impulse beneath, there was giving forth above.
Who really knows?
Who will here proclaim it?
Whence was it produced?
Whence is this creation?
The gods came afterwards, with the creation of this universe.
Who then knows whence it has arisen?
Whence this creation has arisen
- perhaps it formed itself, or perhaps it did not -
the One who looks down on it,
in the highest heaven, only He knows
or perhaps He does not know.

In another portion of *Rg Veda* called *Sukthas,* Cosmic creation is depicted abstractly and contradictorily as Gods sacrificing the *Purusha,* the unmanifest entity, and generating a Brahma who then creates the Universe. Later *Puranic* texts fall more in line with mythological explanations with a trinity of God tasked with creation, preservation and destruction, repeated in cycles. This latter description resonates with physicists who believe in a cyclic Universe. The famous astronomer Carl Sagan, in his public television series "Cosmos" (1980) identified this Hindu philosophy in explaining the Universe as the dance of the God *Nataraja* who creates and destroys the Universe in cycles. Sir Roger Penrose points out to Hindu theological coincidence and "pleasing coincidence" (and is no more than pleasing coincidence).

In an impressive display of intellect, the *Brahma Sutra* (undated, perhaps before 5th century) declared that from theological considerations, there was no beginning: (The following can be found on the Internet web (Jayanarayanan 2013). Also see (Badarayana 1999)).

> "If it be argued that it is not possible (to take Karma into consideration in the beginning), since the fruits of work remain still undifferentiated, then we say, no, since the transmigratory state has no beginning." – (Brahma Sutra 2.1.35)

This was the interpretation of the debater's statement in Chandogya Upanishad (earlier period), which states,

> "There could have been no Karma before creation, in accordance with which a diverse universe could have emerged; for nondifferentiation is emphasized in the text, "O amiable one, in the beginning all this was but Existence, one without a second."-(Chandogya Upanishad 4.2.1)

In an equally adept intellectual argument derived from this, the 7th century sage Sankara declared the Unification of all things in the early Universe and the later appearance of discrimination:

> "It is only after creation that results of work, depending on the diversification into bodies etc. could be possible, and the differentiation into bodies could be possible by depending on the results of work. This will lead to the fallacy of mutual dependence (logical seesaw – the validity of two explanations depending on each other). Thus, well may God become active by depending on the fruits of work after the creation of multiplicity. But before this emergence of diversity it would perforce be without any variety, since the fruits of work bringing about differentiation would be absent."

Hindu Cosmology is highly sophisticated and for many physicists it points the way. Hindu Cosmology is that the Universe is cyclically born out of a Cosmic Egg, the *Brahmaanda* from the *HiraNyagarbha*, the Golden Womb, also the Brahma (not to be confused with *Brahman*, the supreme God). This birth is cyclic in the sense that after every *kalpa* of 4.32 billion years, the Universe is destroyed and recreated. *Rg Veda* (10:129) states *"Darkness there was: at first concealed in darkness this; All was indiscriminate chaos; All that existed then was void and formless: by the great power of Warmth was born that Unit."* *Manu Smriti* (Hindu code for all social classes) states, *'This (Universe) existed in the shape of darkness, unperceived, destitute of distinctive marks, unattainable by reasoning, unknowable and immersed in itself, as it were in a deep sleep. Then the Swayambu (Self Existent) appeared and dispelled the darkness, making this."* All this remains well within a regime that is consistent with modern Cosmology. Other typical mythologies provide further descriptions, such as the churning of the milky ocean with a primordial mountain using a primordial snake, brought out the various things and life forms in the Universe. The sophistication of Hindu understanding is reflected by the statement of self organization of the aspects of the Universe, in the *Mantra Pushpam* of *Yajur Veda*:

Fire is the source of water,
He who knows this,
Becomes established in himself,
Water is the source of fire,
He who knows this,
Becomes established in himself.
He who knows the source of water,
Becomes established in himself,

In this chant, other stanzas have the Sun, the Moon, the stars, air and the clouds and these are similarly related to water and thereby relate to each other in the circular fashion.

One can see that Hindu theists, to begin with, would have no problem with how old the Universe is nor any difficulty with humans not being the first to be created. Together with the incarnation mythology, Hindu Cosmology itself can easily align with Modern Cosmology. The concept of a Universe being born out of a singularity is quite cogent with that born out of an egg and the nothingness before it would also be consistent with Hindu Cosmology. The Big Bang concept is considered to be so aligned with Hindu Cosmology that it is often claimed by Hindu believers that Hinduism had already known that. Having said that, Hindu religious scholars, are not very knowledgeable about Modern Cosmology and the few who might be, do not engage or perhaps, have no quarrel.

Modern Cosmology is a true demonstration of human intellect and enterprise. The telescopes, the observations, the theoretical efforts and the computations are integrated into a cogent picture of how this Universe came about and how it is doing now. It is an amazing story considering how little our mind is and how vast the Universe and its contents are. Yet, at the level of the average person, science gives the confidence that if it can understand this vast problem, it can surely be trusted to provide guidance in the smaller matters of our own lives and our societies.

## REFERENCES

Abrams, N.E. and Primack, J.R; "Essays on Science and Society", *Science*, Vol. 293, No.5536, (2001)p.1769.

Aquinas, *St. T.; Summa Theologica, Great Books of the Western World*, Vol. 19, Encyclopedia Britannica, Chicago, IL (1952) pp.250.

Badarayana, St.; *Brahma Sutras: Text, Word-to-word Meaning, Translation, and Commentary,* Translation and commentary by Swami Sivananda, Islamic Books, Calicut, IN (1999) pp.189-192.

Barbour, I.G.; *Religion and Science: Historical and Contemporary Issues,* Harper Collins Publishers, New York, NY (1997),.

Craig, W.L. and Smith, Q.; *'Theism, Atheism, and Big Bang Cosmology'*, Oxford University Press, Oxford, U.K. (1993).

Davis, T.M. and Lineweaver, C.H.; *"Expanding Confusion: common misconceptions of cosmological horizons and the superluminal expansion of the universe", Publications of the Astronom. Soc. of Australia,* Vol. 21, No.1 (2004) pp.97–109.

McMullin, E.; *"How Should Cosmology Relate to Theology?," The Sciences and Theology in the Twentieth Century*, Editor: Arthur Peacocke, University of Notre Dame Press, Notre Dame, IN (1981).

Fletcher, A.; *"Life, the Universe and Everything: Investigating God and the New Physics",* Lulu.com, Raleigh, NC (2006).

Guth, A.; *The Inflationary Universe-A possible solution to the Horizon and Flatness Problems*, Stanford Linear Accelerator Publication, SLAC-PUB-2576( July 1980).

Halvorson, H. and Kragh, H.,;"*Cosmology and Theology", The Stanford Encyclopedia of Philosophy,* Fall Edition, Editor: Edward N. Zalta (2013).

Hubble, E.; "*A relation between Distance and Radial velocity among Extra-Galactic Nebulae", Proc, National Academy of Sciences,* Vol. 15, (1929) p.168-173.

Jayakumar, R.; *Accelerators, Colliders and Story of High Energy Physics- Charming the Cosmic Snake,* Springer, Berlin, Heidelberg (2011) p.200.

Jayanarayanan, S.; *Chaturmnaya,* Etext: http://lists.advaitavedanta.org/ archives/ chaturamnaya/2013-March/000012.html.

Kamali, S.A.; *"Tahafut Al-Falasifah- The incoherence of Philosophers",* Proc. *Pakistan Philosophical Congress*, Lahore (1963), p.3.

Livio, M.; *Hubble Cleared*, Nature, Vol.479 (10 Nov 2011)p.150.

Pascal, B.; *"Pense'es"* No. 206 (original 1670*),* Translation by A.J. Krailsheimer, Penguin Classics, New York, NY (1995).

Peebles, P.J[a].; *Principles of Physical Cosmology*, Princeton University Press, Princeton, NJ (1993) p. 108.

Peebles, P.J[b].; *Principles of Physical Cosmology*, Princeton University Press, Princeton, NJ (1993) p. 98.

Poe, E.A.; *The Science Fiction of Edgar Allan Poe*, Editor: Harold Beaver, Penguin, London, U.K.(2006) p.401. EText:American Studies at University of Virginia, http://xroads.virginia.edu/~hyper/poe/eureka.html.

Polkinghorne, J.; *Science and Religion in Quest of Truth*, Yale University Press, New haven, CT (2011) p.74.

Rafiabadi, H.N.; *Saints and Saviours of Islam*, Sarup & Sons, New Delhi, India (2005) p.1-34.

Sagan, C., *Contact*, Simon and Schuster, New York, NY (1985).

Stoeger, W.R.; "*Are Anthropic Arguments, involving multiverses and beyond, legitimate?"*, in *Universe Or Multiverse?,* Editor: Bernard Carr, Cambridge University Press, New York, NY (2007).

Tipler, F.J.; *The Physics of Christianity,* Doubleday, New York, NY (2007) p. 92-93.

*Chapter 7*

# RELIGION, SCIENCE AND SCIENTISTS

There are only two ways to live your life. One is as though nothing is a miracle. The other is as though everything is a miracle. – Albert Einstein (Quoted in (Dellinger 1993).

In the introductory chapter of the book, several relationships between science, scientists, society and religion, were mentioned. This chapter explores these relationships in further detail. Einstein's statement on life's miracles has many facets. It is obvious that there are a few people who, through lack of imagination or being disenchanted with life, might lead lives without acknowledging the miracles. The positive attitude of seeing miracles everywhere is a gift that is given by science. Of course, there is the intermediate case of people, for whom some things are miracles or miraculous, whereas other things and happenings are just routinely expected. Most of us fall into this category. The miraculous or the miracle, now, refers to those things or events that are not naturally expected or cannot be explained by sciences. Such things as winning of a lottery or survival of a sick person or finding a soul mate against scientific odds would be considered a miracle and a child growing up every day to become a woman or a man, is a common happening and not a miracle. Deeply religious and spiritual people also see miracles in everything. They see the benevolent hand of the Divine in all that goes on, and to an extent, the lack of scientific understanding helps in believing in God's miracles. A deeply scientific person is spiritual exactly in the same way, but the scientists' awe of the things and happenings and their impression of these things as miracles are increased many fold, as they understand them scientifically. (Some, who are fascinated by the major scientific discoveries that bring great benefit to people, call these discoveries themselves as miracles of science). Just as these every day miracles bring joy to the scientists and the religious, the two groups are also troubled by the same things and happenings. Injustice, disease, ignorance, crime and regressive culture are the things that both adherents fight. But science and religion also have different goals as discussed in the introductory chapter, and may compete for the same space in the society. The topic of science and religion is, of course, a very intense and broad one, with great intellectuals and mighty philosophers engaged in it.

As science and religions evolve, it seems that any conflict between the two is less about what the truth is, but more about which is the way to truth.

## RELIGION'S VIEW OF SCIENCE

> We have a more perfect knowledge of God by grace than by natural reason. Which is proved thus. The knowledge which we have by natural reason contains two things: images derived from the sensible objects; and the natural intelligible light, enabling us to abstract from them intelligible conceptions. Now in both of these, human knowledge is assisted by the revelation of grace. For the intellect's natural light is strengthened by the infusion of gratuitous light.- St. Thomas Aquinas, Summa Theologica, (Aquinas 1981)
>
> The way of devotion is not different from the way of knowledge or Gyaana. When intelligence matures and lodges securely in the mind it becomes wisdom. When wisdom is integrated with life and issues out in action it becomes Bhakti (devotion). Knowledge when it becomes fully mature is Bhakti. If it does not get transformed in to Bhakti such knowledge is a useless tinsel. - Commentary by C. Rajagopalachari, First Indian Governor General of India. (Rajagopalachari 1965).
>
> The statement that "Allah is omniscient" does not justify ignorance- Abu Raihan Al-Baruni in Kitab Al Tafhim quoted in (Nasr1993).
>
> "One of the truly bad effects of religion is that it teaches us that it is a virtue to be satisfied with not understanding." - Richard Dawkins in The God Delusion, (Dawkins 2008)
>
> "We are redeemed in and with the world, not from the world. Part of our task, then, is to articulate a theology of nature, for which we will have to draw from both religious and scientific sources.." - Ian Barbour quoted in (Stratton 2000).

Barbour believes that the two fields are parallel, relying on paradigms, historicity and reliance on a peer review community. In both science and religion, there is a missionary character in the transmission of ideas, and indeed, in both, there is personal fulfillment and a spiritual experience. The main objection religion has with science is that the scientific spiritual experience is not Enlightenment. Hindu wisdom says it plainly – *'Vigyaana asti vipareeta gyaana'* – Scientific knowledge is an inverted knowledge.

But the lack of moral judgment, the lack of attempt to directionally transform the participant in personal ways, the tentativeness (lack of dogma) of any approach is precisely what characterizes science. While scientists may attempt to do so, science does not burden itself or the practitioner with social or personal values and responsibilities. These latter characteristics contrast strongly with religion and even theology. While, scientists and non-believers celebrate these characteristics of science and emphasize that this is the psychologically and socially healthy way, many religious people and theologians condemn science for it. But the trouble is that some influential religious groups do not stop with condemnation, but have actively pursued the dismantling of the scientific enterprise they disagree with, because it collides with their theology. This has, in the past and present, created a backlash against religion from the rationalists.

The story of the fire in a burning bush which spoke to Moses, the most important prophet of Judaism, is an example of where a religion derives its significant authority to influence society. It does not matter very much if it is a true incident or the imagination of an individual who was under the influence of an entheogen -a chemical substance that gives feelings of contact with the divine (See (Shannon 2008)), the influence is real and has shaped history. But, since the advent of Sciences, the Burning Bush is the metaphor for the burning desire to know nature and the flame of rigorous inquiry into nature, set alight by the pioneers, such as Ptolemy, Aristotle, Aryabhatta, Al Jabr, Hippocrates and Avicenna and kept burning by the intense and insatiate minds following them. This too shaped the course of history. To take poetic license, this rigorous inquiry is the God of scientists.

## CHRISTIANITY AND SCIENCE

Western science, as we know it today, began in the womb of Greek polytheism and within a generation, the first philosophers of science had the audacity to move the explanation from mystical to naturalistic causes- essentially moving away from "The Gods did it". While applied science made serious strides, for 2000 years, Aristotelian philosophy held a stranglehold on scientific thought with some basic semi-dogmatic beliefs, such as: The Universe is composed of four elements, earth is the center of all Universe and the gravity is because all objects are trying to reach its center, all matter changes because it reaches for a higher form. The long fingers of Aristotle were slowly removed by physics, chemistry and mathematical discoveries starting from "Atomism" of Democritus culminating in Galileo's celestial observations and Newton's mechanical world view. Ironically, it might have been the resistance of the conservative church to "natural" explanations proposed by Aristotle that might have cracked open the grip. In a matter of a century or so, science proved that matter accelerated only under a force which could be described by a field, matter consisted of atoms which reacted and formed other matter, Earth was orbiting, along with other planets, around the sun which itself is slowly moving at the edge of the milky way galaxy etc.. Newton had great belief in Christianity and believed that the Bible, in particular, the book of Genesis accurately depicted the Truth. Scientists like Copernicus and Galileo believed in Christian God as well and did not intend for their scientific work to be heretic even though the Roman Catholic Church thought so. Galileo is said to have defended his theories from church doctrine, quoting Baronius, a Vatican Cardinal and Librarian, *"the intention of the Holy Ghost is to teach us how to go to heaven, not how heaven goes,"* (Galileo1957). During this time and for several centuries, science was, in some ways, protected and promoted by Calvinist Protestant sentiments. Though there was unease among theologians with scientific constructs in Europe, references to God continued in scientific literatures during the 1800s. For example, the Bridgewater Treatises (Robson 1990), written by several authors, even found ways to harmonize new discoveries with church doctrines. British geologists, discovering inconsistencies with Genesis, still continued bravely attempting reconciliation. While the mainstream scientists were consistently happy with this marriage, fissures began to appear in the middle 1800s, when the radical working class French began advocating a limited role for God, towards achieving some political ends.

Charles Darwin, in the final paragraph of his epic book on natural selection, "*The origin of Species*" (Darwin 1872), states *"There is grandeur in this (evolutionary) view of life, with its several powers, having been originally breathed into a few forms or into one; and that, whilst this planet has gone cycling on according to the fixed law of gravity, from so simple a beginning endless forms most beautiful and most wonderful have been, and are being, evolved."*. This theory of the evolution of life added an entirely new dimension to the debate and to a few it became a belief system with a capital Evolution. As stated in the earlier chapter, Darwin himself started becoming skeptical of the theological foundations of Christianity and started believing that all religions were equal.

Science was marching along in other fields as well. Cave diggings and archeology showed fossil specimens to be hundreds of thousands of years old, contrary to the teaching of the Bible that the earth was only some six thousand years old. In 1863, one of the last fervent theologian geophysicists Sir Charles Lyell in his book T*he Antiquity of Man*, gave in to the weight of evidence against the Bible, but continued to seek a compromise. In Germany, some scientists and philosophers continued to invoke the Devil and God's hand in the findings and in England, parliamentarian Gladstone vainly tried to reinterpret statements of Thomas Huxley (known as Darwin's bulldog) in favor of Genesis. While society was practically benefiting from the scientific discoveries and breeding people's trust in science itself, an expert class of scientists and philosophers and Renee' Descartes and Baruch Spinoza, students of Francis Bacon, empowered itself in discussing the role of science and religion in society, separating the two domains. *The Descent of Man (1874)* by Charles Darwin showed that Man too was an animal descended by the same forces of natural selection. In the US, the ever impressive Princetonian Presbyterian, Charles Hodge, in his book, *"What is Darwinism?"*, argued that if Charles Darwin's theory excluded *the* design argument, it was effectively atheism and could not be reconciled with Biblical Christianity. Asa Gray, the famed botanist, responded that Christianity was compatible with Darwin's science. Both he and many other Christians accepted various forms of theistic evolution, and that Darwin had not excluded the work of the Creator as a primary cause. This demonstrates that even among scientists, there were many with willingness to coexist with concepts of a Supernatural power or even a scriptural God. Churches on the other hand, kept away from debating these topics to a point of resolution and simply adopted a view that (only) scientific notions that promoted atheism and/or against Christian teaching were anathema.

In 1874, John Tyndall, a pantheist and a famous Victorian scientist, addressed the British Association for the Advancement of Science, calling for rationalism and a superior place for scientific truths over religious and non- scientific views and dogmas. The popularity of philosophers like Tyndall, initiated public debate on the conflicts and consonance between science and religion, which has grown ever since. Yet, the idea of an omniscient and omnipotent supernatural power remained. In 1896, the worldwide conflict and debate between science and religion took on new energies with the publication of "T*he History of Warfare of Science with Theology*" by the American author Andrew Dickson White (White 1993), which argues vociferously for throwing out religious dogmas while accepting scientific discoveries, without denying the existence of a great benign power. Following is a quote from his writing:

"The world is finding that the scientific revelation of creation is ever more and more in accordance with worthy conceptions of that great Power working in and through the universe. More and more it is seen that inspiration has never ceased, and that its prophets and priests are not those who work to fit the letter of its older literature to the needs of dogmas and sects, but those, above all others, who patiently, fearlessly, and reverently devote themselves to the search for truth as truth, in the faith that there is a Power in the universe wise enough to make truth-seeking safe and good enough to make truth-telling useful."

(This noteworthy quote has been a beacon. The wonderful feature of Christian theology, not emulated as much in other religions, is that Christian theologians of the recent past noted and acknowledged the scientific discoveries that impacted religious belief and people's morality. Even more remarkably, they diligently study the sciences and then ruminate about and debate the implications to Christian theology. No other theologians from Hinduism or Islam or Buddhism have done such a thorough study of science for theological discussions).

In the 1900s, there were an increasing number of intellectuals coming out against theism. Notable atheists and agnostics included algebraist Emil Artin, astronomer-chemist Francois Arago, popular writer-philosopher Mark Twain, political theorist and a giant of his time, Isaiah Berlin, French philosopher Gilles Deleuze, Australian philosopher John L. Mackie, the premier logician Bertrand Russell, the Austrian-British philosopher Karl Popper, Jawaharlal Nehru, the first Prime Minister of India, inventor Thomas Edison, physicists Pierre Curie and Paul Dirac and so on. Later years in 1900s saw the famous biologist Peter Medawar write wittily and strongly against religion and belief in God, but had this illuminating and humble opinion (Medawar 1988),

"... I believe that a reasonable case can be made for saying, not that we believe in God because He exists but rather that He exists because we believe in Him... Considered as an element of the world, God has the same degree and kind of objective reality as do other products of mind... I regret my disbelief in God and religious answers generally, for I believe it would give satisfaction and comfort to many in need of it if it were possible to discover and propound good scientific and philosophic reasons to believe in God... To abdicate from the rule of reason and substitute for it an authentication of belief by the intentness and degree of conviction with which we hold it can be perilous and destructive... I am a rationalist—something of a period piece nowadays, I admit."

Nehru, despite his great reverence for the theist Mohandas Gandhi, was harsh in his assessment of religion. He stated,

"The spectacle of what is called religion, or at any rate organised religion, in India and elsewhere, has filled me with horror and I have frequently condemned it and wished to make a clean sweep of it. Almost always it seemed to stand for blind belief and reaction, dogma and bigotry, superstition, exploitation and the preservation of vested interests."- Jawaharlal Nehru quoted in (Cardiff 1945).

"Religion merges into mysticism and metaphysics and philosophy. There have been great mystics, attractive figures, who cannot easily be disposed of as self-deluded fools. Yet mysticism (in the narrow sense of the word) irritates me; it appears to be vague and soft and flabby, not a rigorous discipline of the mind but a surrender of mental faculties and a living in a sea of emotional experience. The experience may lead occasionally to some insight into inner and less obvious processes, but it is also likely to lead to self-delusion." (Nehru 1994).

The famous arguments against belief in a personal God and/or against creation by a God were and are usually buttressed by scientific discoveries and successful scientific theories. On the other side of the fence, perhaps in response, the theologians, theists and believers adjusted their arguments and beliefs to conform to the irrefutable scientific discoveries while they kept their objections to other ideas and discoveries that directly contradicted creation by a God and negated faith based approaches. In a radical approach, opposed by the Roman Catholic Church, Pierre Tielhard de Chardin, a Jesuit priest and a paleontologist dismissed traditional interpretations of Genesis on the creation of man. In his book *"The Phenomenon of Man"* (deChardin1959), which was completed in 1930 but posthumously published, he stated that human evolution is an evolution towards complexity, from cell to animal to a human of higher consciousness to an aggregation of consciousness of all individual consciousness to what he called "Omega Point". Some now believe that the modern internet is trending towards such a consciousness. (Frank Tipler at Tulane University has come up with his version of an Omega Point for the future cosmological state in which, expansion of information which has to be consistent with the size of the Universe, will tend to infinity with sentience occupying all of the Universe. This, he says would collapse the Universe to an Omega Point).

This trend has continued into the 21st century. Richard Dawkins, in *"The God Delusion"* (Dawkins 2006), states that in the face of astounding and ground breaking discoveries and theories on cosmology and high energy physics, many protagonists of religious and non-religious theism and Intelligent Design have changed their arguments and present the scriptural texts as symbolic and say that any one taking a literal meaning of the scripture does not understand. In other cases, they use the gaps in science (which are shrinking), to posit God in those gaps. Atheists like Dawkins, Tom Lehrer, Peter Atkins, Christopher Hitchens, Isaac Asimov, Daniel Dennett and James Randi have drawn sharp lines, arguing that faith and religion obstruct growth of an individual and society. In a brilliantly argued response to Francis Collins (former director of National Center for Human Genome Research and a geneticist, who in his book, *"Language of God"* (Collins 2006a) proposed that evolution and Christian theism are compatible), George C. Cunningham, (*"Decoding the Language of God"* (Cunningham 2009)), states that the ideas of Christian miracles, the inconsistency of free will theories with Christianity and the conflict between a loving God and observed "evil" happenings in the world, are demonstrations of an unscientific and illogical basis of Christianity. Steven Weinberg states that this seriously debated issue in the early Church, has become a parody and tells about a mock society (formed by people who are against teaching creationism), asking for equal time in school for teaching flat earth geography (Weinberg 2008). Weinberg believes that there is definitely a tension, if not a conflict, between science and religion and this tension is not due to contradictions between scientific theories and Christian doctrines. He suggests four reasons:

- Before science, religion had gained acceptance and authority from awesome and mysterious phenomena like earthquakes, thunder etc. and this acceptance eroded after more and more scientific explanations became common knowledge. Religious belief was weakened, particularly in more scientifically advanced places and times.

The rational nature of these explanations has also made clear that we have not encountered anything that requires supernatural explanations.

- These explanations have taken away the special place and role of man in the cosmos that Christianity holds. Both the majestic size of the Universe and the theories of Darwin and modern theories of evolution point to the insignificant size and impact of humans on the Universe and also that man's evolution did not follow any special divine intention or intervention.
- The third source of tension is that science seems to take away God's authority and a sense reflected even more in Islam, namely that science ties God's hands.
- Religion relies on authorities such as prophets and scriptures and the inviolability of teachings. While science does accept authorities, it does not do so in guaranteed ways. No authority remains sole authority and a discoverer of any scientific idea is only a forerunner of further ideas and faith in an idea is based on a chain of evidence. Weinberg states, "*If I want to understand some fine point about the general theory of relativity, I might look up a recent paper by an expert in the field. But I would know that the expert might be wrong. One thing I probably would not do is to look up the original papers of Einstein, because today any good graduate student understands general relativity better than Einstein did. We progress. Indeed, in the form in which Einstein described his theory it is today generally regarded as only what is known in the trade as an effective field theory; that is, it is an approximation,..*" (Imagine what would happen if a religious person said that any modern priest knows about Christianity better than a saint.")

Independent of science, atheism had been in the western world for ages. Originally, anti-science, it has now been enabled by advances in physical sciences and Darwin's theory. There are and were great intellectuals and common people, in all walks of life, who are agnostics or atheists. But, this population remains a small fraction of the population of the western world (including Russia), (less than 15% while the world average is about 2%). The fraction of atheists is increasing slowly and steadily, unrelated to science because while scientific interest is modulated by latest science news, atheism is not.

It is important to point out that there has been a continuum of beliefs and convictions between theism, atheism and agnosticism even among scientists, illustrated by utterances and positions of notable scientists: (In fact, this applies even to the question of whether there is a conflict between science and religion). The following quotes illustrate this:

> I'm very astonished that the scientific picture of the real world is very deficient. It gives a lot of factual information, puts all our experience in a magnificently consistent order, but it is ghastly silent about all and sundry that is really near to our heart, that really matters to us. It cannot tell us a word about red and blue, bitter and sweet, physical pain and physical delight, knows nothing of beautiful and ugly, good or bad, God and eternity. Science sometimes pretends to answer questions in these domains, but the answers are very often so silly that we are not inclined to take them seriously- Erwin Schrödinger (Schrödinger 1954).
>
> I shall reexamine the suppositions underlying our belief in science and propose to show that they are more extensive than is usually thought. They will appear to coextend with the entire spiritual foundations of man and to go to the very root of his social existence. Hence I will urge our belief in science should be regarded as a token of much wider convictions.- Michael Polanyi, Hungarian Physical Chemist and polymath, (quoted in (Jha 2002).

> The overwhelming impression is one of order. The more we discover about the universe, the more we find that it is governed by rational laws. . . . You still have to ask the question why does the universe bother to exist? If you like, you can define God to be the answer to the question. _ Stephen Hawking (quoted in (Flew 2008)).
>
> There is a tremendous tradition of distinguished scientists who were and are Christians. I hope that my work is considered sufficiently outstanding to fall into the distinguished among that category. I also hope I have given you enough evidence that you will never again believe that it is impossible to be a scientist and a Christian.- Henry Schaefer (Schaefer 2003a).
>
> "Both of these choices (turning back on Science and turning back on faith) are profoundly dangerous. Both deny truth. Both will diminish the nobility of humankind. Both will be devastating to our future. And both are unnecessary. The God of the Bible is also the God of the genome. He can be worshipped in the cathedral or in the laboratory. His creation is majestic, awesome, intricate and beautiful - and it cannot be at war with itself. Only we imperfect humans can start such battles. And only we can end them."-Francis S. Collins,(Biologist and Director of the Human Genome Project), (Collins 2006b)

In addressing the conflict between science and religious belief, Stephen Jay Gould describes science and religion as each comprising a separate magisterium of human understanding. (The term magisterium borrows a term from the Catholic Church,). According to him, science defines the natural world, and religion the moral world. If each realm is separate, then according to Gould, they are not in conflict. He calls this the principle of non-overlapping magisteria, or NOMA. (Gould 1999).

> "I cannot understand why we idle discussing religion. If we are honest—and scientists have to be—we must admit that religion is a jumble of false assertions, with no basis in reality. The very idea of God is a product of the human imagination. It is quite understandable why primitive people, who were so much more exposed to the overpowering forces of nature than we are today, should have personified these forces in fear and trembling. But nowadays, when we understand so many natural processes, we have no need for such solutions"- Paul Dirac (quoted in (Heisenberg 1971)).
>
> "My own faith, my scientific faith, is that there is nothing the scientific method cannot illuminate and elucidate"- Peter Atkins, in "On Being" (Atkins 2011)
>
> "One of the great achievements of science has been, if not to make it impossible for intelligent people to be religious, then at least to make it possible for them not to be religious. We should not retreat from this accomplishment." (Weinberg 1999)
>
> "The universe we observe has precisely the properties we should expect if there is, at bottom, no design, no purpose, no evil, no good, nothing but blind, pitiless indifference." (Dawkins 1995).

However, many intellectuals such as John Hedley Brooke, (*Science and Religion - Some Historical Perspectives* (Brooke 1991)), believe that the thesis of conflict between science and religion, "*...in its traditional forms, ... has been largely discredited*". The Templeton Foundation has instituted and awarded an annual prize for individuals who show that there is no conflict between science and religion. Several authors have written books and articles with this viewpoint. There have been and are many attempts to reconcile Christianity and science, such as the efforts by intellectuals like Gottfried Wilhelm Leibnitz and Thomas F. Torrance. Recently Pope Benedict appointed the Nobel Prize winning biologist Abel Warner as the President of the Pontifical Academy of Sciences to continue this work. In 2005, the French cardinal Paul Poupard expressed the desire to end the mutual prejudice between religion and science. Elsewhere, the reconciliation is even more of an active process. The State University of New York in Binghamton offers an accredited course on "Evolutionary Religious Studies" - the study of Darwin's theory of evolution from a religious point of view. Even the U.S. National Academy of Sciences got into the act, by letting the Templeton Prize be announced at its venue on March 2013, to a nominee the NAS president himself had nominated.

Notable among the theologically devoted and scientifically minded independent thinkers is the 20$^{th}$ century Canadian Jesuit priest, Bernard Lonergan, who, while being loyal to theological faith, remained an adherent of scientific methods. His famous quote, when speaking of the philosophy and theology of his time, is- *"One entered the rationalist door of abstract right reason, and came out in the all but palpable embrace of authoritarian religion"* (Lonergan 1972). His influence on the Catholic Church, in moving away from the "dangerous" Aristotelian or dogmatic ideas of nature, is evident when we read his writings. His tome and extremely difficult and amazingly profound work, *"Insight"* (Lonergan 1972), is an invitation to theologians to understand the scientific process and apply it to religious thought without abandoning the ethics and faith in religion. He invites readers to the goal of "self appropriation" by obtaining first an objective knowledge through observation and understanding phenomena and then judging the truth of it from metaphysical consideration. He suggests that the final step of this will be a subjective and yet the only true understanding. In this the knower and the knowledge would have a complete and consistent ("isomorphic") structure. His famous quote speaks to his point of view:

> "There is bound to be formed a solid right that is determined to live in a world that no longer exists. There is bound to be formed a scattered left captivated by now this, now that development, exploring now this and now that new possibility. But what will count is a perhaps not numerous center, big enough to be at home in both the old and the new, painstaking enough to work out one by one the transitions to be made, strong enough to refuse half measures and insist on complete solutions even though it has to wait." (Lonergan 1993).

This sentiment is also expressed by H.D. Anthony in the preface to his book, "*Relativity and Religion*".

> "It will no doubt be admitted that philosophy, science, and religion must together ultimately throw light on the vexed question of human personality and man's place in the Universe; but one shrinks from this task of co-ordination, because of the high standard of general knowledge that would be required. The whole question of what is really the essence of Christianity must, on the theoretical side, be related to current philosophical and scientific theories." (Anthony 1927).

In *"Science and Religion in Quest of Truth"*, John Polkinghorne (Polkinghorne 2011) states, *"taking science seriously does not imply a necessary denial of the reality of divine providential action."* He restates strategies to bring modern religion and theology in tune with scientific discoveries, by understanding and reading scripture with *"respect...to the genre of what has been written"*, recognizing symbolism and avoiding *"a simple flat footed literalism."*. He proposes that diligent development of ideas such as theistic evolution (idea that the scientific theory of evolution is true and was set in motion by God) and theodicy (reconciling evidence for evil with a benevolent God) will bring science and religion together.

Alvin Plantinga (Plantinga 2011), in a more radical argument with Dawkins, Dennett and Atkins, uses tools from logic. He questions the very reliability of reasoning as the basis of acceptance of an idea. In arguing that naturalism itself is not consistent with science ideas such as evolution, he questions these authors' definitions of science. He states that the "deep conflict" between science and religion does not exist, but a "deep concord" does. He argues that our quest for knowledge about our world and ourselves is consonant with a God creating us in His image. He states that what science claims as course of nature is essentially God's will. The laws that explain and predict are the description of the constancy and potency of God.

There are many like Polkinghorne, Lennox, Lonergan, Collins and Schaeffer who believe that, with or without some tweaking, religion is not in conflict with science. In their eagerness to reconcile religion with Science, in addition to intellectual inquiry, they demonstrate a desire that these be reconciled because of their belief in both of the domains.

Like Lonergan's, there are other sophisticated theologies that depart radically from orthodox views. An example is the process theology of Alfred North Whitehead and Charles Hartshorne. It postulates its own interpretation of God's attributes. God is temporal as much as the Universe is and God evolves. God co-creates the Universe, co-evolves with the Universe and does not control. Therefore, in this view, God is part of the Universe and very few conflicts with science remain, once the existence of such a God is postulated. This is a highly sophisticated view in which there is no Cosmogony in the sense of creation out of nothing (*ex nihilo*) instead the creative potential (*Potentia*) is always present and creation is out of chaos (in the traditional meaning of the term). There is a good resemblance to modern day cosmology, but the concept uses a more cyclic creation and dissipation of the Universe (and God). Process theology removes many contradictions in theology by stating that God is Omniscient but not Omniprescient (knows everything but not anything in future). This removes the contradiction between omnipotence and omniscience. God has the knowledge of eternal objects (basic laws) as well as, His knowledge is changing with the evolution of the Universe. The Process Philosophy God is not judgmental or passionate. *"God as Cosmic Moralist...God as the Unchanging and Passionless Absolute....God as Controlling Power....God as Sanctioner of the Status Quo....God as Male.... Process theology denies the existence of this God."* (Cobb 1976). One can also interpret Process Philosophy as being science friendly, because it admits fundamental laws without "tying God's hands". Indeed, this is true considering that Process Philosophy was viewed favorably by Heisenberg.

As noted above, this has been a remarkable engagement of scientific and Christian concepts, unparalleled and unemulated in other religions. Perhaps because of this and sadly, much of these discussions have largely excluded an examination of conflict or congruence of religions other than Christianity. Because it is easier to provide a contrast, atheist authors like Richard Dawkins have focused on Christianity, Judaism and post 12$^{th}$ century Islam, and in general, on religions of strong faith. (Francis Collins calls these choices as "straw men" in order to knock down). Hinduism, Buddhism, early Islam, Paganism, Shintoism, Zoroastrian, Mayan and Native American religions that have ethnic, ritualistic, natural phenomena based observances cannot be corralled into a narrow belief system and so most discussions by western atheists have kept away from these religions either for convenience or for scoring points. This has been a disappointing and, in some ways, unscientific aspect of this, otherwise intense, intellectual debate. Some of these religions have had notions of vast spaces and vast scales of time and therefore often would not come into conflict with modern scientific discoveries and theories. The creation mythologies of some of these religions are also highly nuanced and are open to interpretations that fit into such discoveries and theories. Examples of two major religions are given below.

## HINDUISM AND SCIENCE

In Hindu society, it is often claimed, though not necessarily truthfully, that Hindu scriptures and other texts anticipated many of the modern scientific discoveries and inventions. Therefore, Hindu society broadly welcomes and folds in intellectual, rational and scientific advancement. In the day to day understanding of a common religious Hindu, there is usually no conflict in accepting naturalistic explanations that do not involve God. While, many if not most, would see a remote hand of God in all this, the religion itself does not come into direct conflict. However, when it concerns afterlife and personal relationship to God and spirituality, Hindus are deeply religious and would denounce science if it denies that connection. Though there are very few atheists in India, 80% of Indian Hindus believe that Darwin's theory of evolution is an acceptable theory. (British Council Survey presented at World Conference of Science Journalists in London by the British Council as part of its international program *Darwin Now,* 2009). Hinduism, like some other religions, is a way of life, where the polytheist practices are incorporated into monotheism (actually non-duality). Hindus have a history of living with apparently contradictory beliefs and so scientific philosophy is just one more view that can be incorporated in the mix. There is also a deeper reason for this: It is generally admitted by scholars that among all religions, Hindu scriptures come closest to agreement with modern science. Hindu concept of time and space are in line with modern physics. Therefore, as physics and interdisciplinary sciences such as brain research, delve into regions that were originally metaphysical, Hinduism and science find alliances. However, broadly stated, from the viewpoint of Hindu religious scholars, modern science is incomplete and should include examination of spiritual realities, not only physical. There is also a minority of conservatives, who hold fundamentalist views, and believe that the *Vedas* and theist notions are infallible and that modern science is just a fool's errand. At the other end, Hindu atheists and rationalists are very frustrated by Hindu adherence to superstitious beliefs, many of which are traditional fear based, and rituals that are wasteful. These rationalists advocate modernizing Hinduism and a movement towards a true scientific attitude. The current rise in Hindu fundamentalism in Indian society is a worrisome feature for seekers of this goal.

Hindu scriptures and philosophical inquiries are highly layered from common mythological stories to the highest form of inquiry into origins, purpose and lack thereof, and existence of God. The *mimaamsaa*s *(*exegeses) are philosophical investigations into the aspect of *Dharma* (the regulatory mechanisms of the Universe) and *karma*, and corresponding conduct of, and effect on, life forms through critical examination of the *Vedic* texts. It is a telling commentary of the catholic nature of Hinduism that there are Hindu religious leaders who are atheists, even though the population of atheists is only about 5%. This also points out that the urge to believe in a supernatural God among the public is hard to remove by intellectual interpretations. Hindu science flourished in the years preceding $12^{th}$ century with many advances in mathematics, technology, medicine and astronomy instruments. Concepts like atomism (*anu*) and nothingness have been present in ancient Hindu ideas. Time units range from 17 microseconds to 311 trillion years and distance units of about 1.5 cm to 1.5 million meters were reckoned by Hindu philosophers. With this background, it is not surprising that Modern Science is familiar to Hindus and they accept major findings.

With the spirit of questioning and argumentation, and with the inclusiveness of atheist religious texts while pondering the truth in *Vedas,* Hindu scholars generally welcome modern inquiry into science and only protest the excesses of technology for use in creature comforts. Modern theories and discoveries are often enthusiastically incorporated in Hindu religious thought, since the curious coincidence of cutting edge scientific theories with Hindu philosophy, is pleasing to them. This enthusiasm has been reciprocated by physicists of high caliber such as David Bohm, Robert Oppenheimer and Roger Penrose. The 18$^{th}$ century astronomer William Herschel and the 19$^{th}$ century mathematician and inventor of the concept of computers Charles Babbage were influenced by Indian and Hindu logic. Physicists like David Bohm and Fritjof Capra have even founded metaphysical viewpoints based on Hindu and Buddhist ideas, while maintaining a rigorous view of science. This author grew up under the strong influence of Hinduism at intellectual level and it is revealing how positively western scientists view Hindu philosophy, in particular, the *Vedas* (scriptures) and *Upanishads* (also exegeses). Some of this is just seeing that non scientific notions are less prevalent in these scriptures and some of it is just romantic attraction for the East. But, there is legitimate attraction due to the keener inquiry into space, time and cosmology and the intellectual tradition of the questioning method of arriving at the truth. The book description of "*Bhagavath Purana and Biology*" by Jonathan Edlemann, (Oxford University Press, USA (2012), states that the author *"...argues that although Darwinian theory seems to entail a materialistic view of consciousness, the Bhāgavata's views provide an alternative framework for thinking about Darwinian theory. Furthermore, Edelmann argues that objectivity is a hallmark of modern science, and this is an intellectual virtue shared by the Bhāgavata. Lastly, he critiques the view that science and religion have different objects of knowledge (that is, the natural world vs. God), arguing that many Western scientists and theologians have found science helpful in thinking about God in ways similar to that of the Bhāgavata."* (Edelmann 2012). (*Bhagavath Purana* or *Bhagavatam* is a mythological portion of Hinduism and *Bhagawath Gita* contains the most significant of God *Krishna's* teachings).

There is an illustrious class of intellectual thinkers in Hinduism, who transcend religion and are engaged in what might more appropriately be called mysticism. J. Krishnamurti, U.G. Krishnamurti and Swami Vivekananda are examples. For them, the investigation is more about the self than the religion. This investigation which uses scientific logic as it needs, attracts many scientifically oriented people. One of the key questions that is legitimate in this inquiry is that since science is primarily confirmed by observations by the self and many such selves, how does one separate the science from perception. But, while admitting the practicality of scientific discoveries, these thinkers and, to some extent Hindu and other religious philosophers, might take an anti-realist position stating that all that is outside our minds is not real or that the mind is interpreting the inputs from things that are not connected to the mind. Then these philosophers would dismiss science as unreal.

But one has to be very careful with this happy marriage of science and Hinduism, because this is only one face of it and the agreement is more socio political and practical than is necessarily true as a whole. One false view is that Hinduism is well tuned with all sciences and there are no conflicts. It must be realized that at least some of the congruence is due to the scope for nimble explanations of the scriptures. The literarily beautiful texts, which use one of the most advanced and colorful languages with many possible interpretations of a single word, have given this scope. Hinduism is like a crystal with a large number of facets. Depending upon what angle one illuminates it, one sees different colors. In this author's experience, this scope has been misused by so many intellectuals that the true meaning of Hinduism is hard to find. In practice too, objectivity is hard to find in Hindu religious discipline and practice, even among Hindu scholars. The other false view is that Hinduism is coherent and holds clear values for Hindus and therefore Hindus can make religious judgments on science. Modern science has only confused many of Hindus, even highly educated ones, in the context of their religion. What can be said is that Hindus are grateful that they can practice and use science without giving up superstitious or meaningful rituals, because these can always be justified in some form of scriptural-scientific defense. Hindu society lives at the edge of contradictions and this is an apt place for them. Nirad Chaudhuri gives an example of this paradoxical belief system in "Hinduism", *"I would say that the more highly regarded a text is in theory the less it is followed in practice (by the Hindus.)"* (Choudhuri 1979). Despite the fact that Gandhi believed that reason is the main way to truth and that faith will lead to such reason, most Hindus do not appeal to reason in their daily life. A Hindu's life is rich with superstitious beliefs, mythological knowledge as well as practical knowledge and armed with these, they maneuver deftly through life.

At a practical level, Hinduism is eminently useful for Hindus and non-Hindus. Since it does not prescribe faith in one particular God and does not bar most practices which do not harm life, Hindu practices are helpful. Yoga, which has a religious ritualistic basis, is a proven health practice like "zen". The meditational and physical exercises are scientifically proven to be beneficial physiologically, mentally and medically. Science and Yoga are congruent, both in specifics of the method and in the efficacy. Similarly, Hindu vegetarianism is also a scientifically healthy practice. These aspects and the fact that adoption of Hinduism permits integration with scientific thinking, have made it very popular in the West. The important aspects, *Dharma* (righteous acts), *karma* (actions and their fruits to oneself) and reincarnation and liberation from reincarnation are the moral aspects of Hinduism. Happily, though religious institutions would not concede this, Hinduism does not dictate belief and thereby one may, according to Hinduism, practice science at all levels, lead spiritually righteous life and attain liberation. This gives great freedom to most scientists (except when they have to harm animals in medicine or biology). Modern Hinduism is very theistic and would not tolerate science if it were explicitly atheistic. Also to most serious adherents of Hinduism, science that denies traditional rituals and practices, would be frowned upon.

## ISLAM AND SCIENCE

Islamic science absorbed and distilled the Eastern sciences from India and China and this established knowledge greatly benefited the development of western science. (This is sometimes known as the Great Titration (Needham 1969)). Islamic scholars absorbed, criticized and corrected Greek scientific methods and conclusions, again enabling later western sciences. Therefore, in the early Islamic tradition, science was an integral part of scholarly study with no conflict with religious institutions or philosophy. Yet, as in India,

Islamic science started declining. Around 1100, Abu Hamid Al-Gazzhali, a brilliant master of philosophy who critiqued Aristotelian principle (falsafa) and who in his attempt to resolve contradictions between reason and revelations in the scriptures, argued that there cannot be any laws of nature, because that would tie the hands of God. By 1350 CE, Islamic science hit rock bottom. There are a combination of reasons for the interruption and isolation of science by invasions and sectarian and religious conflicts. But, it would seem that the Islamic societies lost interest in science. According to Nobel Laureate Abdus Salam,

> "The very encyclopedic nature of knowledge and science in Islam was now a hindrance in an age of specialization. The wholesome faculty of criticism, by which a young researcher questions what he is taught, re-examines it, and brings forth newer concepts, was no longer tolerated or encouraged. " -(Quote from the speech 'Islam and the West' by Abdus Salam, in Paris at the UNESCO meeting.) (Salam 1994).

There are several Muslim leaders and scholars who tried to bring modernity to Islam that included acceptance of Western sciences. In late 19th century, Jamal ad-Din al Afghani, an Afghan scholar and political activist, who traveled widely working for the unity of Islam, is considered to be one of the founders of Islamic Modernism. He was instrumental in making the Islamic people think about finding the right median between orthodox Islamic philosophy and Western science and technology, which did not always correspond to Islamic values. He stated that Islamic theology (*kalam*) should include modern cosmology (of course, then highly undeveloped). But he was selective in his choice of what he would accept of Western science. He condemned the materialism of Greek thought such as atomism of Democritus and while accepting the principle of natural selection critiqued Darwin's theories that implied that there is no hand of Supernatural power in the functioning of the Universe and which undermined the special role of humans. Afghani saw how science was empowering the Western nations and therefore the Islamic world had to embrace the same.

Martin Heidegger has stated that the distrust of science in the Islamic community is because it was considered a colonial project. Today, people living in Islamic countries welcome the fruits of science and technology, while decrying the decadence that the technology has brought to the Western society. However, there is a great scope for the acceptance of modern science by the Islamic world, since there is an increasing pride in the scientific scholarship of Islamic scholars of the period before the 13th century. There are schools such as the Gulen Movement centered in Turkey which emphasize modern learning including science. All Islamic countries have thriving science and technology efforts. In particular, Iran and Pakistan, with strong believers, have very modern science programs. This speaks to the fact that Islam and science can coexist and energetic debates between science and Islam will bring further illumination.

From Greek scientists like Ptolemy and Aristotle to Aryabhata to Ibn Al-Haitham to Isaac Newton to Paul Rutherford to Albert Einstein to Enrico Fermi to present day scientists like Stephen Hawking and Yoichiro Nambu, science has continued in the spirit of discovery, blind to religious and particular vantage viewpoints. For this reason, the Islamic people, as perhaps other religious populations, believe that scientific inquiry and justifications of physical phenomena take away from the spiritual and ethical needs of a society. Single minded pursuit of science, discarding preconceived or prescribed notions, is the scientific method. It just so happens that in this method of inquiry, theism and values interfere with true practice.

Although unlikely, if religious concepts resulted from a scientific pursuit or are demonstrably consistent with the prevailing science, they would be welcomed into the fold of science. Also, part of the fear of science is unfounded. Just as atheists are as moral as religious people, science has its own spiritual quest and ethics of science is not indifferent to the ethics of society. Indeed, there is a commonality of approaches with some of the world religions. It is likely that the distrust of science is based on a lack of full understanding or a misunderstanding of science and its methods.

Religion has been instrumental in the development of arts, music, architecture, cultures, good social norms and in general, building up of communities. Arguably, in the absence of religion, just humanist tendencies might have given rise to these features of a society. But, a stark fact remains. Scientific development could not have come about through religion, even if it is admitted that religion might permit this development. Some philosophers' arguments notwithstanding, it is not possible to accept that religious principles would permit scientific methods of rigorous inquiry. Despite the intellectual credentials of major religions, the fact that the scriptures have not been revised to include scientific discoveries and disseminated as the more advanced religious knowledge, shows that religions do not follow science's example for advancement. In this sense, religion cannot walk with science.

## THE SCIENTIFIC AND RELIGIOUS METHODS

> "At the heart of science is an essential balance between two seemingly contradictory attitudes--an openness to new ideas, no matter how bizarre or counterintuitive they may be, and the most ruthless skeptical scrutiny of all ideas, old and new. This is how deep truths are winnowed from deep nonsense." (Sagan 1996).
>
> "One cannot bring in the instruments of modern physics without sooner or later introducing its philosophical mentality, and this mentality, as it captures the scientifically trained youth, upsets the old familial and tribal moral loyalties." (Heisenberg 1958).

Francis Bacon held that scientists start with an "Inductive View" by generalizing observations and then formulating a theory directly to fit the data. The theory may include hypotheses, postulates and theorems. In reality, theorists do a lot more- they use other concepts not evident in data and their creativity, imagination, intuition and even bias play a role. (Charles Sanders Peirce, father of a branch of philosophy called Pragmatism, theorizes that the process of "abduction", a feeling precedes deduction and induction in an individual scientist's idea. This "sentient" involvement is said to be the creative process behind science). theory then proceeds along the lines of what Systems Engineering approach uses. First a conceptual entity arises, which is then reviewed. This is followed by intermediate updated versions of the theories, produced using a "deductive view" where experiments are specifically constructed or data are mined to test the theory. When the theory passes the tests, a final mature theory is crafted and accepted by the process of large consensus, which is when all scientists, irrespective of their expertise, start using it as a starting point in their pursuits. But, like in production of the final item in manufacturing, the final design (theory) may be altered or even abandoned, if significant problems arise. The rigor of science comes from the fact that agreement of results of a new experiment or new data is only a necessary condition for the Theory and is not sufficient. If a theory can be falsified by even one single, well designed experiment or an incontrovertible and applicable data, the theory would have to be corrected to bring the data into harmony and if that is not possible, abandoned altogether. (Sometimes, a failure of a theory may be related to the failure in one of the elements of a network of theories and this makes the system both susceptible to infectious errors and also makes falsifiability, a long process). Of course, the above is not the only process; sometimes, experiments and observations lead the science all the way into a theory or no theory is ever used and only empirical data is used (as is frequent in engineering, technology or medicine). In a general sense, the subjectivity in science is to be found only in the collective crafting of the methodology of science by a cultural, social sector of humans. But scientists work hard at this aspect and try to exorcise the method of any subjectivity and, some argue, go overboard in rejecting perfectly good science because they may not conform to the standards of objectivity and consensus.

There is another hidden characteristic of scientific method- this is often called consilience. The reliability of a scientific conclusion is based on the fact that, typically, an observation is made using different branches of sciences producing different tools for observation and analysis. For example, if one were to get telemetry data from a space probe giving pictures of the Moon, the construction of the probe, the viewing telescope, the converter of the light signal to the telemetry signal and the receiving and interpreting system on the earth, all use different branches of science and therefore come together in a consistent way. Now if a similar picture is obtained by an earth-based telescope which uses yet another set of sciences and the two pictures agree, this accounts as proof of the picture being true. The above concept of consilience, popularized by the biologist-humanist E.O. Wilson, is an idealization and in reality, measurements by two different techniques rarely agree 100% and the difference between them may be, for a while, considered to be within the accuracies of the apparatus or may spur a new series of scientific activity directed towards understanding the difference.

The concept of empiricism in science is a key one and using it as the basis of science, the Vienna Circle, started by Ernst Mach, Moritz Schlick, Rudolf Carnap, Otto Neurath and others, define and exclude metaphysical phenomena and explanations from science. Empiricism is the basis for knowledge derived from observations based on sensory experience of the observers. Positivism, championed by August Comte in the 19$^{th}$ century, states that authoritative knowledge can only be obtained by logical and mathematical analysis of sensory experiences or empirical data. The essential argument of the Vienna circles is that while empiricism and positivism can be used to break down an observation into simpler statements, metaphysical statements cannot be broken down. Such metaphysical observations or statements should be discounted. For example, a statement that "The dog caught the ball" can be broken down into simpler empirical statements on the kinetics of the ball, the physical nature of the dog and the ball and so on but a statement such as, "The ghost lives in the underworld" has no empirical parts to it. The Viennese circle would also include many ordinary every day "principles" that cannot be broken down into simpler verifiable parts, in the definition of metaphysics. (See (Carnap 1959)). While, not everyone agrees with the definition of science in these terms, the Viennese Circle manifesto presents one concrete facet of scientific attitudes.

Jesse Thomas (personal communication) believes that Stuart Kauffman's idea of "adjacent possibilities" for biological evolution is applicable to sciences and arts themselves. There is much truth to be found in this. Science too evolves by taking advantage of the environment of current scientific thinking or memes (equivalent of genes, according to Dawkins) and the phrase "survival of the fittest" applies to an extent here. For example, a small breakthrough can open up new vistas and start a spurt of discoveries in one subject. Scientific development, like biological evolution, is partly predictable and partly unpredictable. For these reasons, scientific development can also enter into chaotic regime, for example, circling a solution, finding but not arriving at it, being sensitive to starting points and showing phase transitions.

From the vigorous arguments made by either side on science and religion, one would think that scientists make a rational and/or integral choice on religious matters and science. In this author's experience, this is far from true. It is very likely that most of the intellectually accomplished physicists, working in the physics of cosmology, gravitation, quantum mechanics etc., ponder and readjust their belief or disbelief in a supernatural being to reconcile with their scientific knowledge. Even Newton, who is held up as an example of a man of Christian faith, had to abandon orthodoxy and searched for ancient texts which held additional truths outside the Bible, which would square with the developing view of the operating world. A scientist's view of religion is shaped by his or her own experiences as well, but it is true that at least some scientists have to struggle with religious views.

There is a cardinal rule in the pursuit of science - Never expect or favor a particular result. Do the experiment or investigation objectively without bias and influence on the results. In the realm of religion, there is an astonishing congruence of this creed of the scientists with the Hindu Text *Bhagawath Gita*, (See for example, (Easwaran 1993)) which is purported to be the teachings of Krishna, God incarnate:

> "You have the right to perform your actions, but you are not entitled to the fruits of the actions.
>
> Do not let the fruit be the purpose of your actions, and (with that) you won't be attached to not pursuing your duty." (Bhagawath Gita II: 47).

Among Hindus, this is the most simple and yet the most important teaching, which extols the virtue of leading life with non-attachment to fruits of their actions (karma philosophy). In Christianity too, there is the notion of surrendering the fruits to the will of God, but it is complicated by the notion of free will.

What the religious groups and individuals find most irritating (and some even find it immoral) is that scientists, with strong notions of rational, evidence based truths and physical verifiability or falsifiability, may be intrepid in questioning the fundamental tenets of religion and where the answers to these are given in scriptures. They do not seem to realize that this is essentially the way of science. Others, less fundamental in their beliefs, but still holding to the creation of the Universe and life by a God, question the ability of science to answer these questions. This brings us to the dichotomy between rational practitioners of science and people who believe in God.

Religion defies a formal definition in the sense of a broad application. It is a usually a formalized system of beliefs and rituals. As applied to major religions- Sumerian, Egyptian, Hindu, Christian, Islam, Judaism, Shintoism etc., the religions are mostly theistic (except Buddhism and some sects in Hinduism) and hand down beliefs, traditional practices and morality. To a lesser or greater extent, religions are prescriptive both in duties and punishments and rewards. The method of religion is to have faith in these prescriptive teachings and would include in the prescription, the greater prescription of faith in a Supernatural entity. So, in a rigorous pursuit of religion and its teachings, behavior, acts and life itself has to conform to these principles. In a religion, these principles are fixed according to the scriptures and only very occasionally these dogmas are changed by a religious authority, but even those changes are carefully thought out so as not to be in violation of God's or the Prophet's teachings. After accepting the primary scriptures like Bible without question, further interpretations, dispensations are made in the most intellectually debated and sometimes open environment. This process may resemble that of science. However, in a few religions, particularly the less advanced ones and cults, this privilege of questioning would be reserved only to the leader.

Since religion applies to daily life of a vast number of people, these teachings can be implemented only through a reward and punishment system, dispensed both by the religious institutions and by the supernatural beings. In a sense, the physical and moral practices, based on belief and faith, become the religion and therefore, in successful religions, the religion is a way of life. One might interpret this to mean that religion's goal is to institute a way of life that may include the Glorification of a Supernatural. In the strict application of this modality, scriptures and teachings are more or less fixed, to be obeyed and practiced by the people. But, due to the fact that the implementation of this system, in its rigorous interpretation, is impossible, modern day religion allows latitude in not condemning people for violations, but letting them out with a warning in most cases.

Notably, religion, unlike science, pursues all forces in life in an integrated fashion. The fact that religion addresses all aspects of an individual's life and is highly prescriptive, can be both a comfort and inconvenience to the practicing individual. But, the holistic answers to life's questions on physical and mental health, career, family, society, environment and the Universe, make religions spectacularly successful and in the mind of the accepting and non-skeptical participant, it is an ideal guide and the rewards and punishments, a justification to follow the tenets of the religion. The personal angle motivates all religious practitioners, which is true in other fields as in science, but with religion, this becomes the reason for living itself.

These believers usually have no problem in assigning unexplained phenomena to God's abilities to alter the Universe in large and small ways. This helps them to move on and in most cases it strengthens their faith that something inexplicable happened and therefore there must be a God. There is a circularity in the belief system which gives consistency but without external proof. For example belief in the afterlife, purgatory and concepts of hell and heaven are consistent with believing in ghosts, but there is no palpable evidence for any of it. Similar complimentarity can be seen in the fact that since "God can do it", they pray that "He will". This strong hypothesis and the "wild card" of God allow them free passage through all realms of life and the physical Universe, reconciled to contradictions. There is an evolutionary nature to this faith- it adapts to the irrefutable and new scientific facts that contradict their religious beliefs. As an example, a few centuries ago, the incident of a man being struck down by lightning would have been definitely attributed to an act of punishment by God. But with the present understanding of science of lighting and its randomness, an average religious person would agree with the rationalist view that it was just an accident. So, with developments and understandings gleaned from science, the religious interpretations of creation and daily events, the boundaries of religion (or at least religious practice and belief) are continuously being redrawn to include rationality and new scientific discoveries. After all, one cannot deny the physical objects created by scientific inquiry, the television set, the rockets, the nuclear reactor, the medicines for diseases and the X-ray CAT scanner. This even allows them to admit the existence of an electron which, like God, cannot be seen but only felt. But, for the religious, the boundary for acceptance of science is clearly drawn. The Pew survey of June 2009 finds that, though the American public looks at science very favorably. (About 35% of the population believes that science comes in the way of their religious belief). However, if science were to deny the existence of God or if a scientist were to do so, this would be denounced. For the faithful, a hypothesis, such as God is primary, forms the basis of their understanding, purpose and perhaps their identity. To them, this primary hypothesis has to be affirmed by everything, including scientific discoveries, and evidence not supporting the hypothesis should be ignored or just lived with as anomalies created by God Himself in His wisdom. Some would argue that the God hypothesis is self-knowledge and not a religious dogma. But, even so, this foundational hypothesis does not require the proof or consistency with observations, which science requires.

Figure 7.1. Gods off the hook.

Scientists operate with an opposite methodology. In their ideal approach, they would not like anything to be sacred and be beyond questioning. Typically, scientists work by coming up with their ideas and experimental results, challenge others to repeat and prove it for themselves, bring it to the market place of ideas and compete over what makes more sense in the light of existing physical evidence. They constantly try to take sides providing supporting and opposing theories, views and results. This competition can go on for a long time until finally one idea stands up to the weight of evidence and scientific logic. Others are corrected so that they become consistent with the latest scientific understanding. Even so, any assumption or hypothesis is continuously assaulted to test its applicability. It must be noted that hypotheses even such as those of Einstein, one of the greatest minds, are challenged even today. It is interesting that when two theories give same or similar results, one may be discarded in favor of the other because general applicability and a sounder reasoning are valued more. A classic example is the hypothesis of imponderable ether by H.A. Lorentz and the hypothesis of constancy of light speed in Special Theory of relativity by Albert Einstein. Lorentz ether hypothesis and Lorentz contraction (transformations) fully explained the constancy of light speed in all direction of earth, but Ernst Mach and Einstein thought that this was not general enough and that - aether was just a trick not a solution. So, Einstein believed that hypothesizing that speed of light is constant, was more proper. One may find a parallel in the rational view of scientists in the matter of the existence of God. The detachment from the desired end result, the constant error checking and examination of assumptions make scientists take pride in their approach, as opposed to the approach based on faith. While individual scientists are just humans with flaws such as any and may not adhere to the scientific methods at one time or another, be deceitful etc., this internal examination tries to weed out any build up of bad science. It is also true that science would be the first to admit that it has blind spots, but would maintain that such blind spots would probably be revealed in time. On the aspect of spiritualism, science can claim as much spiritual experience as a deeply religious devotee would. It is true that the average scientist, even an avowed atheist, remains in awe of the Universe and the life forces that he or she is investigating.

Science, unlike any other calling, is born in awe, inspiration and curiosity, nurtured by thirst for knowledge of Nature, untainted by a value or belief system. Ideally, scientific work is carried out without fear of treading into new territories or fear of stepping on other's feet and finally realized in the understanding of more than what was sought. There is a serious anti-symmetry between science and religion in this regard. To be accepted as good science, an idea, however trivial or profound it is, has to cross enormous barriers of evidence, logic and agreement with other rational fields it impacts. The arrival of a theory is not its proof, but the path it arrived by has to be illuminated thoroughly and is up for scrutiny any time, even far into the future. So, it is reasonable that scientists believe that accepted knowledge of science stands on much firmer ground than the revealed or "realized" knowledge of religion or metaphysics. To a great extent, this confidence is the reason why Stephen Hawking, Steven Weinberg, Richard Dawkins, Peter Atkins and others feel that the discussion of God is unnecessary or is counterproductive.

Both the religious and scientists operate from a knowledge base and use their respective methodologies to understand the world around them. It is fair to say that since religion is extremely familiar to most people, the scientifically knowledgeable or inclined people understand at least some of the foundational knowledge in religion and its methods. But the reverse is perhaps not true, except for the highly philosophically inclined individuals seen above. The religious understand that scientists do not accept religious faith based knowledge. But, there is confusion in the mind of people who are against scientific methods, because they do not fully understand it. One example is that when a scientist says that there are no ghosts, she means that there is no proper evidence for ghosts, ghosts have not been and cannot be observed and that any such claims have not been verifiable. She is also liable to act accordingly unless and until she has evidence. But the religious, who are more familiar with faith based methodology, think that the scientist "believes" that there are no ghosts and it is a type of dogmatic disbelief. It is unfamiliar to them that the scientist would be the first to jump on to a scientific inquiry if an incontrovertible scientific basis (theory or observation) for the existence of ghosts, is found. However, it must be admitted that this gulf is not easy to bridge. One interpretation that anti-science people might make is that science is dogmatic about its "beliefs" and "disbeliefs".

To illustrate further, if science were to "finally" accept something that religion or tradition has been saying all along (there are very few examples of this, such as in traditional medicine, social practice and *ex-nihilo* Creation of the Universe), this is considered to be "proof" that non-scientific methods are well ahead of science. But what is not noted in this thinking is that this argument cherry picks the topics. Science arrives at its finding, even if it starts with a creative mental process of a scientist or an accidental observation, by the process of rational inquiry and makes sure that the finding is consistent with all the other things it knows. Until, this consistency with known science is proved, the scientific jury is still out. Therefore, the particular congruence with religious or traditional (non scientific) finding, was arrived at by a robust independent process, explains more than just the concept and provides proof so that the finding is sound. In other words, in cases where this congruence between the scientific and the non scientific occurs, one should be thankful to the scientific finding, because it confers robustness to the finding. A clear example is the rather mystifying agreement on the evolution of mammals, between science and Hindu mythology. While wondering at this agreement, it should not be lost that science is what has made the finding a proper one rather than a belief unrelated to other natural phenomena.

For the religious, perhaps because of the success of science in affecting people's lives directly at material level, endorsement by a scientific discovery provides a major support. This support is rarely questioned and as science does, this support will not be examined for the scope and extent of its applicability and how that support might invalidate some other tenet of religion. The religious gladly accept support from science, without endorsing its methods and its primary rule that everything including explanations on the divine should not be inconsistent with all other scientific findings.

## PUBLIC, SCIENCE AND SCIENTISTS

Whatever the field of endeavor, science has earned the respect and no field will go unscathed of criticism if there is no real or proposed, consistency with well-known scientific discoveries and notions. As Susan Gallagher and Roger Lundin state (*"Literature Through the Eyes of Faith"*, Harper Collins), *"...their (sciences') undeniable usefulness in helping us organize, analyze, and manipulate facts has given them an unprecedented importance in modern society."* (Quoted in (Schaefer 2003b)). The abandonment, by a broad public, of the literal interpretation of the book of Genesis, is an example. The number of scientists in the world, who believe in God or a god, is estimated to be less than 4% and is dropping and the number of atheists is increasing as science marches on. This states that, religion, which has much to offer, at least by way of psychological comfort and cultural progress, has to reconcile its methods, rituals and theology, so as not to be in conflict with science. As stated above, religions should update scriptures, lose concept of sanctity of text and make these scriptures consistent with scientific findings. Explicitly stated, the Bible, the *Vedas*, the *Torah* should be rewritten to include modern understanding. The orthodox would believe that such an effort would be heretic, but modern society is too far advanced to believe in a conceptual basis of heresy.

Figure 7.2. Enlightenment.

It is quite clear that science, in intent, proceeds independently of religious and social influence. Social priorities might push one field or aspect of science over the other, but the science of the topic itself is internally consistent. Also, individual scientists may proceed to pursue science in subjective or even personal goal oriented ways. But, the much tested peer review process, the requirement of rigorous proof with minimum uncertainties and inconsistencies, keeps science from the clutches of dogma and arbitrariness. Scientific endeavors such as in psychology, psychiatry, social sciences, economics etc. are not (yet) seen to be adequately advanced to delve broadly into individual and societal issues and find well accepted solutions, but these sciences are progressing in that direction. The fact that rigorous science cannot and will not take shortcuts to examine and analyze these domains is both a testament to its integrity and the reason for the perceived indifference of science to a broad range of topics.

There are three consequences of the above situation with regard to the public's perception of science: (1) The public trusts scientists with their results and knowledge where matters of societal needs are concerned- medicines, electronic devices, communications, transportation, energy etc. (2) The strong theists are firm that science is incapable of solving the true mystery of the Universe and life that has such ordered complexity. Some theists see science as feeding atheism and thereby endorsing a moral ambiguity or even immorality in society. (The average person in United States thinks that atheists are more likely to commit serious crimes, which of course, is nonsense and is opposite of the facts). They also believe that science comes in the way of acquiring 'true knowledge' - the realization of God and purpose of life. (3) For the broader public, which is not 100% committed to theologies either, science and scientists are, more or less, opaque. While there can be excitement on a discovery such as Higgs Boson, there is not a true understanding of why scientists think the way they do or how they arrive at conclusions. For some, science is like the work of a high saint, trying to understand the mind of God. For most others, it is just the activity of a parochial group of workers attached to a particular discipline of work, not unlike farming. The latter opinion is confirmed by a good number of so called scientists who are not dedicated to science but are in it for power or money or just to make a living. But fortunately, there are people who see that scientific discoveries contribute to true knowledge and the scientific profession is unlike any other, not only because it requires a good brain, but also because it shines the light on the highest human aspirations. This group forms the core of the more enlightened people and will guide the future.

But, as John W. Judd said,

> "For it is not only by their intellectual greatness that we are impressed. Every man of science is proud, and justly proud, of the grandeur of character, the unexampled generosity the modesty and simplicity which distinguished these pioneers in a great cause. It is unfortunately true, that the votaries of science—like the cultivators of art and literature—have sometimes so far forgotten their high vocation, as to have been more careful about the priority of their personal claims than of the purity of their own motives—they have sometimes, it must be sadly admitted, allowed self-interest to obscure the interests of science".-Judd, J. W., T*he coming of evolution: The story of a great revolution in science.* (Judd 1910).

Scientists are people too. While they take pride in their pursuit, the competition it engenders and the potential accolades for discoveries can corrupt scientists. In addition, the funds that the scientific enterprises attract and the power it bestows on the leaders, has politicized the process more so than in the past. It is a shame that in many countries, the science managers are the more visible people garnering credit and presenting the face of science, rather than the scientists in the trenches.

Scientists are not a monolithic group representing uniformity of values and morals and are members of their profession or calling, much like any other. Many, if not most scientists, hold independent and non-integrated views on science, religion and faith in a supernatural being or phenomenon. In fact, many scientists do not labor on the issue of conflict between science and religion and simply make their own smaller version of a world view, where opinions, nurtured beliefs, skepticism and rationalism can coexist. (Though, as noted before, these scientists-believers are becoming fewer and fewer). Many scientists, particularly engineers and technologists, paradoxically, may not reflect on the conflict between their conceptions and perceptions. Many even have emotional, philosophical and personal reasons to pursue religious rituals, which do not make scientific sense. This is because, like most humans, scientists are also bundles of contradictions. But, perhaps because of the training, when it comes to their science, good scientists keep their biases, faith and irrationality away, and try to keep their science side untainted by religious considerations. Part of the reason is, as stated before, many so called scientists are in it just for making a living and have less commitment to principles in science. This is increasingly true in a world, where science is funded by external sources and does not demand significant personal sacrifices. In truth, the scientific methodology itself is designed taking into account the weakness in scientists as erring and feeling humans with blind spots and biases. The methodology is developed to be self-correcting to the extent humans can devise such a method. This tradition starts with Socrates in Greece, Confucius in China and Krishna in India and continues until today.

In a curious schizophrenic attitude, much of the public, while enjoying the technological fruits in full measure, can disregard the very science that made that technology possible. Somewhere there must be a misguided logic in their mind that tells them that the two are not connected and may consider religion as equal or even a better guide to science itself. The process of internal squabbling and wild chaotic process of discoveries in science is unknown to them. They do not see the integrity of the process and the overcoming of the challenge in reasoning out the most elusive concepts. As a result, they do not truly appreciate the work of the scientists and the validity of scientific truths, even if they admire them. They do not understand why scientists have such confidence in scientific discoveries, which are well beyond average intellectual understanding and most importantly, cannot be observed or felt in day to day life. To illustrate, they do not understand why scientists "know" there are electrons when one cannot see them. This inability to relate to science leads to confusion and this is one of the reasons for conflict between science and religion. For example, the lay person is likely to draw a false analogy that "just as the electron is there and cannot be seen, so is God", which is due to a lack of understanding of the weight of scientific proof for the electron. (The discovery of electron by J.J. Thomson was the result of an actual experiment and provided answers to many scientific questions. See, *"Accelerators, Colliders and the Story of High Energy Physics",* by Raghavan Jayakumar (Jayakumar 2012).

In a most important book, "Quark and the Jaguar", the Nobel Prize winning physicist Murray Gell-Mann, (Gell-Mann 1994) states that religion, mythology and superstition form a schema that people develop in order to understand and navigate around in their lives. Often, this becomes an agreed form of schema for the society. In other words, a world view is formed which helps individuals to go on with their lives without understanding a lot. (By the way, this is much the way, our brain and even our vision works. (See, for example, *"How the Mind Works"*, By Steven Pinker (Pinker 1997)). Yet, as sciences have developed, people have adapted to the new knowledge that is accepted and have changed their schema. Now, in scientifically advanced countries with a large science educated population, the superstition and adherence to religion on the basis of fear of God striking us down, is on the decline and scientific solutions are accepted for daily life. Yet, science, as the first guide to more significant questions in life, society and Government, is still far behind religion. The religious schema still dominates the scene.

## Author's Views on Science and Religion

Hopefully, the breadth, depth and intensity of this debate are evident from the above paragraphs. This author will add his two bits of practical observations to this mighty collection. The view on the conflict between science and Christianity is evolving because the Christian view is evolving. (Any reconciliation between science and religion because of some specific aspect of science or a recent discovery is not relevant for this discussion). Undoubtedly, there is a conflict between science and the fundamentalist and literal interpretation of the Bible. Sophisticated theological arguments apart, the public can either accept that the Universe is over ten billion years old and abandon the literal acceptance of the Bible, or withdraw into a small group of like- minded fundamentalists who decry this or all scientific development. The same is true about accepting the science of human evolution and many other scientific concepts and precepts. Individuals do not have a choice of accepting one scientific discovery and not accepting the other, nor can they accept arbitrary points in the chain of development of a particular aspect of science. This is because science is a rigorous discipline and each link is a strong link and links in different fields are also connected. (See concept of consilience above). A fundamental flaw in the calculation of properties of electrons will devastate all aspects of physics and other fields. As stated before, one finding that incontrovertibly contradicts evolution would bring down major fields in biology and medicine. Therefore, the only way a Christian believer can deal with science without losing faith, is to believe that the Bible is symbolic, take poetic and dramatic license and that the history in it is only for the purpose of moral and religious lessons, unconnected to natural phenomena, where there is conflict.

There can be a conflict in the minds of lay people, between science and religion when it comes to topics in which science has evidence based answers and when people ignore or are urged not to follow the scientific advice or the scientific methods in dealing with such topics. Often, these conflicts are not in ponderous or philosophical or mystical issues, but simple day to day topics, such as whether to treat a child with science based medicine or prayer or a traditional way and how much faith or how much skeptical analysis to apply in day to day situations. In such situations, it is important for religious authorities to impress on people to trust science because it has never let people down in the long term, even though it may make mistakes in the short term. Understanding the process of science, the emphasis of this book, is the key to resolving this conflict. In simple words, where there is a clear conflict, a person must choose science to be the guide.

One very coherent area of religion and in many ways, common to all religions (except for Buddhism), is afterlife (life after death). The ideas of how a dear one would be received in Heaven by angels, the dedication with which rituals are performed for the dead, the reverence for the dead, the compassion for the dying and all the superstitions associated with treating the dead, permeate the consciousness of many religious and semi religious individuals and societies. These ideas and practices are usually very consistently rooted in tradition and religion and there is very little conflict in dealing with the dead religiously. In fact, even religious persons, who could start a riot on other occasions, would be respectful of the other religious rites for the dead. It cannot be denied that, in a sense, often dead are treated better than the living. This strong and abiding idea of afterlife has no basis in science and this is a matter of a strong philosophical difference with serious consequence to the living (and the dead for the religious people). Science, an existential philosophy, would emphasize proper and humane treatment of the living. Every individual must examine her or his view on this intellectually and spiritually, and must choose sides. Unlike some people like the author, people fear death and religions give hope with the rewards of afterlife, but with increasingly unorthodox life styles, a believer is likely to be mortified about the punishment of hell than be energized by the hopes of heaven. A scientific understanding of life and its mechanisms provides a perspective on these untestable and unverifiable hopes and fears. When examined with a scientific attitude one often comes to the conclusion that we reap the fruits of our actions in our existing lives. The fact that some do not get just rewards or others do not get just punishments, is nature acting in a chaotic fashion, and is because nature does not conform to human concepts of justice.

Most people who have strong faith in religion and/or traditional values and methods that are not supported by science, cherry pick scientific topics and applications both for their use and for denouncing them. Following situations are examples:

- A person would use modern conveniences of airplanes, cars, cell phones and television, admire the exploration of outer planets and see the photographs of Saturn's moons, happily chat about the discovery of the God Particle and talk about the identification of the absconding criminal using DNA mapping. But the same person might claim that science is clueless about the significant astrological effects of planets on individual lives or science is wrong when it denies that God created miracle of a weeping Madonna statue, or science does not and cannot understand the importance of starting some work at the auspicious time of day appointed by the scriptures. This type of discrepancy can be seen in Francis Collins' book *"Language of God"*.
- Most would rush the loved family member to a modern Western medicine hospital or clinic when the health condition is critical or serious, keep consulting modern medicine and taking modern medications for serious illnesses and use modern instruments and laboratories for diagnosis of health condition. But, on the side, these same persons would denounce modern Western medicine as clueless about health and wellness and claim that traditional medicine, including naturopathic medicine, has all the answers and is infallible.
- Zealous people cherry pick science to quote them as proof of the validity of religious or traditional values and at the same time, cherry pick science ideas for denouncing when it does not support the religious or traditional view.

Science gives evidence that unscientific prejudice is immoral and certain tendencies and behaviors are destructive and self –destructive. What lessons one draws for personal life from science, is not dictated and is, in a true sense, a self-realization. Then each scientific discovery becomes a life lesson, enlightenment and a guide to the future. In many cases, such lessons are demonstrable and are moral in the traditional sense. As stated before, science is a whole enterprise. Its edifice may have different wings and parts to it, but all parts of sciences are connected through pathways and all parts are built from the same brick and mortar of evidence and reasoning. One cannot choose to use part of science wholeheartedly and deny the truth of the other part completely. (One can and should have healthy and logical doubts). The above examples are clear cases of hypocrisy not unnatural or uncommon to humans in other areas. Therefore, a proper scientific- spiritual evaluation of one's approach to life and one's world view is always healthy. Such introspection is strengthened by bringing science into one's life and even making it a way of life. This capability is strengthened by knowledge such as the one provided by this book. Scientists have a role in bringing the crucial scientific knowledge to the public domain in an accessible user-friendly way.

This individual instinctive knowledge can only be fostered and strengthened by public discussion and setting of standards on what is good science. Rudolf Carnap and Otto Neurath believed that science should be evident at public level using a public language. They believed that this is what would make it distinct from metaphysics in a politically significant way. In that spirit, science should be testable in front of the public and only such a public science would form the body of science. (Galison 1996). Obviously, it is impossible to demonstrate the nitty gritty of gauge theory and its application to the discovery of the Z particle to the public, but the public can be informed about the big questions and the results, inviting them to probe further. In this context, the recent growth of science in private nurseries of Universities and laboratories, well beyond the grasp of the public, has caused political issues. A well argued article by Nicolas Kristof in the New York Times (Kristof 2014) states that instead of making scientific and academic information more accessible, institutions and academics believe that engagement with public is a frivolous distraction. The turgid prose dissuades people from learning academic findings. Kristof quotes Jill Laporte, *"a great, heaping mountain of exquisite knowledge surrounded by a vast moat of dreadful prose."* Modern development on the ideas of unity of sciences has reached ultra-philosophical levels and have stopped being understandable to even scientists. This trend is against the concept of practicing science in public and also moves away from a holistic understanding of sciences by the public. Scientists have to work to refocus their language so that the public can be engaged. If jargon in each field is any gauge (a glaring example is the information technology field), scientists and technologists are moving away from this. This does not bode well for the popularity of science. Perhaps, the scientific institutions can learn from religious ones in bringing their message to the public.

## REFERENCES


Anthony, H.D.; *Relativity and Religion*, London University Press, London, U.K.(1927) p.vii.

Aquinas, St. T; *Summa Theologica, Question 12, Article 13,* Christian Classics (1981) http://www.ccel.org/ccel/aquinas/summa.FP_Q12_A13.html.

Atkins, P.; *On Being,* Oxford University Press, Oxford, U.K. (2011) p.104.

Brooke, J.H.; *Science and Religion - Some Historical Perspectives, Cambridge University Press, Cambridge, U.K.* (1991).

Cardiff, I.D.; *What Great Men Think of Religion*, The Christopher publishing house, Boston, MA (1945). EText: Free thought Almanac http://freethoughtalmanac. com/?p=3922.

Carnap, R.; *"Elimination of Metaphysics by Logical Analysis of Language"* in *Logical Positivism,* Editor: A. J. Ayer, ed.,. Glencoe, Ill.: The Free Press, (1959) pp. 60-81.

Chaudhuri, N.; *Hinduism*, Oxford University Press(1979), p. 30.

Cobb, J.B., Griffin, D.R.; *Process Theology: An Introductory Exposition,* Westminster Press, Philadelphia (1976), pp. 8-9.

Collins, F[a].; *The Language of God: A Scientist Presents Evidence for Belief,* Chapter Four, Free Press, New York, NY (2006).

Collins, F[b].; *The Language of God: A Scientist Presents Evidence for Belief,* Free Press, New York, NY (2006) p.211.

Cunningham, G.C.[a], *Decoding the Language of God*, Prometheus Books, Amherst, NY (2009) pp.58-77.

Darwin, C.; *The Origin of Species by Means of Natural Selection, or the Preservation of Favoured Races in the Struggle for Life* (6th edition), John Murray, London, U.K. (1872) p.235. Ebook: On the Origin of Species, 1st Edition, Project Gutenberg, http://www.gutenberg.org/catalog/world/readfile?pageno=235&fk_files=3274399.

Dawkins, R.; *River Out of Eden,* Basic Books, New York (1995) p.133.

Dawkins, R.; *The God Delusion,* Marine Books, Houghton Mifflin, Boston (2008) p.152.

Dawkins, R.; *The God Delusion,* Marine Books, Houghton Mifflin, Boston (2008) p.151, p.286.

de Chardin, T. *"The Phenomenon of Man"*, Harper and Brothers, New York, NY (1959).

Dellinger, D.T; *Yale to Jail: The Life Story of a Moral Dissenter, Wipf & Stock Publishers, Eugene, OR* (1993) p. 418.

Easwaran, E., *Bhagawat Gita,* Nilgiri Press, Tomales, CA (1993).

Edelmann, J.; *Bhagavath Purana and Biology*, Oxford University Press, New York, NY (2012).

Flew, A.; "*There is a God",* Harper Collins College Division, New York, NY (2008) p.97.

Galileo, G.; Discoveries and opinions of Galileo, Translator and Editor: S.Drake, Anchor, New York, NY(1957) p.186.

Galison, P.; *The Disunity of Science: Boundaries, Contexts, and Power,* Editors: by Peter Galison and David J. Stump, Stanford University Press,(1996)).p.16, p.160.

Gell-Mann, M.; *The Quark and the Jaguar*, W.H. Freeman and company, New York, NY(1994) p.17, p.243.

Gould, S.J.; *Rock of Ages,* Ballantine Books, New York, NY(1999) Etext: Unofficial Stephen Jay Gould Archive, http://www.stephenjaygould.org/library/ gould_noma.html.

Heisenberg, W.; *Physics and Philosophy: The Revolution in Modern Science,* Harper and Row Publishers, New York, NY (1958) p.2.

Heisenberg, W.; *Physics and Beyond: Encounters and Conversations,* Harper and Row, New York, NY(1971) p.85-86.

Jayakumar, R.; *Accelerators, Colliders and Story of High Energy Physics- Charming the Cosmic Snake,* Springer, Berlin, Heidelberg (2011) p.8.

Jha, S. R.; *Reconsidering Michael Polanyi's Philosophy,* University of Pittsburg Press, Pittsburg, PA (2002) p.26.

Judd, J. W.; 1910. *The coming of evolution: The story of a great revolution in science*, Cambridge University Press, Cambridge, U.K.(1910) p.3.

Kristof, N.; *"Professors, We Need You!", New York Times Sunday Review*, 15 Feb 2014. EText: http://www.nytimes.com/2014/02/16/opinion/sunday/kristof-professors-we-need-you.html?_r=0.

Lonergan, B.; "*Dimension of Being" in Collected Works of Bernard Lonergan,* University of Toronto Press, Toronto, Canada (1993) p. 245.

Lonergan, B.; *"Insight"*, Editors: Frederick E. Crowe and Robert M. Doran, University of Toronto Press, Toronto, Canada (1972) p.4-6.

Medawar, P.; "*The question of the existence of God"* in *The limits of science,* Oxford University Press, New York, NY(1988) p.96.

Nasr, S.H.; *An introduction to Islamic cosmological doctrines: conceptions of nature and methods used for its study by the Ikhwān al-Ṣafā, al-Bīrūnī, and Ibn Sīnā*, 2nd edition, Revised. State University of New York Press, Albany, NY (1993) p.111.

Needham, J.; *The Grand Titration: Science and Society in East and West*, Allen & Unwin, Crows Nest, Australia(1969).

Nehru, J.; *Discovery of India*, 1946, Oxford University Press, Delhi, India(1994) p.26. EBook: Slideshare, http://www.slideshare.net/sanjaybhatt0330/discovery-ofindiabyjawaharlalnehru.

Pinker, S.; *How the Mind Works*, W.H. Norton and Co., New York (1997).

Plantinga, A.; *Where the Conflict Really Lies*, Oxford University Press, Oxford, U.K.(2011) p.46-47.

Polkinghorne, J.; *Science and Religion in Quest of Truth*, Yale University Press, New haven, CT (2011) p. 113.

Rajagopalachari, C.; *"Bhaja Govindam"*, M. S. Subbulakshmi, CD, Label- saregama (1965).

Robson, J.M.;*"The Fiat and the Finger of God: The Bridgewater Treatises,"* in *Victorian Faith in Crisis: Essays on Continuity and Change in Nineteenth-Century Religious Belief,* Editor: R. J. Helmstadter and B. Lightman, Macmillan, London (1990).

Sagan, C.; *The Demon-Haunted World: Science as a Candle in the Dark*, Balantine Books., New York, NY (1996) p.304.

Salam, A; speech on *"Islam and the West"* in Paris at the UNESCO House on April 27, 1984, *Renaissance of Science in Arab and Islamic Countries*, Editors: Dalafi, H.R and Hassan., M.H.A.; Singapore: World Scientific Press (1994) p.39.

Schaefer, H.F. III[a]; *"Science and Christianity Conflict or Coherence?"*, The Apollos Trust, (2003) p.35.

Schaefer, H.F. III[b]; *Science and Christianity Conflict or Coherence?"*, The Apollos Trust, (2003) p.7.

Schrödinger, Erwin, Physicist 1954. *Nature and the Greeks,* Vol.II., Cambridge University Press, New York, NY (1954) p.93.

Shannon, B.; "*Biblical Entheogens: a Speculative Hypothesis"*, Time and Mind: The Journal of Archaeology Consciousness and Culture, Vol. 1, Issue 1, (2008 )pp. 51-74.

Stratton, *S.B.; Coherence, Consonance and Conversation: The Quest of Theology, Philosophy and Natural Science for a Unified World-View*, University Press of America, Lanham, MD (2000) p.111.

Weinberg, S.; *Without God*, New York Review of Books, Vol 55, No. 14 (2008)p.1.

Weinberg, S.; Address at the Conference on Cosmic Questions, American Association for the Advancement of Science (AAAS), *Dialog on Science, Ethics and Religion*, Washington, D.C. (14-16 April 1999). EText: http://www.counterbalance.org/cq-wein/nocon-frame.html.

White, A.D.; *The History of Warfare of Science with Theology,* Prometheus books, Amherst, NY(1993) p. 247.

*Chapter 8*

# SCIENCE FOR LIFE

We are the same people as those of Gobekli Tepe. Only, we are separated by eras of cultural, religious, scientific and technological development. Our fundamental aspirations remain the same. One of our primary commonalities with the people of Gobekli Tepe is for us to succeed individually and as a society. Success is defined in terms of material, intellectual, emotional and spiritual achievements. Science is one of the crucial tools for the success of individuals and society. But science has to be understood, appreciated and applied appropriately by individuals comprising the society. This, at first, might seem like an impractical statement. How can individuals who may or may not have scientific training do this? We act from our knowledge including memories and also from certain inclinations which we may call attitudes. So, acting with a scientific attitude is what is suggested by using science as a tool in life. A scientific attitude is cultivated through (a) knowledge of science, (b) knowledge of the scientific method of reasoning. Hindu epistemology states that we acquire knowledge by empirical evidence (*pratyaksha*), deductive reasoning (*anumaana*), comparison and inductive reasoning (*upamaana*), postulation (*arthapatti*), non-apprehension or unawareness of something (*anupalabdhi*) and verbal testimony of a trusted one (*sabda*). This is applicable to the learning of science. The introduction to the topics in science presented to the readers in the previous chapters would be very helpful in this journey in scientific knowledge. One might think that the knowledge of all sciences is too deep and vast for empirical, deductive and non-apprehension knowledge. So, some of the public's and individual's knowledge must come from their own learning and some from the testimony of trusted scientists and groups. One can then apply this knowledge to make a proper judgment of situations and for problem solving. In addition to this, one has the greater benefit of wisdom from the understanding of science.

An average individual is more interested in the overall science rather than the details and knowing what that means to one's life. In subsequent sections the following aspects to integrate science into one's life, are discussed: (1) Seeing sciences as a whole discipline and being able to identify non science, (2) scientific curiosity and being aware of science in daily life, understanding where science can and does apply and where it stops, (3) developing a scientific attitude, (4) having a clear understanding where one's beliefs lie and how these beliefs affect one's life.

## SEEING SCIENCES AS A WHOLE DISCIPLINE

The topic of Unity of Sciences is a vast one. At the level of an average science enthusiast, there is a hunger to see all sciences to be placed on the same basis. The experience of such an enthusiast is, in many ways, to the contrary. There is a sense in which biology, psychology and physics seem to have some common basis, but this basis is not evident, in general. Science involves topics within branches within disciplines, different methods and approaches and even a set of values to guide the methodology. For example, the Gravitational theory needs to be unified with quantum mechanics. But can psychology be unified with biology and biology with physics and so on? Can the methods of physics, biology and social sciences be unified?

Descartes proposed that sciences would be unified through the language of geometry and expressions of algebra. An aspect of this is coming true. Physics descriptions of space time, gravitation, crystalline characteristics and molecular biological functions include structures and topologies. A similar description is possible in chaotic phenomena. In principle, many relationships, such as networks and phase diagrams can be considered to be topological descriptions. The question is whether such descriptions are complete.

An alternative way of Neurath, is to see all sciences as being interconnected to make the whole of science without necessarily having the same language or starting point. We certainly see evidence of this type of relationship between chemistry, biology, psychology and chaos. The concepts of algorithms and complexity also provide the interconnection. Whether this complex network of science is a satisfying edifice, is a matter of debate. It certainly takes one away from the "reductive unity", where sciences would be expected to unify under some reduced set of fundamental components and laws. Ernest Rutherford's statement, *"all of science is physics or is stamp collecting"*, reflects this thinking. This going away from reductive unity is perhaps more satisfying to theologians who see diversity and interconnectivity as God's work. As stated above, unification does not mean reducibility of one branch into another. Stuart Kauffman actually believes that biology is not reducible to physics, because biology operates in the fuzzy area of "adjacent possibilities" and remote from other infinite number of possibilities.

The other way is to return to the concept of falsifiability as the principle of science and that which separates science from non science, as proposed by Karl Popper. Though Popper did not state that falsifiability of theories is a unifying principle, it could well be. But what complicates this argument is how to unify empirical science that is without theories, which is what we find in much of medicine, psychology and social sciences.

Hopefully, the topical chapters have given a complete sense of what constitutes science and the scientific methodology. The success of scientific enterprise in propelling society is evidence that the body of knowledge created, even in the last century alone, has a unique characteristic that is not present in any other field. There are many ways of viewing the different branches of science as a single human enterprise. Except for the Cartesian theorists who see mind and matter as distinct, most would agree that all sciences arise in a logical fashion, from laws of physics. The fundamental particles interacting through fundamental forces constitute every aspect of matter and non- matter and give rise to phenomena from black holes to viruses. In that sense, a scientific description is, in principle, possible from all sciences. At the beginning, it is important to see this unity of sciences. But, even with a tremendous human effort, such physics based descriptions of complex objects and phenomena are practically impossible. The magnitude and complexity of natural objects and phenomena and the uncertainties due to chaos and quantum effects, prohibit such an enterprise. Therefore, it is necessary to understand why we "know" that medicine, biology, chemistry, psychiatry and physics are sciences. For many of us, this is due to formal education. But, the "epistemological" question of how we obtain knowledge has a clear and general answer for sciences. One answer is the common basis of the process of knowledge- namely one does not go around making up things without concrete observations and logic. Since small initial errors can take one far away from the truth (see Sensitivity to initial conditions in a previous chapter), a scientific process includes consistency checks in logical deductions and this too unites the sciences. One might start from calculating the mass of an electron from first principles, but it must be checked with an experiment on the measurement of mass and this mass should be consistent with light emitted from an atom. In the same way, one might start from a description of bacterial DNA, but it must be checked with the expression of the DNA in coding of proteins and finally, the symptoms should be consistent with such expression of the DNA. This process has to be followed rigorously and deviations explained, corrected or researched further. There is another check that can be made of the interconnectivity and integration of sciences. One can take a small piece of a complex phenomenon and explain it through or in association with another science. If a chemical affinity between an antigen and antibody is seen, a specific molecular bond can be tested through chemistry, which, in turn, can refer to physics principles. In life, one can make such checks at one's own level of knowledge. For example, if the water bill for the last period shows a very large increase in water consumption and one finds a water leak in the garden, such a visible leak can be checked and fixed and wait for the next bill to confirm that the leak has been fixed. But one can (should) do a simple calculation of the size of the leak from this local leakage, for 24 hours a day 7 days a week and check against the total loss of water to see if this fix would solve the problem. If it does not explain the change, one has to go look for additional leaks. This is essentially the scientific method. A very nice book on this topic of estimating various things (back of the envelope calculations) is "Consider a Spherical Cow" by John Harte (University Science Books, (1988)).

A superficial but frequently perceived commonality is the generation of technology or technical sciences. For example, some see the first splitting of the atom as the same as the first discovery of the DNA. But we persist. What is the connection that we unconsciously make between the knowledge of cosmology and knowledge of a biological organism? But we don't make this connection with say, knowledge of literature, arts or cooking, in the general sense. Rudolf Carnap stated that all sciences are characterized by logical statements which are put together by many experiences with experimental verifications. A simple application of this statement can be made the following way: The visceral connection we make among sciences is based on our instincts and understanding of our environment. This is both instinctive and intellectual. We make sense of the world around us and respond with an approximate scientific understanding. In dodging a falling branch, we react with an instinctive knowledge of kinematics, in eating fruit we act with instinctive knowledge of its nutritional value and in protecting and rearing our children we act in an instinctive understanding of the commonality of genes. In understanding the meaning of a father and mother and grouping them as parents despite their significant difference and roles, in understanding different nutritional items as foods despite significant differences in taste and consistency and in grouping activities as play and fun despite their tremendous variation, we exhibit the capability of discerning unity and commonality. We recognize science as the reliable field that provides objective information, whatever the details and variations are. In other words, very often, we know science when we see it. The fostering of instinctive understanding helps in developing a scientific attitude.

## Scientific Curiosity and Awareness and Recognition of Non-Science

Awareness of science is exhibited by the fact that when an individual encounters a routine process, he or she would wonder what physical phenomena are active in the process. This was mentioned in relation to persons being cognizant about the physical process associated with an electric light. With that type of understanding, a prosaic action of lighting the room becomes a miraculous process of illumination, pun intended. That is only the beginning. An inquiry and even a semblance of knowledge in these areas lead to an exciting life, where everything becomes magic (in a performance sense), secrets of which one can actually learn. Once one knows the scientific processes behind everyday objects, one can then never look at them the same way again. This awareness would inform the person of the reasons and mechanisms behind all the things that are personal and non- personal. This awareness of how things work would also bring a non- religious understanding of even complex phenomena. One need not admire a butterfly metamorphosed from a caterpillar and assign it to God in one step. One can know the full details of how this happens and then, if still unsatisfied and one wants to, assign it to God's doing. But scientific confidence would actually assign it to as yet undiscovered scientific processes. Contrary to the uneducated opinions that science takes away nature's beauty, scientific understanding enhances the experience and increases the respect for even inanimate objects. The amazement at the butterfly would not be reduced one bit; instead it would be enhanced by the amazement at the natural mechanisms and steps involved in the metamorphosis. Similarly, when we look at the sky and marvel at it, we would know so much more about it and be amazed all the more. One can also see readily the practical use of such awareness. Selecting the right product for use and its operation and maintenance become easier, when we know the science behind it. When we are sick, we are able to judge how serious it is from the knowledge of the mechanisms and symptoms and this helps us prevent such sickness. We can more effectively participate in our own treatment and would be less vexed or anxious about the progress of our cure. A simple case in point: if one gets wet in rain, seemingly catches a cold immediately, one can conclude that it is not due to a virus that we caught in the rain, because the virus would have an incubation time of a couple of days. Instead, it is due to either an allergy or lowered immunity to a virus, already present in the body. (The distinction is important). In a more serious area, an awareness of the cosmology lets us know that planets are too far away and their influence too miniscule to have any definite influence on our lives and if there is any effect at all (extremely unlikely and science has not found any such effect or possibility of such an effect), it might be very occasional and due to some chaotic phenomenon. This knowledge of cosmic phenomena and interactions would take away the superstitious belief that astrology is a science to be relied upon.

An analogy for knowledge is the volume contained in a sphere. If we say that we know what is inside the sphere is our knowledge, its surface is the edge of what we don't know. As we increase the knowledge, we come to know more about things that we don't know and so we become more aware of our own ignorance. This increased knowledge of our ignorance also increases our amazement at natural objects and phenomena. This increased awareness increases our curiosity to know how things are the way they are. Curiosity is a primary sign of intelligent beings wanting to learn about the world they live in. Curiosity, in turn, is accompanied by the desire to find answers that are concrete and satisfying. Science helps in finding these concrete answers and in cases where science does not have the answers, the lack of an answer itself would be revealing. For this reason, one sees that intelligent and curious people are attracted to science. Curiosity is both the driver for knowledge and the result of the knowledge. It is also the result of the awareness of our ignorance. In that sense, curiosity is humility. This is why curiosity, which overcomes shyness or pride, is highly celebrated. It is good to ask why, what, how and when, to the point of tiring oneself and others out. Children exhibit this unbounded curiosity between the ages of two and four and as we grow, this curiosity and eagerness to know reduces.

Curiosity is closely linked with imagination which enables one to form mental images of things that are not firmly established. Together, knowledge, curiosity and imagination provide the brick and mortar of creativity. Scientific knowledge, in particular, roots this creativity in rational ideas. Curiosity permits us to explore the edge of this knowledge. Imagination lets our minds take flight, combining the two. In addition, scientific and mathematical concepts make abstract concepts more accessible to a person. A good education system fosters all three, so must an individual. It is sometimes assumed that awareness of practical phenomena and scientific knowledge limit imagination. This is far from true. This is clearly demonstrated by the fact that most scientists working in fundamental areas like physics, chemistry, biology and mathematics have greater imagination. This fact is also illustrated by the fact that many scientists also show creativity in other fields, such as music and other arts.

## Discriminating between Science, Pseudoscience and Non- Science

One of the keys if not the key need for an individual is to recognize what is not science. Many people are unable to make this discrimination and therefore, are often in a quandary. The rather evident examples of prayer, astrology and some types of medicines have been given. It is worth repeating these and giving more definite characteristics, in order to identify information and advice that is not scientific. A friend of this author had a question about the prolific claims on wearing magnetic objects on one's body and their efficacy in preventing various diseases like heart disease and cancer. This comes under the category of pseudo science. How indeed would we even have an idea whether this claim is scientific?

We saw that when a significant scientific claim is made and there is a potential scientific basis behind it, it usually eventually gathers momentum, even if it is not accepted initially. There are many instances of once refuted scientific claims now being accepted with rigorous proof. A major example of this is the Big Bang Theory. Geoffrey Marcy proposed detecting planets in other parts of the cosmos from dips in star light intensity and though this was not taken seriously then, now this is one of the standard methods. Some pseudoscientific notions, not accepted by science, claim the same status. On the other hand, spectacular claims of cold fusion were viewed with skepticism and remain so today. This is so, because though mechanism proposed for cold fusion is possible, the physics and the quantitative and empirical observations are on shaky ground. Spectacular claims of discovery, which are not scientific, wind down and disappear over a period. The problem of misinformation in general and pseudoscience in particular, has been made worse by internet, where authentic looking information is provided to the gullible consumer. The assessment of the legitimacy of a scientific claim or notion requires some work. The following are guidelines in determining the scientific validity of a claim:

- Is the information confirmed by an authentic publication or a scientifically recognized source? This question can have answers at many levels. It could be a school science teacher or a general scientist or a specialist in that very field. One should use a source that is appropriate for the depth of confirmation needed. One has to be careful when anonymous sources such as those on the internet, are used, because many people have confident but wrong ideas and opinions and some mislead deliberately with an agenda.
- Is it anecdotal or has a proper scientific test been done? An example is a miracle cure for cancer. If this were an established cure, science would have noticed it and done trials on it. A doctor would know about it or there would be scientific publications. If it is a new cure anecdotally proved, remain skeptical. (It is often argued that the scientific establishment is blocking such cures for profit. Only in exceptional cases, this may be true). If scientists deny the claims or cannot confirm it, then make the effort to see if the scientific rationale holds up. In general, one must be extremely skeptical about anecdotal evidence, unless it is overwhelming and shared by a large number of people from different sections of the society.
- Is it statistically probable or would it be grouped under miracles? If it is statistically improbable, remain skeptical. A qualification is needed here. Winning a big lottery

with 6 numbers is statistically improbable and yet someone does win. In this case, the win is according to statistical laws. One could check the probability of an event occurring and if the claim is that it happens more often than what statistics would predict, then it is to be discounted. If it is only according to statistical probability then again, it is not a miracle. The statistical verification is necessary even when a case can be proven without doubt. In most cases, the cause and effect have to be proven conclusively. An example can be taken from an episode of the television series "House". The attending doctor insists on trying a treatment not well supported or confirmed by science, on a patient. Unexpectedly, the patient is cured. But the team of doctors decides not to tell the attending doctor about it, because "he just got lucky" and next time he might kill a patient, meaning that the patient might have been cured by something else.

- Are the circumstances and conditions consistent? In the above stated example, one could catch a cold from being in the rain, but it would take a couple of days to exhibit symptoms. If a person was cured by the ritual of exorcism, were his or her behaviors consistent with a psychiatric illness which might be cured by the shock therapy of exorcism? Are the conditions so unusual that it is hard to compare with normal states? In this case, make sure to consult a specialist before drawing conclusions.
- There is a powerful concept in the proof of mathematical theorems and physics concepts: necessary and sufficient. When one states that if A is true then B is true, this is an incomplete statement. What needs to be examined is, is it sufficient for A to be true for B to be true? In this case, getting A is a way to get B. In such a case, there may be other ways too, such as if C is true then also B is true. But if it is necessary that A be true for B to be true, then one must have A to get B. But if this is only necessary, then perhaps A being true alone is not enough and something else is needed in addition for B to be true. If it is necessary and sufficient that A be true for B to be true then the relationship between A and B is cause and effect. For a person to get high scores in an examination, it is necessary and sufficient to study hard and know all answers to the questions that are likely to appear in the exam. Often, in life, we are deceived by not taking into account this concept of "necessary and sufficient". This necessary and sufficient condition ensures that the facts are complete and one is not making decisions based on half truths. Ascertaining this is part of the scientific attitude.
- Do the magnitudes of the causes and effects make sense? It may be that a well-known scientific phenomenon is being misused to ascribe inconsistent effects. A very good illustration is that of magnet therapy to cure various diseases. Here is why it does not work:

  First, at a gross quantitative level, many patients have been placed in MRI magnets which have an intense magnetic field of 20,000 gauss, nearly uniformly over the body and 20 to 40,000 times the earth magnetic field of about 0.5 to 1 gauss. Government agency scientists have looked for an effect of an MRI and do not see any (Formica 2004). Only patients with metal implants or pacemakers are not permitted for MRI, since an additional alternating field is applied in the procedure. In contrast, a magnet therapy gadget has about 1000 gauss at the surface of the object of say 1 cm thickness and drops off as the cube of the distance. So, 1 cm below the skin, the field is only about 125 gauss. 10 cm away from the magnetic object, the field is 1 gauss, about the same as the earth magnetic field. There are ways to intensify the fields locally with some shapes, but still the field is small. The inquirer must then ask, if 20,000 gauss over 10 million c.c. does not make any difference, even allowing for the fact that it is applied for only 30 minutes, then why would the 100 gauss field over 1 c.c. volume make a difference? The conclusion is that magnet therapy is a case of pseudo science. However, a more scientific inquiry into this is warranted if only from a sense of curiosity, because we know that birds and other life forms do sense even the weak earth magnetic field. So, one might question if any of our faculties are actually affected by the magnets.

One can ask if power lines and cell phones can be safe. A power line carries up to 500 kilovolts of alternating electricity. A review of studies on harmful effect of power-frequency fields found that 46% of studies found no effects, 22% found some evidence for DNA damage and 32% were inconclusive. (Vijayalaxmi 2005). The electric field at the ground level from a 500 kV line can be as much as 10 kV/m. These can induce up to 0.01A/m$^2$ of current in the brain. This should be compared with the currents of up to 10-150 A/m$^2$, in the body. (Yoon 2003), (Ahlbom 1993). Therefore, it is unlikely that power lines cause cancer. On the other hand, the jury is out on the cell phone effect. The conclusion is that alternating electric fields do not seem to cause cancer or mental confusion as far has been determined, but it is good to look for any further developments.

- The cases of medical cures by faith healing, superstition such as, walking under a ladder is bad luck, ghosts and materialization of objects out of thin air are clearly non science because the concepts are not rooted in a scientific hypothesis, postulates and theories that are confirmed by scientific observations in a scientifically controlled environment. The lack of any foundational science, derivable from natural laws, is what characterizes these claims. A symptom of non -science is the irreproducibility of any claim. Often non- science parades as special powers of individuals, saints and swamis. Non- science is also characterized by violating natural laws of cause and effect, conservation of mass, momentum and energy, principles of thermodynamics and proposing phenomena, fields and forces that have not been observed scientifically.
- It is common to explain using analogies. Science is not exempt in providing simple analogies for complex questions. But this is fraught with errors and danger of misuse. The analogy may not be valid for reasons of magnitude or character of the phenomena. A charlatan might misuse the analogy of cancer treatment with radiation for his suggestion of standing in sunlight for several hours as a cancer cure. Similarly misuse of terminology is also not an uncommon misuse of science. The term "The God Particle" for the Higgs Boson is one of them, though not intentional. Another good example is the use of the phrase, "It is only a theory" implying that it is not true in practice, while the results of the theory may be confirmed in practice.

## DEVELOPING A SCIENTIFIC ATTITUDE

A scientific attitude is the position that all phenomena, objects and events come about from natural causes, which are often explained by science. The attitude also implies that where there is no scientific explanation, it would be eventually found using scientific methodology. Therefore, the attitude comes mainly from the conviction that scientifically reasoned, empirically proven and consistent answers are the ones that are correct. It also acknowledges that such answers may not be easy to come by and often have to be teased out. An abiding interest in science and its methodology are the source of a scientific attitude. The attitude also enables a person to seek out the difference between science, pseudoscience and non science.

The key part of a scientific attitude is the reasoning that science uses. The development and use of this reasoning power makes a person rational and able to make sense of the world, without taking short cuts. It also enables the individual to identify answers that are genuine and those that are incorrect in part or whole. Deductive reasoning is a top down reasoning that starts from a general principle and logically goes to the specific to test its truth. For example, a simple affirmative reasoning is: The Sun is very hot, it emits infrared rays, infrared rays warm me and when I stand in the Sun I am hot. A deductive reasoning of negative kind is as follows: Mango trees do not grow well in cold climates. My area has a cold climate, I cannot grow mango trees. A reasoning which is in conflict with the result requires checking: I see my plant is not growing. All plants need water, good soil and sunlight. My plant gets water, good soil and sunlight. My plant should grow well. This conflict in the end of reasoning points to some problem. It could be an error in one of the statements (maybe the soil does not have an important nutrient) or some factor is missing (may be it is too cold). Deductive reasoning is the root of testing falsifiability. Karl Popper, who believed that all life is problem solving, showed that falsifiability has to be a fundamental method of checking out something. If A is true then B has to be true, then, if I can show that B is not true then A cannot be true. While the idea of deductive reasoning is correct, in practice, deductive reasoning requires a good amount of knowledge. Otherwise, it would lead to wrong or conflicting answers. An example of this is the climate change effects. We have come to know that accumulation of carbon dioxide in the atmosphere causes the earth to warm up since the earth's radiation to space is blanketed. While the earth is going through such warming, a wrong deductive reasoning would say that winters would also be warmer. In reality, climate is a complex phenomenon involving multiple interacting systems and therefore, actually in many places winters can be colder in some years, when the global warming is taking place. Deductive reasoning should be employed after acquiring as much information about the situation as possible and the answers should be cross checked with some other way of reasoning or an observation.

The other reasoning method is inductive reasoning. This type of reasoning is built into our survival instincts where we tend to derive a conclusion for one situation from observations in other situations. So a good inductive reasoning leads to a high probability of the answer being right. This reasoning is necessary because we may not have the ability to reason out something from fundamentals. We see luscious berries on a bush and our experience says that such bushes have good berries. We decide that it is very edible. Though that has a high probability of being right, there is a chance that the bush is not exactly the same kind as we know and the berry is poisonous. So inductive reasoning has to be employed with caution. In some cases, one errs on the safe side when reasoning inductively. There are similar looking bushes with similar looking berries and it is safer not to eat them. Another example is when parents tell children not to trust strangers even though the probability that a stranger has bad intentions is small. Therefore, here the reasoning has to take into account the risk of one action vs another.

The objectivity, which the scientific attitude engenders, helps in obtaining correct solutions, particularly for personal problems. A scientific attitude is helpful in observing our environment and deriving the right conclusions, in understanding roots of a problem and potential solutions and making a choice in action. If one's otherwise normal child is not doing well in school, understanding the root of it is important. Even if it is only because the child is not working hard enough, there are underlying reasons. To tease it out requires a scientific attitude, which can discriminate between emotional and rational reasoning. A scientific attitude brings about harmony in life, compassion for all life and a morality in which doing harm to another is a last resort. There is an impression that people who approach situations scientifically, are cold hearted. This is rarely the case. One has only to test people with scientific attitudes in a situation which requires empathetic or sympathetic response. Taking the example of sickness with pain, when a doctor is unconcerned about the pain, it is only because she knows it is temporary and her priority would be more about treating the underlying cause than to attend to her pain. Equally, a teacher with a scientific attitude would be kind to a student who is behind in studies, because he understands the underlying disability.

Individuals with scientific attitudes are the best ingredients for a healthy society. From arriving at solutions for public health and safety and understanding material needs and their limits set by natural cycles to Governmental involvement in material, economic and social progress, such individuals collectively would make the right decisions. The larger the number of people with scientific approach to solutions, the greater will be the accuracy. It is also likely that people with scientific attitudes are likely to make better decisions on war and peace situations because the curiosity and the ability to question deeply would make them come to more proper conclusions.

## Skepticism, Faith and Cultural Traditions

Science cannot be based on beliefs and faith. This is a difficult topic. The goal of this book is to help the readers to bring science into their lives as a part of it. This essentially means that belief and faith cannot be a large driver in their decisions. Yet, belief in a system and faith in people and oneself are important in our daily life, peace of mind and harmony in society. The line where belief and faith must stop and science takes over is an individual decision. But, if faith and belief say one thing and science another, it is a clear indication that a scientific approach is necessary.

Faith takes many forms. Faith in the existence of God is faith in the religious definition. Faith in a person is trust that is either developed over a period or based on an instinctive feeling. But faith may apply in a situation where one just believes that the basic work has been done and trusts that he or she can proceed from that point. We drive on roads, eat foods, ride on airplanes, put our money in a bank and send our children to school based on this trust.

This type of faith is based on experience and assurances from responsible people who also have vested interest in our safety and benefit. This faith is also the basis for many scientists not having to work out everything that they need as a starting point. But such a faith is a choice. These are testable and verifiable situations and one can be skeptical about these at the cost of expediency if one chooses. Religious belief does not give this choice.

But, at all points, some skepticism is also healthy. When carrying out the activities described above, it is healthy to be cautious and understand the situation before proceeding. This assessment comes from being armed with knowledge. When a politician says that he has a plan for greater prosperity in the country, it is good to find out the detailed proposals and assess the chances of his plans being realistic or just a vague promise or a lie to get elected.

As we saw, skepticism in science is required when a new idea, not concretely established in scientific understanding, is proposed. Many of the religious, mystical and metaphysical ideas are of this form. So, faith in such cases must be reposed only after considerable introspection and from the vantage point of scientific knowledge and scientific attitude. It is understandable that people believe that religious faith is very personal. But a scientific attitude is also very personal and shared in the same way as religious faith is, but with a rational process. Harmonizing one's spiritual attitudes with science brings harmony in life. Many of the discussions on science and religions point to many ways of doing this. If one has not thought about this issue at all, then it is necessary to take the first step by assessing where one is, on this aspect. A deeply spiritual matter is a matter of personal faith, wherever it arises from. A scientific attitude makes one aware of the physical consequences of such a faith and whether it is causing any harm to oneself or others. This awareness and discrimination can ensure the balanced evaluation of such a faith.

We follow many traditional customs based on culture, which themselves might have risen out of faith. But these might be so rooted in one's culture that they bring peace of mind and joy and harmonize one's life. Playing or singing devotional musical compositions can still remain the high art form, as would any other art form that originates from worship. Appreciation of the wisdom in the mysticism of ancient civilizations, which might be irrational, teaches lessons that one cannot derive from modern civilizations. Irrational rituals that are performed fully knowing that these are only for personal satisfaction and the satisfaction of those whom you care about, are also personal choices, as long as one does not associate any superstitious belief with them. Wearing a tie to a job interview and wearing a wedding ring are such rituals. These do not harm anyone, but do make one happy. Irrational as these are, these connect us to our past and keep us from becoming too arrogant about our intelligence brought on by modernization.

***

Sciences are a human discipline and in that sense, nurtured and shaped according to human understanding as well as human applications. It is obvious that science can be very useful for all individuals at various levels. One does not need the science of cosmology to do the work of say, a bank employee or the home maker. But knowledge of cosmology builds a sense of how things work and when one makes various links in the reasoning in different topics, one develops one's own world view, which is what would be used to carry out the daily duties and tackle crises. One cannot deny that wondering at the beauty of the stars at night and contemplating their nature, has a significant influence on how one feels in the morning, how a sense of proportion develops and how one feels about the problems of the day. In the same way, knowing how the macrophages work helps you to understand what it means when you are sick and your white blood cell count is up. You realize your body is fighting back and this gives the sense of comfort and confidence that nature works the way it is supposed to. You are one step closer to knowing yourself.

Science is a good friend in that it does not lie to you. When it does not have an answer, it tells you so and does not make one up. When it does have an answer, it also warns you that no answer is 100% correct. Science is a true guide and its wisdom will continue to increase and future generations, whatever their predicament or prosperity, will increasingly benefit from it provided they continue scientific pursuits in the same spirit that has characterized modern science in the past few centuries.

## REFERENCES

Ahlbom, *A.* et.al*.; "A metastudy of the effect of power lines on children shows no significant effect", Electromagnetic fields and childhood cancer*, Lancet, Vol. 343 (1993)pp.1295-1296.

Formica, D. and Silvestri, S.; *"Biological effects of exposure to magnetic resonance imaging: an overview". Biomed Eng Online*, Vol. 3, No. 11 (2004).

Harte, John; *"Consider a Spherical Cow"* University Science Books, Mill Valley, CA (1988).

Vijayalaxmi and Obe, G.; *"Controversial cytogenetic observations in mammalian somatic cells exposed to extremely low frequency electromagnetic radiation: A review and future research recommendations", Bioelectromag,* Vol. 26(2005)pp.412-430.

Yoon, R.S.; *Biological applications of current density imaging*, Ph.D. thesis, Electrical and Computer Engineering, University of Toronto, ON, Canada (2003).

# APPENDIX

## PRIMER FOR MATHEMATICAL AND PHYSICAL CONCEPTS

- X/Y or $\frac{X}{Y}$ : X divided by Y
- $X^n$: X multiplied by itself n times
- $X^{-n} = 1/X^n$; $(X+a)^{-n} = 1/(X+a)^n$
- If a quantity Y is dependent on a quantity X, then Y is said to be a function of X, written as Y=f(X), where f is the function. For example, if Y changes as the square of X then $Y=X^2$.
- A curve or a plot or graph (say of Y versus X) shows dependence of one quantity on another or how one quantity changes (varies) when the other quantity changes. Usually, the dependent quantity is on the vertical axis and the independent quantity on the horizontal axis.
- Some examples of functions: Y function is shown as a function of X in curves in figure i.1.

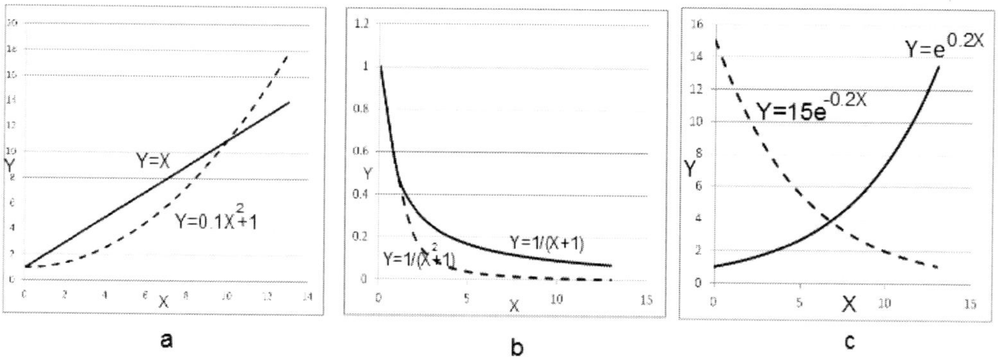


Figure i.1. Different functional variations. In figure c, e represents a mathematical constant approximately equal to 2.71828. Often e is also written as exp.

Sometimes, this type of plot shows a third quantity as a function of the other two. A route map is an example. (See events and world lines in Chapter 5).

- The following (figure i.2) is the example of vehicles A and B and how distance changes with time. Horizontal line shows increase of time with increasing length along the line from the reference point O. The vertical distance shows increasing distance with increasing length from point O. Positive distance is vertically up and negative distance is down from the origin O. Vehicle A is moving at constant speed (covers the same distance in the same time interval) and vehicle B is accelerating (increasing speed with increasing time and therefore increasing distance for the same duration of time interval).

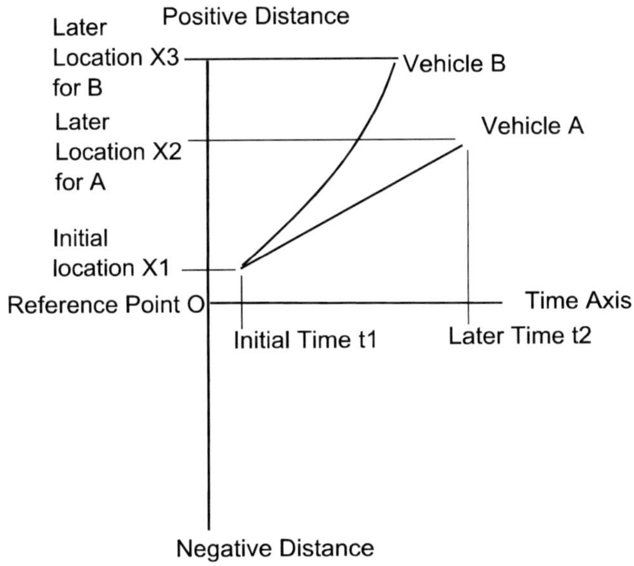


Figure i.2. Distance and velocity for steadily moving and accelerating objects.

- dX or δX or ΔX – a small piece or portion of X such as if X is a distance, dX is a small distance within X. (See figure i.3)

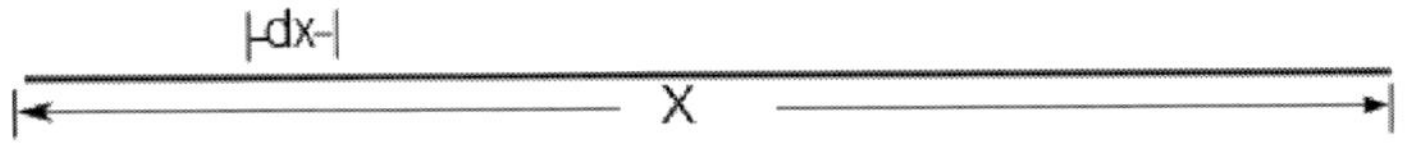


Figure i.3. Differential length.

- If a quantity Y changes (varies) as X varies, the slope at a point is the rate of change of y with x at that point. In the graph i.4 (which is similar to vehicle B above), dX is the change in the horizontal direction in X value at the point P and dY is the change in value of Y in the positive direction and -dY at the point P' is the change in Y in the negative direction.
- dY/dX is then the slope of the curve. (See figure i.4). It is positive (increasing Y with increasing X) for point P in upper curve and is negative at the point P' in the lower curve (decreasing or increasingly negative Y). The slope is also the rate of change of Y with X. (The rate at which y changes when x changes by a given amount).

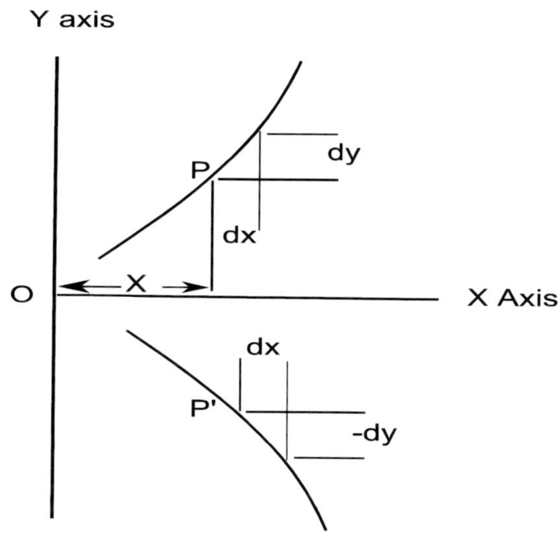


Figure i.4. Gradient or Slope.

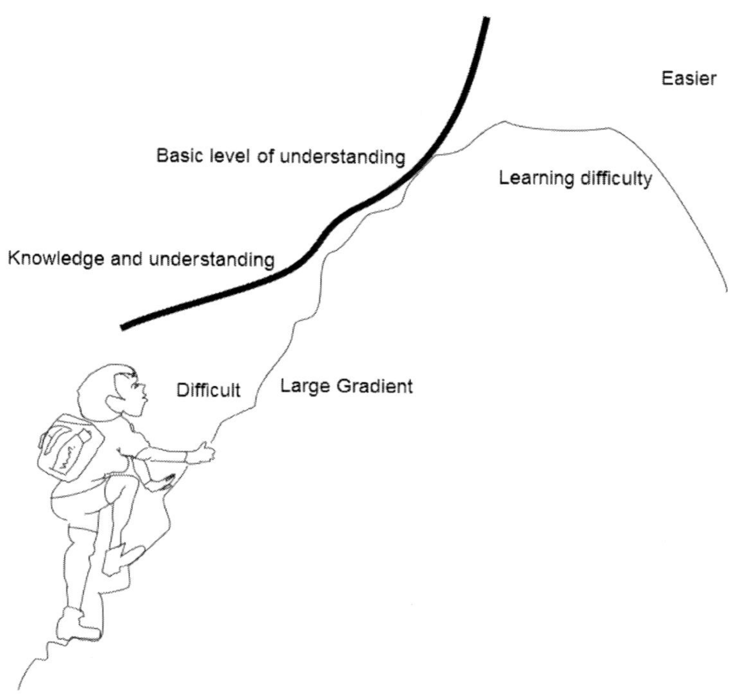


Figure i.5. Easy and difficult slopes. dY/dX is also called the derivative of Y with respect to X.

- For the distance versus time curve in figure i.2, the slope dY/dX (small distance covered in a small time) at any point in time is the speed of the vehicle. For both A and B the velocity is positive. For the vehicle corresponding to the lower curve in figure i.4, the velocity would be negative.
- If the curves represent a hill, the upper curve would be an uphill and the slope is the gradient and is positive and the lower curve, for downhill, would have negative gradient. Therefore the gradient is also the slope.

- The slope or the gradient or the derivative of Y with respect to X can itself be plotted as it varies with X and the slope of this curve (slope of the slope or gradient of the gradient) is called the second derivative of Y with respect to X and is written as $d^2Y/dX^2$.
- If a quantity is dependent on more than one parameter such as Y depends on X and t, a partial derivative of Y with respect to X is the rate of change of Y with respect to X with t held constant and is written as $\frac{\partial Y}{\partial X}$ and the rate of change of Y with respect to t with x held constant and is written as $\frac{\partial Y}{\partial t}$. The symbol $\partial$ is also used for higher partial derivatives such as $\frac{\partial^2 Y}{\partial X^2}$ is the second partial derivative of Y with respect to X.
- Calculation of the derivative is called differentiation. There are derived rules for differentiating a function. For example, if $Y=X^n$, then differentiating Y (calculation of the derivative of Y with respect to X) is given by, $dY/dX = nX^{n-1}$. There are similar rules for all functions.
- The inverse process of differentiation is integration. Taking the above example, if $Z=nX^{n-1}$, then the integral of Z with X $= \int Z\, dX = X^n$. An integral is the sum of little pieces of a function. Integrals are also readily available for different functions or can be calculated according to a rule.

Here, as in general learning, once one achieves a basic level of understanding learning becomes easier and the understanding and ability continue to increase. (See figure i.5). Once this primer is understood, the equations will not be intimidating and the reader can understand even more than what is given in the text.

# INDEX

## #



## A



## B

## C



## D



## E

## I



## J



## K



## L



## M

## N



## O



## P

## Q



## R



## S